SAT Chemistry

Sharon A. Wynne, MS

XAMonline

To obtain permission(s) to use the material from this work for any purpose including workshops or seminars, please submit a written request to:

XAMonline, Inc.
21 Orient Avenue
Melrose, MA 02176
Toll Free: 1-800-301-4647
Email: info@xamonline.com
Web: www.xamonline.com
Fax: 1-617-583-5552

Library of Congress Cataloging-in-Publication Data
Wynne, Sharon

SAT Chemistry/Sharon Wynne
ISBN: 978-1-60787-568-0

1. SAT 2. Study Guides 3. Chemistry

Disclaimer:

The opinions expressed in this publication are the sole works of XAMonline and were created independently from the College Board, or other testing affiliates. Between the time of publication and printing, specific test standards as well as testing formats and website information may change that are not included in part or in whole within this product. XAMonline develops sample test questions, and they reflect similar content as on real tests; however, they are not former tests. XAMonline assembles content that aligns with test standards but makes no claims nor guarantees candidates a passing score.

Cover photo provided by ©Can Stock Photo Inc./robwilson39; ©Can Stock Photo Inc./labamba; ©Can Stock Photo Inc./karenr; ©Can Stock Photo Inc./radiantskies; ©Can Stock Photo Inc./Bialasiewicz

Printed in the United States of America

SAT Chemistry
ISBN: 978-1-60787-568-0

Table of Contents

Introduction: SAT Chemistry

What Are The SAT Subject Area Tests?

The SAT Subject Area tests are designed to test your knowledge and problem-solving ability in a specific subject. The College Board for SATs administers 20 subject area tests. For a list of all the subject areas go to www.sat.collegeboard.org. The SAT Subject tests give you a chance to show your college admissions evaluators that you are knowledgeable in a specific area. In particular, if you intend to major or minor in a subject area, then doing well on the subject area test showcases your credentials for the college.

Why Take The SAT Chemistry Test?

The SAT Subject test is an opportunity to show the colleges you are applying to that you understand a specific subject area. It is not required that you take the SAT Subject Area tests for admission to college. However, if you have taken chemistry in high school, are comfortable with the subject matter, and did reasonably well in the coursework, then it might be an easy test to take to add to your achievements.

While most colleges do not REQUIRE the Subject Area tests, several of them CONSIDER them for admissions and placement. Sometimes, you might be able to waive an introductory course, depending on your scores on the SAT test. In some cases, the SAT Subject Area test is considered as a replacement for the ACT test. So, as you get ready in your junior year to take the SAT, look at the colleges you might be interested in applying to and see if there is a benefit to taking the SAT chemistry test to highlight your knowledge.

How Is The SAT Chemistry Exam Administered?

The SAT subject area tests are administered by the College Board, the same organization that administers the general SAT exams. In the 2015-2016 year, the SAT Chemistry test will be given six times: October, November, December, January, February, and June. So, if you are a junior, you should plan on taking the general SAT first, and then at the next date window, taking the SAT Chemistry test. So, for example, if you take the SAT in October, then take the Chemistry test in November. As mentioned earlier, there is some similarity in the math subject matter for both the SAT Math test and the SAT Chemistry subject test. So, if you can, take the Chemistry test while the SAT material is still fresh.

You will find registration information online at www.satcollegeboard.org. From the website, you can register online or by mail. In both cases, you will need to upload a digital photo of yourself as identification. You can find more about the photo requirements at https://sat.collegeboard.org/register/photo-requirements. The photo will become part of your exam admission ticket. The current fees can be found at https://collegereadiness.collegeboard.org/sat/register/fees. Four registration score reports are included in the fee; additional reports will need to be paid for separately.

Overview Of The SAT Chemistry Exam

The SAT Chemistry exam tests your knowledge of college preparatory chemistry. It assumes you have had at least one year of introductory college-preparatory chemistry, a one-year course in algebra, and experience in the laboratory. The focus of the test is on problem solving, so you need to have your thinking hat on, and have an understanding of the concepts and their application to the question.

In the 2015-2016 testing year, the Chemistry Subject SAT is offered six times. You cannot take the SAT and the SAT subject tests on the same date, so plan to take the Chemistry test in the next test date offering after you take the SAT.

The test consists of 85 multiple-choice questions and you will have 60 minutes to complete the test. A periodic table indicating the atomic numbers and masses of elements is provided for all test administrations. Problem solving requires simple numerical calculations. The metric system of units is used. **Calculators are NOT permitted.**

Three types of questions are used in the Chemistry Subject Test: classification questions, relationship analysis questions, and standard multiple choice questions, each with five answer choices. Relationship analysis questions use a different format, and you will mark your answers to them in a separate section of the answer sheet, as explained in the directions.

In a relationship analysis question, each question consists of two statements. The left-hand column contains a statement labeled I and the right-hand column contains a statement labeled II. For each question, determine whether statement I is true or false and whether statement II is true or false and fill in the corresponding T or F circles on your answer sheet. THEN fill in circle CE only if statement II is TRUE AND a correct explanation of the TRUE statement I.

Areas Tested In Chemistry

The SAT Subject Area for Chemistry covers material taught in most high school chemistry courses. You may find that you don't know the answer to every single question, as you are not expected to have learned every topic on the test. But, as you prepare for the test, use the topic lists below to study, and ensure that you are knowledgeable about all the topics. Below, you can see the specific areas tested, and the percentage of questions from that area.

1. **Structure of matter (25%)**
   Atomic structure, quantum numbers and energy levels (orbitals), electron configurations, periodic trends; molecular structure, including Lewis structures, three-dimensional molecular shapes, polarity; bonding, including ionic, covalent, and metallic bonds, relationships of bonding to properties

and structures; intermolecular forces such as hydrogen bonding, dipole-dipole forces, dispersion (London) forces

2. **States of matter – 16%**
   Gases, including the kinetic molecular theory, gas law relationships, molar volumes, density, and stoichiometry; liquids and solids, including intermolecular forces in liquids and solids, types of solids, phase changes, and phase diagrams; solutions, including molarity and percent by mass concentrations, solution preparation and stoichiometry, factors affecting solubility of solids, liquids, and gases, qualitative aspects of colligative properties

3. **Reaction types -14%**
   Acids and bases, including Brønsted-Lowry theory, strong and weak acids and bases, pH, titrations, indicators; oxidation-reduction, including recognition of oxidation-reduction reactions, combustion, oxidation numbers, use of activity series; precipitation, including basic solubility rules

4. **Stoichiometry 14%**
   Mole concept, including molar mass, Avogadro's number, empirical and molecular formulas; chemical equations, including the balancing of equations, stoichiometric calculations, percent yield, and limiting reactants

5. **Equilibrium and reaction rates 5%**
   Equilibrium systems, including factors affecting position of equilibrium (Le Châtelier's principle) in gaseous and aqueous systems, equilibrium constants, and equilibrium expressions; rates of reactions, including factors affecting reaction rates, potential energy diagrams, activation energies

6. **Thermochemistry 6%**
   Including conservation of energy, calorimetry, and specific heats, enthalpy (heat) changes associated with phase changes and chemical reactions, heating and cooling curves, entropy

7. **Descriptive chemistry 12%**
   Including common elements, nomenclature of ions and compounds, periodic trends in chemical and physical properties of the elements, reactivity of elements and prediction of products of chemical reactions, examples of simple organic compounds and compounds of environmental concern

8. **Laboratory 8%**
   Including knowledge of laboratory equipment, measurements, procedures, observations, safety, calculations, data analysis, interpretation of graphical data, drawing conclusions from observations and data

How The SAT Chemistry Test Is Scored

The highest possible score for Chemistry is 800. The score range is from 200 to 800, and increases in 20-point intervals. You will receive 1 point for each question you answer correctly. It's important to understand the ramifications of guessing on this test. If you answer a question incorrectly, then points are subtracted, based on the following scale:

¼ point subtracted for each 5-choice question
1/3 point subtracted for each 4-choice question
½ point subtracted for each 3-choice question
0 points subtracted for questions you don't answer

Most of the test is 5-choice questions so you will loose ¼ point for each wrong answer.

Adding up the number of correct answers – then subtracting (0.25) times the number of wrong answers – gives the raw score.

The raw scores are then extrapolated to the 200 – 800 point scale. This ensures that different tests and scores of other students do not affect your score. It also means that your score is not graded on a curve (where the highest scorer gets 800, and the remaining get scores relative to that).

Three types of scores are reported for the test. (1) Your score on the 200 – 800 point range. (2) The average score based on the most recent tests in that subject area. (3) The percentile score – for example, for your score of 700, if your percentile is 85, then it means that you did better than 85% of the students taking that test.

In the year 2014, you can see the percentile and corresponding scores for the Chemistry test.[1] So, 51 percent of the students scored below 640 and 49 percent of the students scored above 640.

800	99 percentile
700	74 percentile
640	51 percentile
560	24 percentile

Should You Guess the Answer on the Test?

Now that you know that your score is penalized for incorrect guessing, you should evaluate on a question-by-question basis, as to whether you should guess the answer. If you have some understanding of the question being asked, and the subject matter, and can discard 1-2 of the multiple-choice responses, then take an EDUCATED guess. It's probably not worth losing points by WILD GUESSING. The best option is to mark that question on the test sheet, and keep moving forward on the questions, and go back to it, once you've finished the test.

1 Percentile Scores for SAT Subject Area tests administered in 2014. https://secure-media.collegeboard.org/digitalServices/pdf/sat/sat-percentile-ranks-subject-tests-2014.pdf

Things to Bring on Test Day

Prepare your test day material the day before. The tests usually start in the morning, so don't wait until the last minute to find your gear. Here's what you will need at a minimum:

1. A printed out copy of your Admission Ticket
2. Photo identification
3. No. 2 pencils and an eraser
4. A watch, so you know how much time is remaining

You cannot bring the following into the testing room:

1. Any form of cell phone, tablet, or computer device
2. No iPods or music devices
3. No cameras or recording devices

Testing Tips:

1. **Get smart, play dumb.** Sometimes a question is just a question. No one is out to trick you, so don't assume that the test writer is looking for something other than what was asked. Stick to the question as written and don't overanalyze.

2. **Do a double take.** Read test questions and answer choices at least twice because it's easy to miss something, to transpose a word or some letters. If you have no idea what the correct answer is, skip it and come back later if there's time.

3. **Turn it on its ear.** The syntax of a question can often provide a clue, so make things interesting and turn the question into a statement to see if it changes the meaning or relates better (or worse) to the answer choices.

4. **Get out your magnifying glass.** Look for hidden clues in the questions because it's difficult to write a multiple-choice question without giving away part of the answer in the options presented. In most questions you can readily eliminate one or two potential answers, increasing your chances of answering correctly to 50/50, which will help out if you've skipped a question and gone back to it (see tip #2).

5. **Call it intuition.** Often your first instinct is correct. If you've been studying the content you've likely absorbed something and have subconsciously retained the knowledge. On questions you're not sure about trust your instincts because a first impression is usually correct.

6. **Graffiti.** Sometimes it's a good idea to mark your answers directly on the test booklet and go back to fill in the optical scan sheet later. You don't get extra points for perfectly blackened ovals. If you choose to manage your test this way, be sure not to mismark your answers when you transcribe to the scan sheet.

7. **Become a clock-watcher.** You have a set amount of time to answer the questions. Don't get bogged down laboring over a question you're not sure about when there are ten others you could answer more readily. If you choose to follow the advice of tip #6, be sure you leave time near the end to go back and fill in the scan sheet.

Accommodations for Students with Disabilities

If you have a documented disability, you may be eligible for special accommodations for the test. The SAT College Board's website describes the requirements to obtain the approvals for Services for Students with Disabilities (SSD). Look up the current requirements and steps to register at: https://sat.collegeboard.org/register/for-students-with-disabilities

In order to obtain the SSD approvals, you will need to file a request with the SAT College Board, well before you intend to take the test. The approvals take about seven weeks to process, and documentation of the student's disability and need for specific accommodations is required. The deadlines for SSD approvals are much earlier than the SAT Subject Area deadline. If you intend to take the SAT (and the SAT Subject Area) in the fall of your junior year, then the College Board recommends applying for your accommodation approvals in the spring of your sophomore year. Keep in mind, that if you were approved for special accommodations by the College Board for the SAT, then you might not have to reapply for the Math Subject Level 1 test. But, it is always helpful to confirm this with the College Board.

Once you are approved for special accommodation for test-taking, the accommodation will be noted on your admission ticket. If your accommodation request is not approved, then you must take the test as a standard test-taker.

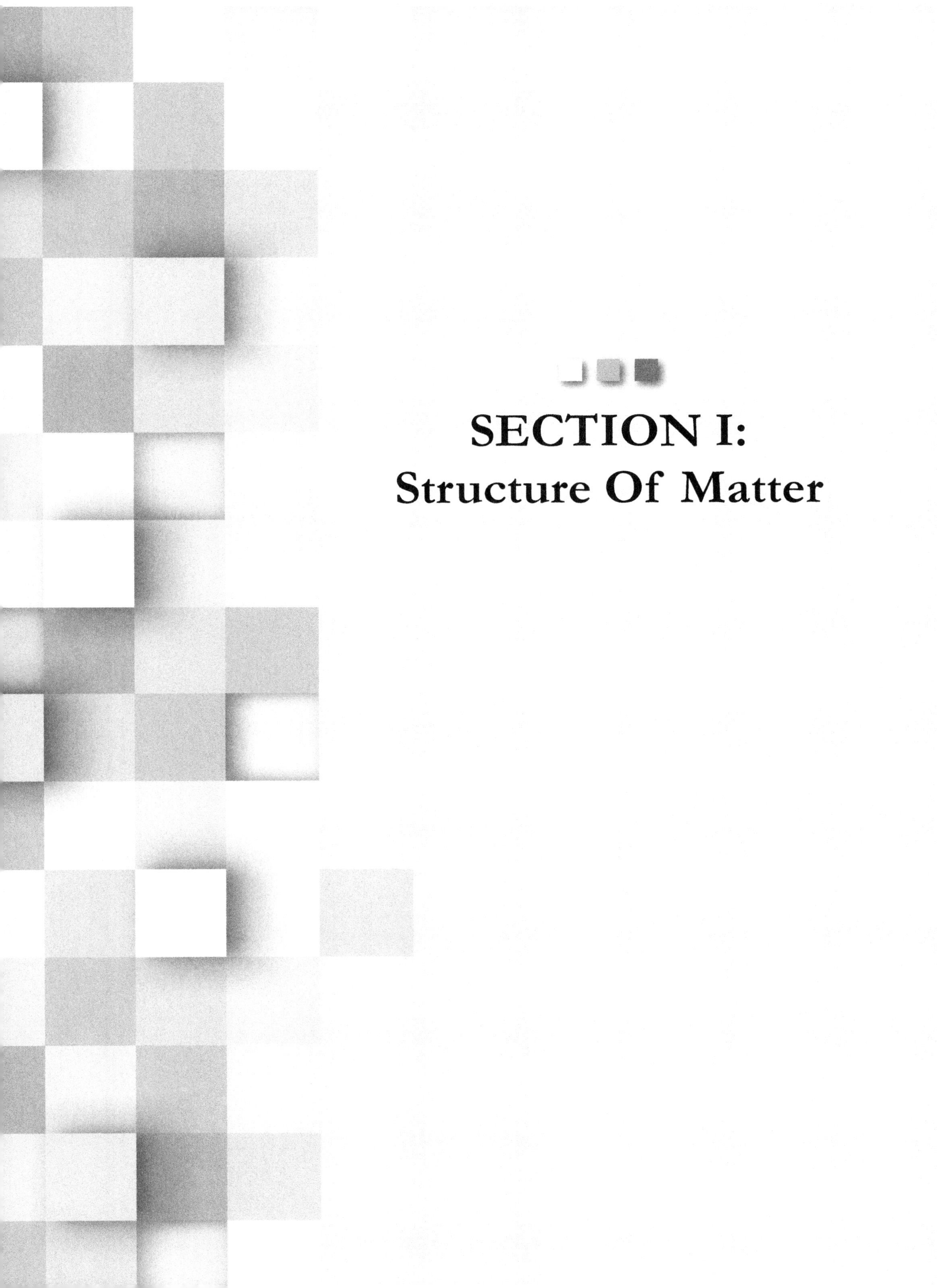

SECTION I: Structure Of Matter

Chapter 1: Atomic Structure

1.1 Experimental evidence of atomic structure

Prior to the late 1800s, atoms were thought to be small, spherical and indivisible particles that made up matter. However, with the discovery of electricity and the investigations that followed, this view of the atom changed.

Dalton's 1808 AD Elements							
	wt.		wt.		wt.		wt.
Hydrogen	1	Sulphur	13	Strontium	46	Lead	95
Azote	5	Magnesia	20	Barytes	68	Silver	100
Carbon	5	Lime	23	Iron	38	Gold	100
Oxygen	7	Soda	28	Zinc	56	Platina	140
Phosphorus	9	Potash	42	Copper	56	Mercury	167

The idea that pure materials called elements existed and that these elements were composed of fundamentally indivisible units called atoms was proposed by ancient philosophers even though they had little evidence. Modern atomic theory is credited to the work of **John Dalton** published in 1803-1807. Observations made by him and others about the composition, properties, and reactions of many compounds led him to develop the following postulates:

1. Each element is composed of small particles called atoms.
2. All atoms of a given element are identical in mass and other properties.
3. Atoms of different elements have different masses and differ in other properties.
4. Atoms of an element are not created, destroyed, or changed into a different type of atom by chemical reactions.
5. Compounds form when atoms of more than one element combine.
6. In a given compound, the relative number and kinds of atoms are constant.

Dalton determined and published the known relative masses of a number of different atoms. He also formulated the law of partial pressures. Dalton's work focused on the ability of atoms to arrange themselves into molecules and to rearrange themselves via chemical reactions, but he did not investigate the composition of atoms themselves. **Dalton's model of the atom** was a tiny, indivisible, indestructible **particle** of a certain mass, size, and chemical behavior, but Dalton did not deny the possibility that atoms might have a substructure.

Joseph John Thomson, often known as **J. J. Thomson**, was the first to examine this substructure. In the mid-1800s, scientists had studied a form of radiation called "cathode rays" or "electrons" that originated from the negative electrode (cathode) when electrical current was forced through an evacuated tube. Thomson determined in 1897 that **electrons have mass**, and because many different cathode materials release electrons, Thomson proposed that the **electron is a subatomic particle**. **Thomson's model of the atom** was a uniformly positive particle with electrons contained in the interior. This has been called the "plum-pudding" model of the atom where the pudding represents the uniform sphere of positive electricity and the bits of plum represent electrons.

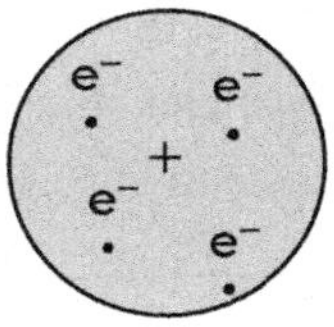

Max Planck determined in 1900 that **energy is transferred by radiation in exact multiples of a discrete unit of energy called a quantum**. Quanta of energy are extremely small, and may be found from the frequency of the radiation, f, using the equation:

$$E = hf$$

where h is Planck's constant and hf is a quantum of energy.

Ernest Rutherford studied atomic structure in 1910-1911 by firing a beam of alpha particles at thin layers of gold leaf. According to Thomson's model, the path of an alpha particle should be deflected only slightly if it struck an atom, but Rutherford observed some alpha particles bouncing almost backwards, suggesting that **nearly all the mass of an atom is contained in a small positively charged nucleus. Rutherford's model of the atom** was an analogy to the sun and the planets. A small positively charged nucleus is surrounded by circling electrons and mostly by empty space.

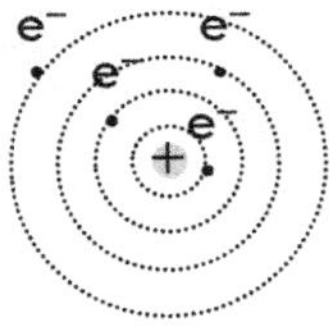

Niels Bohr incorporated Planck's quantum concept into Rutherford's model of the atom in 1913 to explain the **discrete frequencies of radiation emitted and absorbed by atoms with one electron** (H, He^+, and Li^{2+}). This electron is attracted to the positive nucleus and is closest to the nucleus at the **ground state** of the atom.

When the electron absorbs energy, it moves into an orbit further from the nucleus and the atom is said to be in an electronically **excited state**. If sufficient energy is absorbed, the electron separates from the nucleus entirely, and the atom is ionized:

$$H \rightarrow H^+ + e^-$$

The energy required for ionization from the ground state is called the atom's ionization energy. The discrete frequencies of radiation emitted and absorbed by the atom correspond (using Planck's constant) to discrete energies and in turn to discrete distances from the nucleus. **Bohr's model of the atom** was a small positively charged nucleus surrounded mostly by empty space and by electrons orbiting at certain discrete distances ("shells") corresponding to discrete energy levels.

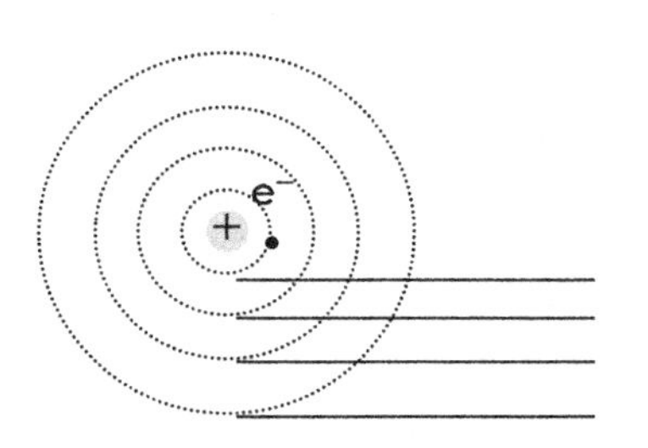

Bohr's model of the atom didn't quite fit experimental observations for atoms other than hydrogen. He was, however, on the right track. De Broglie was the first to suggest that matter behaved like a wave. Until then, waves had wave properties of wavelength, frequency and amplitude while matter had matter properties like mass and volume. De Broglie's suggestion was quite unique and interesting to scientists.

1.2 Energy levels

Depending on the experiment, radiation appears to have wave-like or particle-like traits. In 1923-1924, Louis de Broglie applied this **wave/particle duality to all matter with momentum**. The discrete distances from the nucleus described by Bohr corresponded to permissible distances where standing waves could exist. **De Broglie's model of the atom** described electrons as **matter waves in standing wave orbits** around the nucleus. The first three standing waves corresponding to the first three discrete distances are shown in the figure.

The realization that both matter and radiation interact as waves led **Werner Heisenberg** to the conclusion in 1927 that the act of observation and measurement requires the interaction of one wave with another. This interaction results in an **inherent uncertainty** in the location and momentum of particles observed. This limitation in the measurement of phenomena at the subatomic level is known as the **Heisenberg uncertainty principle**, and it applies to the location and momentum of electrons in an atom.

When **Erwin Schrödinger** studied the atom in 1925, he replaced the idea of precise orbits with regions in space called **orbitals** where electrons were likely to be found. **Schrödinger's model of the atom** is a mathematical formulation of quantum mechanics that describes the electron density of orbitals. The **Schrödinger equation** describes the **probability** that an electron will be in a given region of space, a quantity known as **electron density or** Ψ^2.

This is the atomic model that has been in use from shortly after it was introduced up to the present. The diagrams below are surfaces of constant Ψ^2 found by solving the Schrödinger equation for the hydrogen atom $1s$, $2p_z$ and $3d_0$ orbitals.

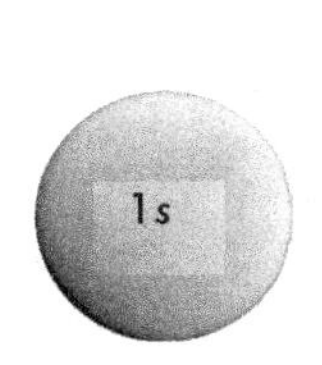

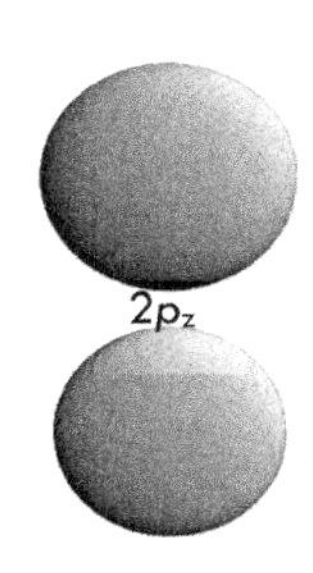

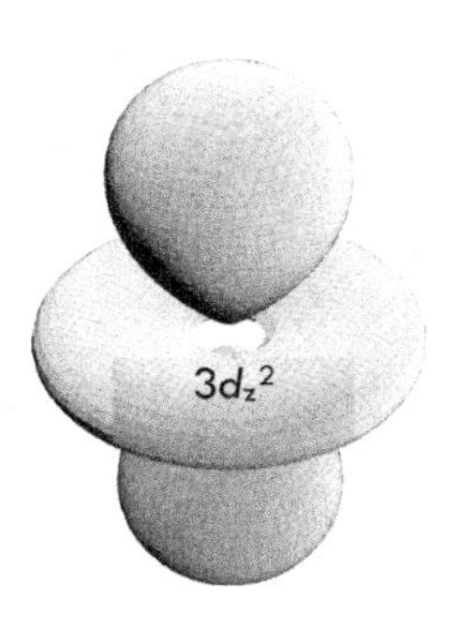

This model explains the movement of electrons to higher energy levels when exposed to energy. It also explains the movement of electrons to lower energy levels when the source of energy has disappeared. Accompanying this drop in energy level is the emission of electromagnetic radiation (for example, visible light is one possibility). In other words, radiation results from the movement of matter. Using Plank's equation, the frequency of that electromagnetic radiation (EMR) can be determined.

Wolfgang Pauli helped develop quantum mechanics in the 1920s by developing the concept of spin and the **Pauli exclusion principle**, which states that if two electrons occupy the same orbital, they must have different spin (intrinsic angular momentum). This principle has been generalized to other quantum particles.

Friedrich Hund determined a set of **rules to determine the ground state** of a multi-electron atom in the 1920s. One of these rules is called **Hund's Rule** in introductory chemistry courses, and describes the order in which electrons fill orbitals and their spin.

Atomic Theory

Probably one of the best examples of the progressive development of science would be the development of atomic theory. The ancient Greeks debated over the continuous nature of matter. Two schools of thought had emerged: matter was either continuous or matter was not continuous. The continuous idea was promoted by Aristotle, and due to his high regard among scholars, that was the idea that flourished. However, there was no effort made by the Greeks to prove or disprove this idea. During the dark ages, alchemists started experimenting and keeping records of their results, sending science on a pathway of experimentation and discovery. Robert Boyle and his famous J-tube experiment in 1661, gave the first experimental evidence for the existence of atoms. He even used words similar to Democritus to describe the results saying that the air consisted of atoms with a void between them. By increasing the pressure inside the J-tube, some of the void was squeezed out, decreasing the volume. Slowly, experimental evidence, including the work of Antoine Lavoisier and Joseph Priestley, began to mount, and in 1803, John Dalton proposed the Modern Atomic Theory which contained 5 basic postulates about the nature and behavior of matter.

Ben Franklin's discovery of electricity in 1746 sent scientists to work to understand it. J.J. Thomson investigated a cathode ray tube and identified the negatively charged particle in the cathode ray in 1897. His work was closely followed by Robert Millikan who discovered that the electron had a -1 charge in his oil drop experiment.

Experiments were under way to understand how electricity and matter interact when the discoveries of x-rays and radioactivity were announced. Scientists trying to understand radioactivity experimented day and night. Ernest Rutherford was one of them. He tried to understand the nature of radioactivity and classified it into three basic types. While trying to find out more about radioactivity, he conducted his gold foil experiment that ultimately provided greater insight into the subatomic nature of the atom, by discovering the nucleus. He also identified the proton present in the nucleus.

Rutherford's graduate student, Neils Bohr, made slight alterations to Rutherford's model of the atom, to account for his experimental results. These changes helped Rutherford's model stand up to classical physics. However, scientists looking for other patterns and information proposed more changes to the planetary model of the atom. These changes fit with spectroscopy experiments and the quantum mechanical view of the atom was formed. About the same time that this new theory was emerging, in 1932, James Chadwick, a collaborator of Rutherford's, announced the discovery of the neutron within the nucleus. This discovery ultimately led to the discovery of fission and the development of the nuclear bomb.

1.3 Quantum numbers

The quantum-mechanical solutions from the Schrödinger Equation utilize three quantum numbers (*n*, *l*, and *ml*) to describe an orbital and a fourth (*ms*) to describe an electron in an orbital. This model is useful for understanding the frequencies of radiation emitted and absorbed by atoms and chemical properties of atoms.

The **principal quantum number** *n* may have positive integer values (1, 2, 3, …). *n* is a measure of the **distance** of an orbital from the nucleus, and orbitals with the same value of *n* are said to be in the same **shell**. This is analogous to the Bohr model of the atom. Each shell may contain up to $2n^2$ electrons. The highest quantum number *n* that any shell of an element has is the same as the number of the row of the periodic table in which the element is found.

The **azimuthal quantum number** *l* may have integer values from 0 to n-1. *l* describes the angular momentum of an orbital. This determines the orbital's **shape**. Orbitals with the same value of *n* and *l* are in the same **subshell**, and each subshell may contain up to 4*l* + 2 electrons. Subshells are usually referred to by the principle quantum number followed by a letter corresponding to *l* as shown in the following table:

Azimuthal quantum number *l*	0	1	2	3	4
Subshell designation	*s*	*p*	*d*	*f*	*g*

Therefore, *s* subshells may have $4 \times 0 + 2 = 2$ electrons, and *p* subshells may have $4 \times 1 + 2 = 6$ electrons. Helium is in the first row of the periodic table, and has only a single *s* subshell with two electrons. This subshell would be notated as 1*s*.

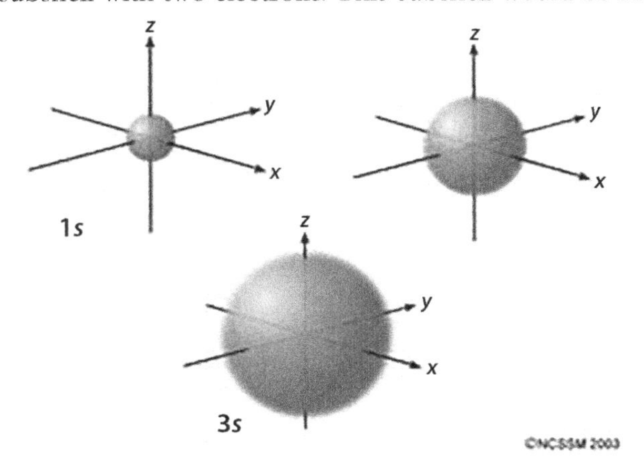

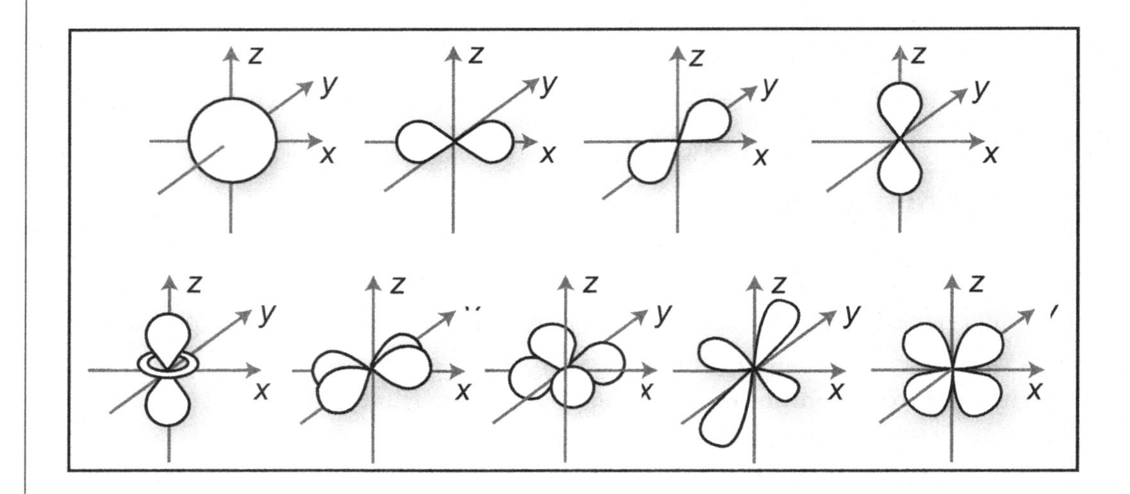

z
y
x

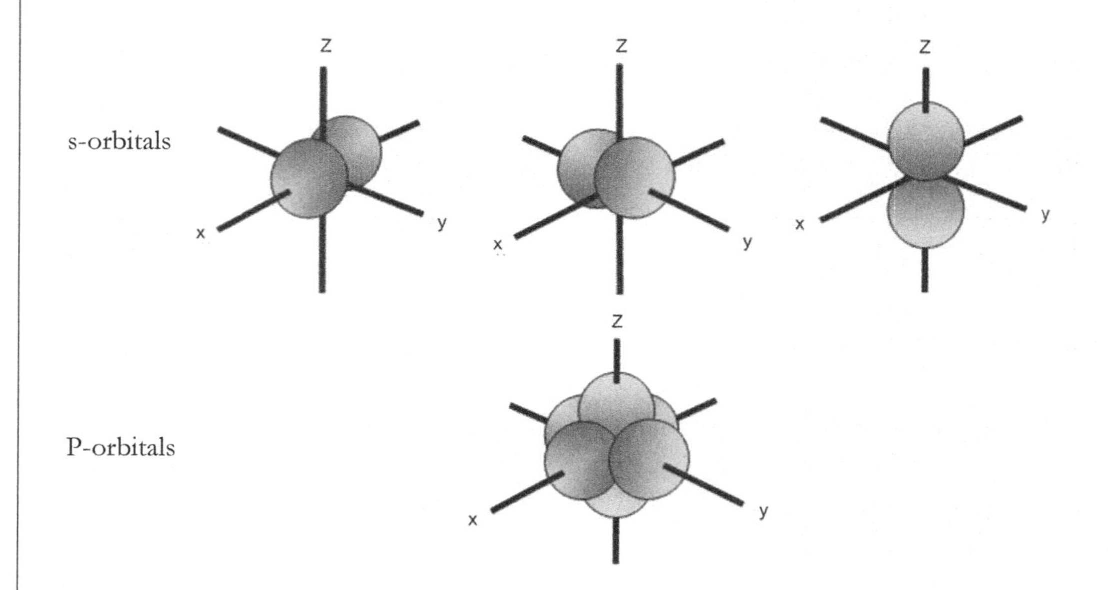

s-orbitals
P-orbitals
z
x
y

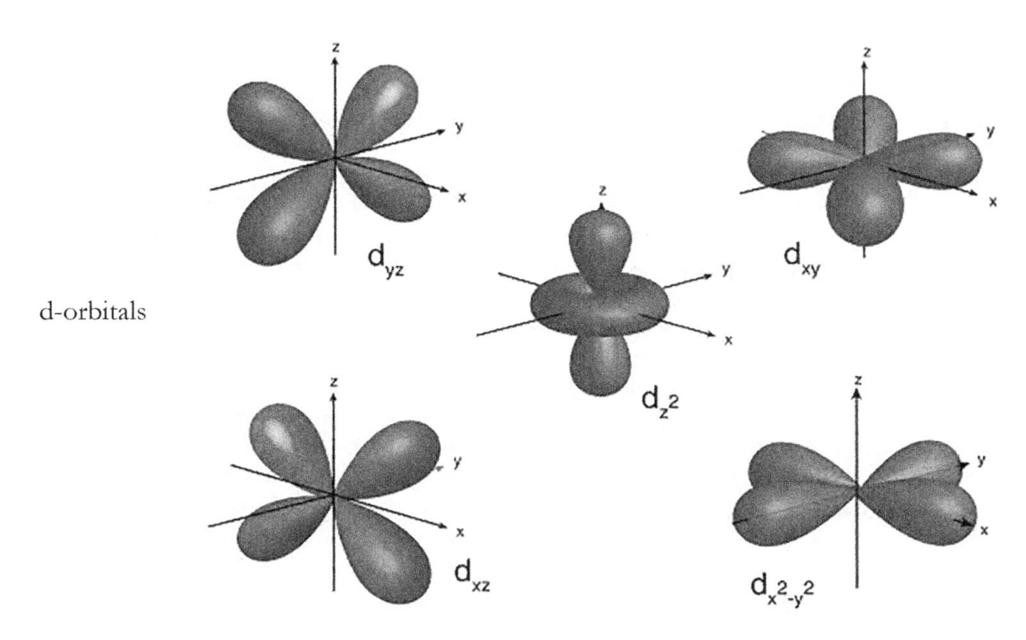

d-orbitals

The **magnetic quantum number** *ml* or *m* may have integer values from –*l* to *l*. *ml* is a measure of how an individual orbital responds to an external magnetic field, and it often describes an orbital's orientation. In any given shell, there is only one *s* orbital with an *m* value of 0. If *p* orbitals are present in a shell there will always be 3 *p* orbitals with values of -1, 0, and 1 (usually referred to as x, y, and z to denote 3-dimensional orientation). A subscript—either the value of *ml* or a function of the *x*-, *y*-, and *z*-axes—is used to designate a specific orbital within a subshell. For example, *n* = 3, *l* = 2, and *ml* = 0 would be shown as the $3d_0$ orbital. Each orbital may hold up to two electrons.

The spin quantum number *ms* or *s* has one of two possible values: –1/2 or +1/2. *ms* differentiates between the two possible electrons occupying an orbital. Electrons moving through a magnet behave as if they were tiny magnets themselves spinning on their axes in either a clockwise or counterclockwise direction. These two spins may be described as *ms* = –1/2 and +1/2 or as down and up.

The **Pauli exclusion principle** states that **no two electrons in an atom may have the same set of four quantum numbers**.

The following table summarizes the relationship among *n*, *l*, and *ml* through *n*=3:

n	*l*	Subshell	*ml*	Orbitals in subshell	Maximum number of electrons in subshell
1	0	1*s*	0	1	2
2	0	2*s*	0	1	2
	1	2*p*	–1, 0, 1	3	6
3	0	3*s*	0	1	2
	1	3*p*	–1, 0, 1	3	6
	2	3*d*	–2, –1, 0, 1, 2	5	10

Subshell energy levels

In single-electron atoms (H, He^+, and Li^{2+}) above the ground state, subshells within a shell are all at the same energy level, and an orbital's energy level is only determined by n. However, in all other atoms, multiple electrons repel each other. Electrons in orbitals closer to the nucleus create a screening or **shielding effect** on electrons further away from the nucleus, preventing them from receiving the full attractive force of the nucleus. **In multi-electron atoms, both n and l determine the energy level of an orbital.** In the absence of a magnetic field, **orbitals in the same subshell with different ml all have the same energy** and are said to be **degenerate orbitals.**

The following list orders subshells by increasing energy level:

$1s < 2s < 2p < 3s < 3p < 4s < 3d < 4p < 5s < 4d < 5p < 6s < 4f < 5d < 6p < 7s < 5f < \ldots$

This list may be constructed by arranging the subshells according to n and l and drawing diagonal arrows as shown below:

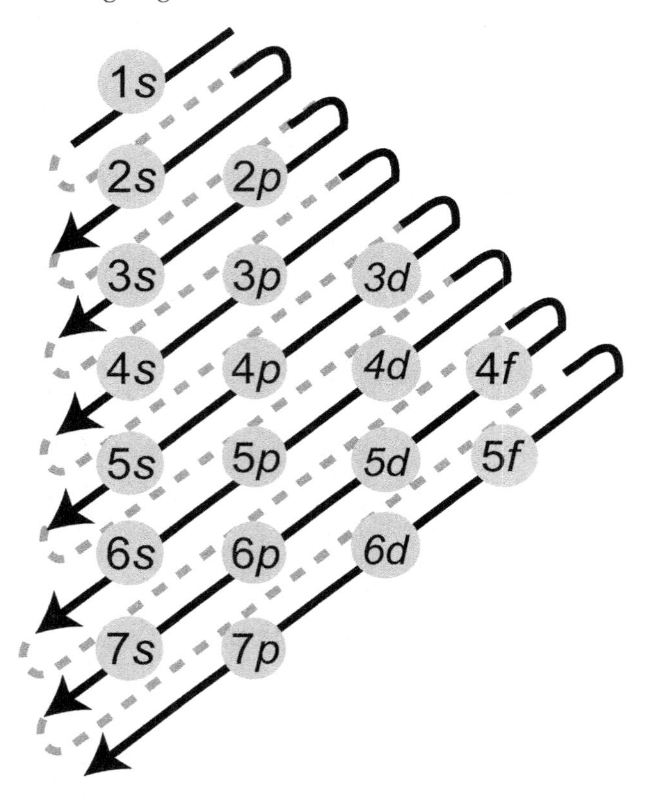

1.4 Electron Configurations

Electron arrangements (also called electron shell structures) in an atom may be represented using three methods: an **electron configuration**, an **orbital diagram**, or an **energy level diagram**. All three methods require knowledge of the subshells occupied by electrons in a certain atom. The **Aufbau principle** or **building-up rule** states that **electrons at ground state fill orbitals starting at the lowest available energy levels.**

An **electron configuration** is a **list of subshells** with superscripts representing the **number of electrons** in each subshell. For example, an atom of boron has 5 electrons. According to the Aufbau principle, two will fill the $1s$ subshell, two will fill the higher energy $2s$ subshell, and one will occupy the $2p$ subshell which has an even higher energy. The electron configuration of boron is $1s^2 2s^2 2p^1$. Similarly, the electron configuration of a vanadium atom with 23 electrons is:

$$1s^2 2s^2 2p^6 3s^2 3p^6 4s^2 3d^3.$$

Configurations are also written with their principle quantum numbers together:

$$1s^2 2s^2 2p^6 3s^2 3p^6 3d^3 4s^2.$$

Electron configurations are often written to emphasize the outermost electrons. This is done by writing the symbol in brackets for the element with a full p subshell from the previous shell and adding the **outer electron configuration** onto that configuration. The element with the last full p subshell will always be a noble gas from the right-most column of the periodic table. For the vanadium example, the element with the last full p subshell has the configuration $1s^2 2s^2 2p^6 3s^2 3p^6$. This is $_{18}Ar$. The configuration of vanadium may then be written as $[Ar]4s^2 3d^3$ where $4s^2 3d^3$ is the outer electron configuration.

Electron arrangements may also be written by noting the number of electrons in each shell. For vanadium, this would be:

2, 8, 11, 2.

Orbital diagrams assign electrons to individual orbitals so the energy state of individual electrons may be found. This way of depicting electron arrangements is most like the bus-boarding analogy, and shows how electrons occupy orbitals within a subshell. **Hund's rule** states that **before any two electrons occupy the same orbital, other orbitals in that subshell must first contain one electron each with parallel spins**. Electrons with up and down spins are shown by arrows, and these are placed in lines of orbitals (represented as boxes or dashes) according to Hund's rule, the Aufbau principle, and the Pauli exclusion principle. Below is the orbital diagram for vanadium:

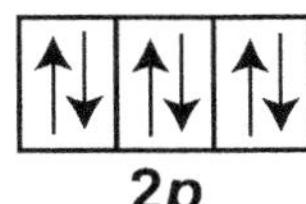
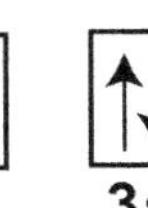
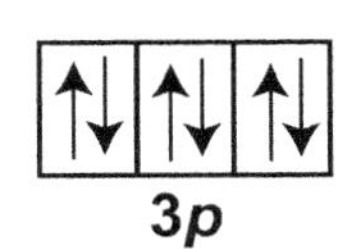
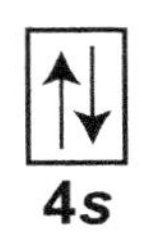
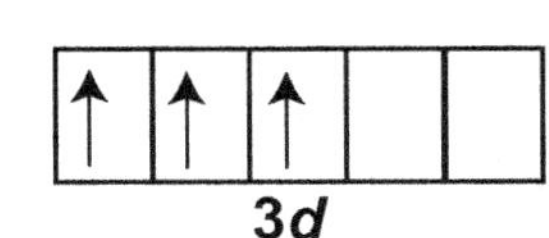

An **energy level diagram** is an orbital diagram that shows subshells with higher energy levels higher up on the page. The energy level diagram of vanadium is:

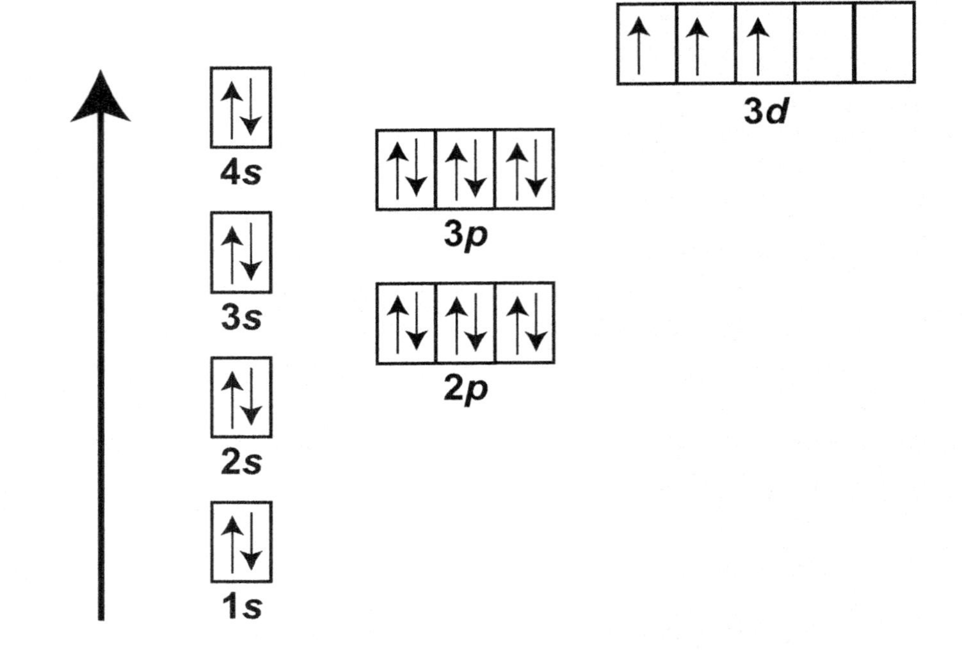

Valence shell electron arrangements and the periodic table

Electrons in the **outermost shell** are called **valence shell electrons**. For example, the electron configuration of Se is [Ar]$4s^2 3d^{10} 4p^4$, and its valence shell electron configuration is $4s^2 4p^4$.

The **periodic table** may be used to write down the electron configuration of any element. The table may be divided up into **blocks corresponding to the subshell** designation of the most recent orbital to be filled by the building-up rule.

Elements in the s- and p-blocks are known as **main-group elements**. The d-block elements are called **transition metals.** The f-block elements are called **inner transition metals**.

The maximum number of electrons in each subshell (2, 6, 10, or 14) determines the number of elements in each block, and the order of energy levels for the subshells creates the pattern of blocks. These blocks also usually correspond to the value of l for the **outermost electron** of the atom. This has important consequences for the physical and chemical properties of the elements. The outermost shell or valence shell principle quantum number (for example, 4 for Se) is also the period number for the element in the table.

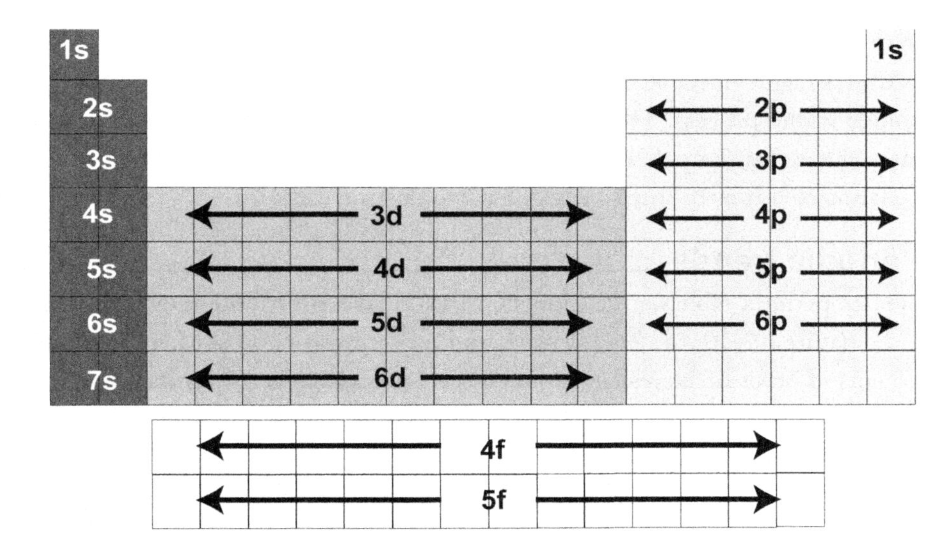

Atoms in the d- and f-blocks often have unexpected electron arrangements that cannot be explained using simple rules. Some heavy atoms have unknown electron configurations because the number of different frequencies of radiation emitted and absorbed by these atoms is very large.

Nonmetals gain electrons or share electrons to achieve stable configurations and **metals lose electrons** to achieve them.

Valence electrons are the electrons involved in chemical reactions, and valence electron configuration is responsible for chemical activity. The octet rule states that when there are less than four valence electrons, the atom will tend to lose electrons to go back to the previous energy level that was full. When the outer energy level is more than half full (more than 4 valence electrons), the atom will tend to gain electrons to complete the valence energy level.

For example, aluminum has 13 electrons with a configuration of $1s^2$,$2s^2$, $2p^6$,$3s^2$, $3p^1$. There are only three electrons in the third energy level, and the octet rule states that there should be eight. Aluminum can gain 5 more electrons to fill the 3 p orbital or lose the $3s^2$, and $3p^1$ electrons. It takes less energy to lose the three electrons. When these electrons are lost, an Al^{+3} ion forms with an electron configuration of $1s^2$, $2s^2$, $2p^6$, which has a complete octet in the outer shell.

With this in mind, you can notice that the families with valence configurations s^1, s^2 or s^2, p^1 tend to lose electrons to form ionic compounds. Families with valence configurations s^2, p^2; s^2, p^3; s^2, p^4; s^2, p^5 tend to gain electrons to form ionic compounds or share electrons to form molecules held together covalently.

Members of the noble gas family have an s^2, p^6 configuration and therefore a complete octet in their valence shells. This causes them to be unreactive. However, the heavier noble gases form a number of compounds with oxygen and fluorine, such as KrF_2 and XeO_4. Remember that atoms with electrons in the "d" or "f' orbitals have s^1 or s^2 in their valence configuration.

This information helps us to understand some other elemental characteristics. For example, the elements in the alkali metal family are not found as free elements in nature.

This is due to their high chemical activity. The one valence electron in the outermost energy level is highly unstable, and it tends to be easily lost, forming many different ionic compounds in the process. The halogen family consists of molecules, rather than free elements. The almost complete valence energy level is easily completed by two halogen atoms sharing electrons to form a covalently bonded molecule.

1.5 Periodic trends

The first periodic table was developed in 1869 by **Dmitri Mendeleev** several decades before the nature of electron energy states in the atom was known. Mendeleev arranged the elements in order of increasing atomic mass into **columns of similar physical and chemical properties**. He then boldly **predicted the existence and the properties of undiscovered elements** to fill the gaps in his table. These interpolations were initially treated with skepticism until three of Mendeleev's theoretical elements were discovered and were found to have the properties he predicted. It is the correlation with properties—not with electron arrangements—that have placed the periodic table at the beginning of most chemistry texts.

In the modern periodic table, **the elements in a column are known as a group,** and groups are numbered from 1 to 18. Older numbering styles used Roman numerals and letters. These elements, known as a family, have similar properties due to their similar outermost electron arrangements. Each family or column has a group name. **A row of the periodic table is known as a period**, and periods of the known elements are numbered from 1 to 7. The lanthanoids are all in period 6, and the actinoids are all in period 7.

Today, the 115 named elements are organized into the Periodic Table. The design of this table provides much information about a particular element or group of elements.

The identity of an **element** depends on the **number of protons** in the nucleus of the atom. This value is called the **atomic number** and it is sometimes written as a subscript before the symbol for the corresponding element. Atoms and ions of a given element that differ in number of neutrons have a different mass and are called **isotopes**. A nucleus with a specified number of protons and neutrons is called a **nuclide**, and a nuclear particle, either a proton or neutron, may be called a **nucleon**. The total number of nucleons is called the **mass number** and may be written as a superscript before the atomic symbol. The **mass number** of an atom is the total number of nucleons it contains. This is the **sum of the number of protons and neutrons** in the nucleus.

$^{14}_{6}C$ represents an atom of carbon with 6 protons and 8 neutrons.

Different isotopes have different natural abundances and have different nuclear properties, but an atom's chemical properties are almost entirely due to electrons.

The number of neutrons may be found by subtracting the atomic number from the mass number. For example, uranium-235 has 235 – 92 = 143 neutrons because it has 235 nucleons and 92 protons.

Atomic mass is a mass relative to carbon-12, and ^{12}C has an assigned value of exactly 12 u. This means that **the mass number of an atom is usually a good guess at the**

atom's atomic mass, and for ^{12}C it is an exact value and not a guess at all. For example, uranium-238 has an atomic mass of 238.05 u.

But if ^{12}C is exactly 12 u, why does carbon have an atomic mass of 12.011 u in the periodic table? This is because 1.1% of carbon on Earth exists as the stable isotope carbon-13. This carbon is heavier by 1 u and contributes the additional 0.011 g to the expected mass of a mole of carbon atoms that we'd find anywhere on this planet. In general, the **atomic mass** of an element is the average of the atomic masses of all stable isotopes of that element weighted by their abundance on Earth.

Atomic mass of element X = (Fraction of X as isotope A) (Atomic mass of isotope A) +
(Fraction of X as isotope B) (Atomic mass of isotope B) +
(Fraction of X as isotope C) (Atomic mass of isotope C) +

Chlorine, for example, exists on Earth in two stable isotopes: ^{35}Cl and ^{37}Cl. 75.76% of the chlorine is ^{35}Cl which has an atomic mass of 34.969 u and 24.24% is ^{37}Cl which has an atomic mass of 36.966 u.

Elements with no stable isotopes are often listed in tables of atomic masses with a number in brackets. This value is the mass number of the isotope with the longest half-life. A list of isotopic compositions and atomic masses for all natural isotopes is at http://physics.nist.gov/cgi-bin/Compositions/stand_alone.pl.

The most metallic element is francium at the bottom left of the table. The most nonmetallic element is fluorine. The metallic character of elements within a group increases with period number. This means that **within a column, the more metallic elements are at the bottom**. The metallic character of elements within a period decreases with group number. This means that **within a row, the more metallic elements are on the left**. Among the main group atoms, **elements diagonal to each other** as indicated by the dashed arrows **have similar properties** because they have a similar metallic character. The noble gases are nonmetals, but they are an exception to the diagonal rule.

Physical properties relating to metallic character are summarized in the following table:

Element	Electrical/ thermal conductivity	Malleable/ ductile as solids?	Lustrous?	Melting point of oxides, hydrides, and halides
Metals	High	Yes	Yes	High
Metalloids	Intermediate. Altered by dopants (semiconductors)	No (brittle)	Varies	Varies (oxides). Low (hydrides, halides)
Nonmetals	Low (insulators)	No	No	Low

Malleable materials can **be beaten into sheets. Ductile** materials can **be pulled into wires. Lustrous** materials **have a shine**. Oxides, hydrides, and halides are compounds with O, H, and halogens respectively. Measures of intermolecular attractions other than melting point are also higher for metal oxides, hydrides, and halides than for the nonmetal compounds. A dopant is a small quantity of an intentionally added impurity. The controlled movement of electrons in doped silicon semiconductors carries digital information in computer circuitry.

The **size of an atom** is not an exact distance, due to of the probabilistic nature of electron density, but we may compare radii among different atoms using some standard. The sizes of neutral atoms increase with period number and decrease with group number. This trend is similar to the trend described above for metallic character. The smallest atom is helium.

Atomic radius decreases left to right due to the electrons being pulled in tighter to the nucleus with the addition of each electron and proton. The atomic radius increases going down the periodic table due to the addition of another level of electrons with each row.

Group names, melting point, density, and properties of compounds

Groups 1, 2, 17, and 18 are often identified with the group names shown on the table below.

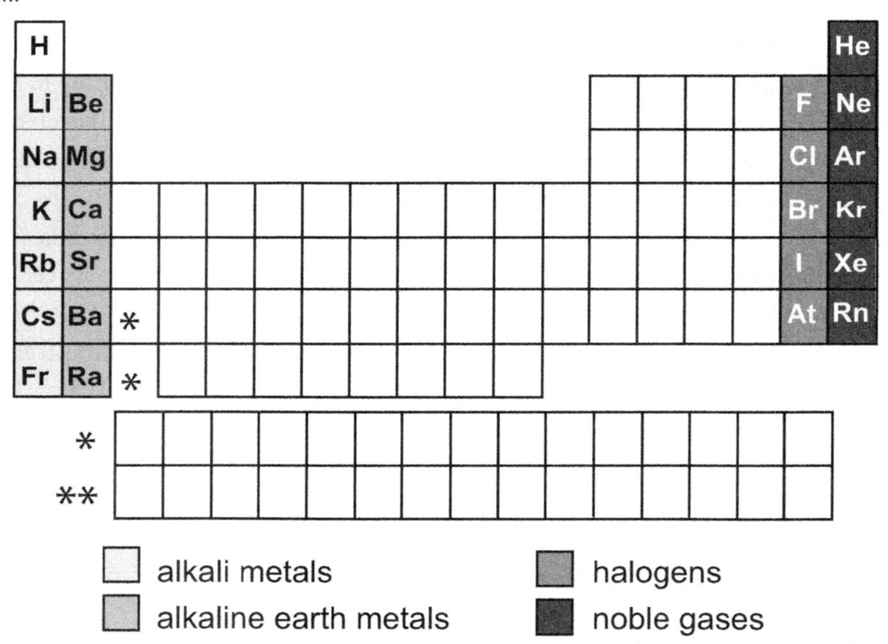

Seven elements are found as **diatomic molecules**: **(H_2, N_2, O_2, and the halogens: F_2, Cl_2, Br_2, and I_2)**. Mnemonic devices to remember the diatomic elements are: "$Br_2I_2N_2Cl_2H_2O_2F_2$" (pronounced "Brinklehof" and "**H**ave **N**o **F**ear **O**f **I**ce **C**old **B**eer." Another way to remember them is by using the Rule of Sevens: 7 of them which form a 7 on the Periodic Table and 4 of the 7 are from Group 7. These molecules are attracted to one another using **weak London dispersion forces**.

Note that **hydrogen** is *not* an alkali metal. Hydrogen is a colorless gas and is the most abundant element in the universe, but H_2 is very rare in the atmosphere because it is light enough to escape gravity and reach outer space. Hydrogen atoms form more compounds than any other element.

Alkali metals are shiny, soft, metallic solids. They have **low melting points and low densities** compared with other metals because they have a weaker metallic bond. Measures of intermolecular attractions including their **melting points decrease further down the periodic table due to weaker metallic bonds** as the size of atoms increases.

Alkaline earth metals (group 2 elements) are grey, metallic solids. They are harder, denser, and have a higher melting point than the alkali metals but values for these properties are still low compared to most of the transition metals. Measures of metallic bond strength, such as melting points, do not follow a simple trend down the periodic table.

Halogens (group 17 elements) have an irritating odor. Unlike the metallic bonds between alkali metals, **London forces between halogen molecules increase in strength further down the periodic table**. London forces make Br_2 a liquid and I_2 a solid at 25 °C. The lighter halogens are gases.

Noble gases (group 18 elements) have no color or odor and exist as **individual gas atoms** that experience London forces. These attractions also increase with period number.

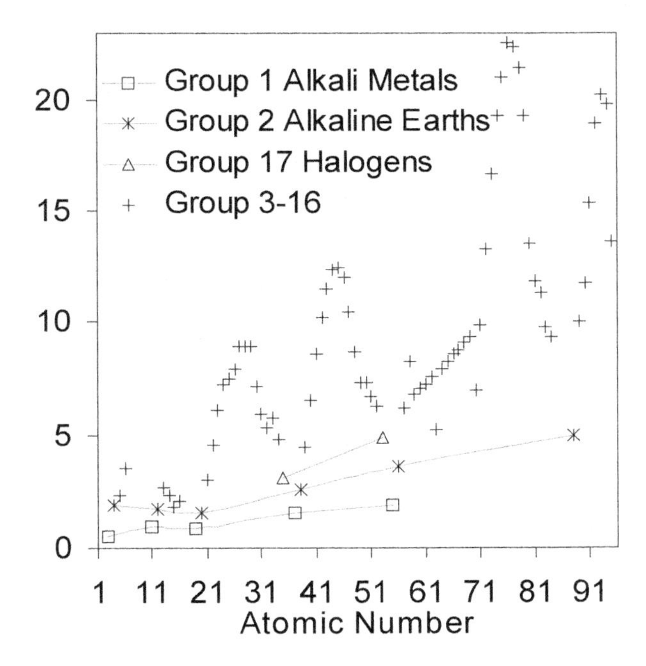

The known **densities** of liquid and solid elements at room temperature are shown below. **Intermolecular forces contribute to density** by bringing nuclei closer to each other, so the periodicity is similar to trends for melting point. These group-to-group differences are superimposed on a general trend for **density** to **increase with period number** because heavier nuclei make the material denser.

Trends among properties of **compounds** may often be deduced from **trends among their atoms**, but caution must be used. For example, the densities of three potassium halides are:

2.0 g/cm^3 for KCl 2.7 g/cm^3 for KBr 3.1 g/cm^3 for KI.

We would expect this trend for increasing atomic mass within a group. We might also expect the density of KF to be less than 2.0 g/cm^3, but it is actually 2.5 g/cm^3 due to a change in crystal lattice structure.

Physics of electrons and stability of electron configurations

For an isolated atom, the **most stable system of valence electrons is a filled set of orbitals**. For the main group elements, this corresponds to group 18 (ns^2np^6 and $1s^2$ for helium), and, to a lesser extent, group 2 (ns^2).

The next most stable state is a set of degenerate half-filled orbitals. These occur in group 15 (ns^2np^3). The least stable valence electron configuration is a single electron with no other electrons in similar orbitals. This occurs in group 1 (ns^1) and to a lesser extent in group 13 (ns^2np^1).

An atom's first **ionization energy** is the **energy required to remove one electron** by the reaction $M(g) \rightarrow M^+(g) + e^-$. Periodicity for ionization energy is in the opposite direction from the trend for atomic radius. The most metallic atoms have electrons further from the nucleus, and these are easier to remove. Factors that affect ionization energy are:

1. Nuclear charge – the larger the nuclear charge, the greater the IE
2. Shielding effect – the greater the shielding effect, the lower the IE
3. Radius – the greater the distance between the nucleus and the outer electrons of an atom, the lower the IE

Sublevel – an electron from a sublevel that is more than half full requires additional energy to be removed

An atom's **electron affinity** is the energy released when one electron is added by the reaction $M(g) + e^- \rightarrow M^-(g)$. A large negative number for the exothermic reaction indicates a high electron affinity. Halogens have the highest electron affinities.

Trends in **ionization energy and electron affinity** within a period reflect the **stability of valence electron configurations**. A stable system requires more energy to change and releases less when changed. Note the peaks in stability for groups 2, 13, and 16 to the right.

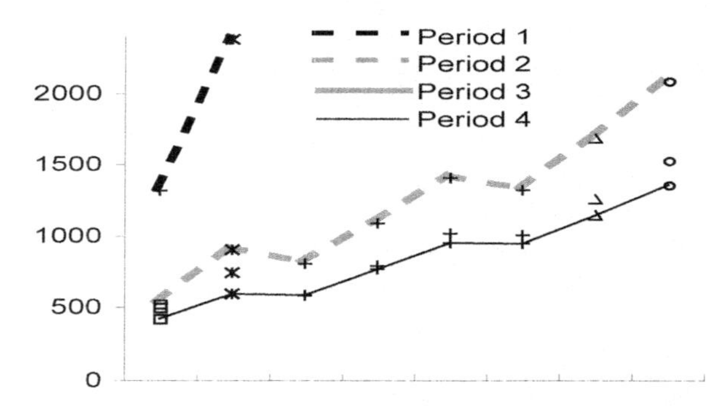

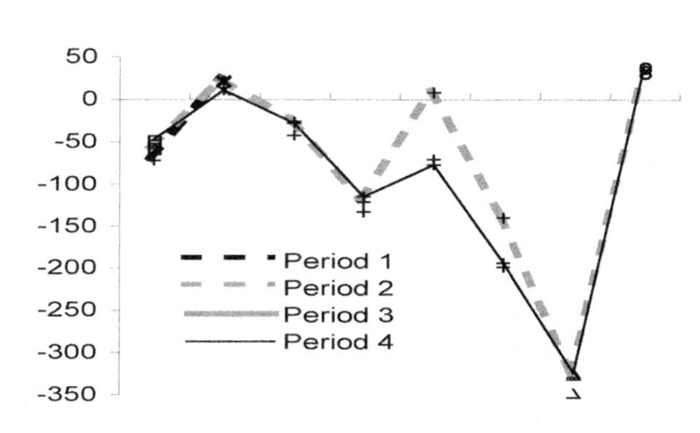

Electronegativity and reactivity series

Much of chemistry consists of atoms **bonding** to achieve stable valence electron configurations. **Nonmetals gain electrons or share electrons** to achieve these configurations and **metals lose electrons** to achieve them.

Qualitative group trends

When cut by a knife, the exposed surface of an **alkali metal or alkaline earth metal** quickly turns into an oxide. These elements **do not occur in nature as free metals**. Instead, they react with many other elements to form white or grey water-soluble salts. With some exceptions, the oxides of group 1 elements have the formula M_2O, their hydrides are MH, and their halides are MX (for example, NaCl). The oxides of group 2 elements have the formula MO, their hydrides are MH_2, and their halides are MX_2, for example, $CaCl_2$.

Halogens combine with each other and also form a wide variety of oxides. They combine with hydrogen to form HX gases, and these compounds are also commonly used as acids (hydrofluoric, hydrochloric, etc.) in aqueous solution. Halogens form salts with metals by gaining electrons to become X^- ions. Astatine is an exception to many of these properties because it is an artificial metalloid.

Noble gases are **nearly chemically inert**. The heavier noble gases form a number of compounds with oxygen and fluorine such as KrF_2 and XeO_4

Electronegativity measures the ability of an atom to attract electrons in a chemical bond. The most metallic elements have the lowest electronegativity. The most nonmetallic have the highest electronegativity.

The most powerful chemical reactions between elements occur between metals and non-metals. This also corresponds to reactions between the most and least electronegative elements. In a reaction with a metal, the most reactive chemicals are the **most electronegative elements** or compounds containing those elements. In a reaction with a nonmetal, the most reactive chemicals are the **least electronegative elements** or compounds containing them. The reactivity of elements may be described by a **reactivity series**: an ordered list with chemicals that react strongly at one end and nonreactive chemicals at the other. The following reactivity series is for metals reacting with oxygen:

Metal	K	Na	Ca	Mg	Al	Zn	Fe	Pb	Cu	Hg	Ag	Au
Reaction with O_2	Burns violently		Burns rapidly					Oxidizes slowly				No reaction

Copper, silver, and gold (group 11) are known as the **noble metals** or **coinage metals** because they are very unreactive. The impact of electronegativity on chemical bonding is discussed in detail below.

Fluorine is the most electronegative element in the periodic table, and the elements decrease in electronegativity going from right to left and from top to bottom with the halogens being the most reactive (electronegative) group.

Valence and oxidation numbers

The term **valence** is often used to describe the number of atoms that may react to form a compound with a given atom by sharing, removing, or losing **valence electrons**. A more useful term is **oxidation number**. The **oxidation number of an ion is its charge**. The oxidation number of an atom sharing its electrons is **the charge it would have if the bonding were ionic**. There are four rules for determining oxidation number:

1. The oxidation number of an element (i.e., a Cl atom in Cl_2) is zero because the electrons in the bond are shared equally.
2. In a compound, the more electronegative atoms are assigned negative oxidation numbers, and the less electronegative atoms are assigned positive oxidation numbers equal to the number of shared electron-pair bonds. For example, hydrogen may have an oxidation number of –1 when bonded to a less electronegative element or +1 when bonded to a more electronegative element. Oxygen almost always has an oxidation number of –2. Fluorine always has an oxidation number of –1 (except in F_2).
3. The oxidation numbers in a compound must add up to zero, and the sum of oxidation numbers in a polyatomic ion must equal the overall charge of the ion.
4. The charge on a polyatomic ion is equal to the sum of the oxidation numbers for the species present in the ion. For example, the sulfate ion, SO_4^{2-}, has a total charge of -2. This comes from adding the -2 oxidation number for 4 oxygen (total -8) and the +6 oxidation number for sulfur.

Example: What is the oxidation number of nitrogen in the nitrate ion, NO_3^-?

Solution: Oxygen has the oxidation number of –2 (rule 2), and the sum of the oxidation numbers must be –1 (rule 3). The oxidation number for N may be found by solving for *x* in the equation $x + 3 \times (-2) = -1$. The oxidation number of N in NO_3^- is +5.

There is a **periodicity in oxidation numbers** as shown in the following table, for examples of oxides with the maximum oxidation number. Remember that an element may occur in different compounds in several different oxidation states.

Group	1	2	13	14	15	16	17	18
Oxide with maximum oxidation number	Li_2O	BeO	B_2O_3	CO_2	N_2O_5		Cl_2O_7	XeO_4
	Na_2O	MgO	Al_2O_3	SiO_2	P_2O_5	SO_3	Br_2O_7	
Oxidation number	+1	+2	+3	+4	+5	+6	+7	+8

They are called "oxidation numbers" because oxygen was the element of choice for reacting with materials when modern chemistry began, and as a result, Mendeleev arranged his first table to look similar to this one.

Acidity/alkalinity of oxides

Metal oxides form basic solutions in water because the ionic bonds break apart and the O^{2-} ion reacts to form hydroxide ions:

metal oxide $\longrightarrow$ metal cation (aq) + $O^{2-}(aq)$ and $O^{2-}(aq) + H_2O(l) \longrightarrow 2\ OH^-(aq)$

Ionic oxides containing a large cation with a low charge (Rb_2O, for example) are most soluble and form the strongest bases.

Covalent oxides form acidic solutions in water by reacting with water. For example:

$$SO_3(l) + H_2O(l) \rightarrow H_2SO_4(aq) \rightarrow H^+(aq) + HSO_4^-(aq)$$

$$Cl_2O_7(l) + H_2O(l) \rightarrow 2HClO_4(aq) \rightarrow 2H^+(aq) + 2ClO_4^-(aq)$$

Covalent oxides at high oxidation states and high electronegativities form the strongest acids. Acids and bases are discussed in **Chapter 7.** Note that the periodic trends for acid and base strength of the oxide of an element follow the same pattern as electronegativity, in that acidity decreases with increasing period number and increases with increasing group number.

http://jcrystal.com/steffenweber/JAVA/jpt/jpt.html contains an applet of the periodic table and trends.

http://www.webelements.com is an on-line reference for information on the elements.

http://www.uky.edu/Projects/Chemcomics/ has comic book pages for each element.

Summary

A summary of periodic trends is shown below. The properties tend to decrease or increase as shown depending on a given element's proximity to fluorine in the table.

Metallic character ⇓	Ionization energy ⇑
Ionic character of halides ⇓	Covalent character of halides ⇑
Atomic radius ⇓	Electroneativity ⇑
Alkalinity of oxides ⇓	Acidity of oxides ⇑

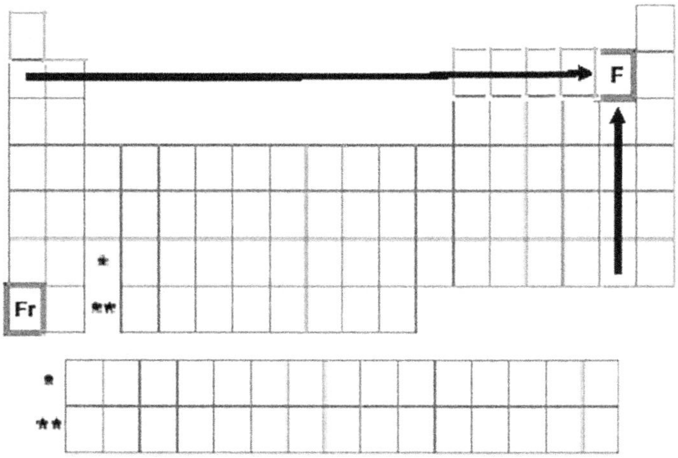

*Note: Memorize the trends below that are in italics; everything else increases top ⟶ bottom, left ⟶ right

Group Trends (top to bottom):

Increases: atomic radius, nuclear charge, ionic size, shielding effect, atomic number (number of protons, number of electrons), covalent character of halides, acidity of oxides

Decreases: ionization energy, electron affinity, electronegativity, ionic character of halides, alkalinity of oxides

Period Trends (left to right):

Increases: electronegativity, nuclear charge, ionization energy, electron affinity, atomic number (number of protons, number of electrons), acidity of oxides

Decreases all the way across: atomic radius, metallic character, ionic character of halides, alkalinity of oxides

Decreases through cations (positive) and again through anions: ionic size

Stays the same left to right: shielding effect

Chapter 2: Molecular Structure

2.1 Lewis structures

Lewis dot structures are used to keep track of each atom's valence electrons in a molecule. Drawing Lewis structures is a three-step process:

1. Determine the number of valence shell electrons for each atom. If the compound is an anion, add the charge of the ion to the total electron count because anions have "extra" electrons. If the compound is a cation, subtract the charge of the ion.
2. Write the symbols for each atom and show how the atoms within a molecule are bound to each other.
3. Draw a single bond (one pair of electron dots or a line) between each pair of connected atoms. Place the remaining electrons around the atoms as unshared pairs. If every atom has an octet of electrons except H atoms with two electrons, the Lewis structure is complete. Shared electrons count towards both atoms. If there are too few electron pairs to do this, draw multiple bonds (two or three pairs of electron dots between the atoms) until an octet is around each atom (except H atoms with two). If there are too many electron pairs to complete the octets with single bonds then the octet rule is broken for this compound.

Example: Draw the Lewis structure of HCN.
Solution:

1. From the periodic table, we know that each atom contributes the following number of electrons: H : 1, C : 4, N : 5. Summing these numbers, we see that this molecule will have a total of 10 valence electrons (because it has no net charge we do not add to or subtract from this number).
2. The atoms are connected with C at the center and will be drawn as: H C N. Having H as the central atom is impossible because H has one valence electron and will always only have a single bond to one other atom. If N were the central atom then the formula would probably be written as HNC.

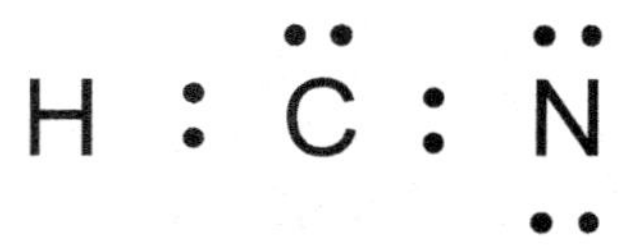

H : C : N (with lone pairs on C and N)

3. Connecting the atoms with 10 electrons in single bonds gives the structure above. H has two electrons to fill its valence subshells, but C and N only have six each. A triple bond (below) between these atoms fulfills the octet rule for C and N and is the correct Lewis structure.

H : C:::N:

Resonance

O_2 contains a total of 12 valence electrons and the following Lewis structure:

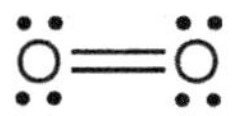

O=O

Ozone, O_3, has a total of 18 valence electrons, and two Lewis structures are possible for this molecule

O=O–O ⟷ O–O=O

Equivalent Lewis structures are called **resonance forms**. A double-headed arrow is used to indicate resonance. The actual molecule does not have a double bond on one bond and a single bond on the other. The **molecular structure is in an average state between the resonance forms**.

Benzene (C_6H_6) has the following resonance forms:

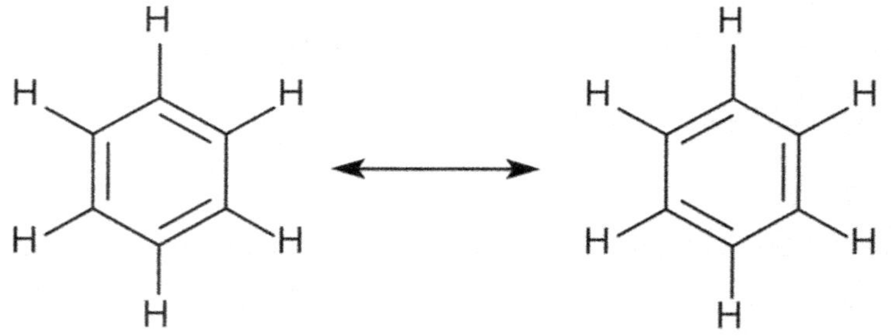

H H H H H H ⟷ H H H H H H

Each carbon atom in benzene bonds to three atoms, so their electrons are in three sp^2 orbitals and one p orbital as we've seen for C_2H_4. The p atomic orbitals combine to form molecular orbitals with delocalized electrons as shown in the pi bonding molecular orbital below to the right. (The electrons are, however, not truly in pi bonds; in a benzene ring, they are in a delocalized pi system.)

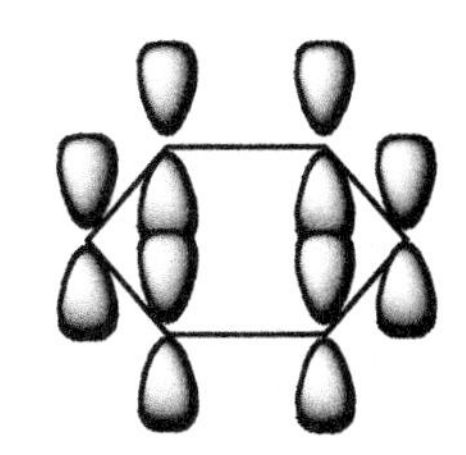

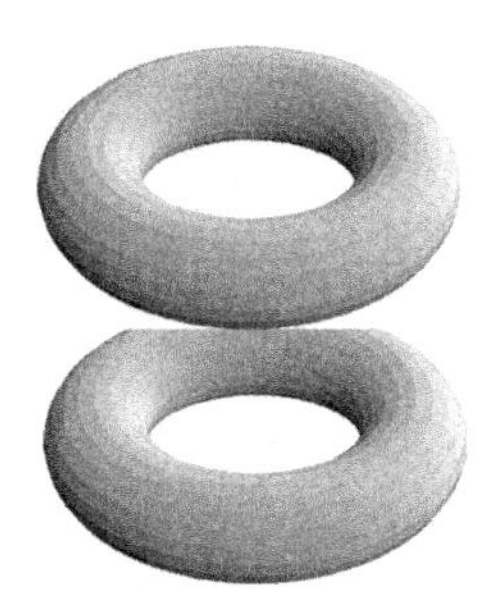

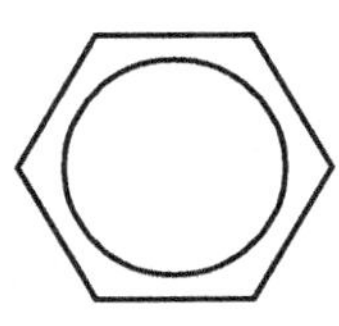

Aromatic molecules are often drawn with a circle in the center of their benzene rings (shown to the left) to represent the delocalized electrons. The atoms of a benzene molecule are all located in the same plane. This is in contrast to molecules that contain only sigma bonds as shown to the right for cyclohexane, C_6H_{12}.

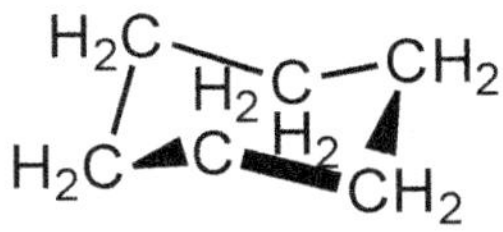

To select the most probable Lewis dot structure for a compound or molecule that follows the octet rule, review the structures and compare to the method for constructing Lewis dot structures given above.

Example: Which of the electron-dot structures given below for nitrous oxide (laughing gas), N_2O, is/are acceptable?

```
   ..   ..  ..
I. : N : : N : O :
              ..

     ..    ..
II. : N : N : : O :

            ..
III. : N : : : N : : O :
```

Solution: Both nitrogen and oxygen follow the octet rule so the Lewis structure should show each atom in the molecule with 8 electrons, either unshared or shared. Upon examination, only choice I provides each atom in the molecule with 8 electrons. Choice II has only 6 electrons around each of the nitrogen atoms and choice III has 10 electrons around the center nitrogen atom.

2. 2 Three-dimensional molecular shapes

Molecular geometry describes the specific three dimensional arrangements of atoms in molecules. It is also defined as the positions of the atomic nuclei in a molecule. Different experimental techniques such as X-ray crystallography can be used to tell us where the atoms are located in a molecule. Also, by using advanced techniques, very complicated structures for proteins, enzymes, DNA, and RNA have been determined. Molecular geometry is associated with a wide area of chemistry including chemistry of vision, taste, smell and odors, drug reactions and enzyme controlled reactions.

Molecular geometry is related to the specific orientation of bonding atoms. Correct molecular geometry determination depends upon careful analysis of electron distributions in orbitals. In addition, the simple writing of Lewis diagrams will also provide important clues for the determination of molecular geometry.

Molecular geometry is predicted using the valence-shell electron-pair repulsion or VSEPR model. VSEPR uses the fact that electron pairs repel each other. Imagine you are one of two pairs of electrons in bonds around a central atom. You want to be as far away from the other electron pair as possible, so you will be on one side of the atom and the other pair will be on the other side. There is a straight line (or a 180° angle) between you and the other electron pair on the other side of the nucleus. In general, electron pairs lie at the largest possible angles from each other.

2.3 Valence Shell Electron Pair Repulsion (VSEPR) theory

The VSEPR theory states that the electron pairs around a central atom arrange themselves in a manner that allows them to be as far apart as possible from each other.

The outermost electron-occupied shell of an atom that holds the electrons involved in bonding is called the valence shell. In a covalent bond, two electrons are shared between two atoms. However, in polyatomic molecules, several atoms are bonded to a central atom using two or more electron pairs. The repulsion of negative electron pairs either in bonds or as lone pairs causes them to stay apart from each other as much as possible.

This idea of electron pair repulsion may be better understood if we imagine them as several inflated balloons tied at their necks. Each balloon, representing an electron pair, will try to minimize the crowding and stay away from the others. Based on the VSEPR theory, the molecular geometry about the central atom as other atoms are added can be determined systematically.

Lewis diagrams provide information about what atoms are bonded to each other and the electron pairs involved.

Molecules can then be divided into two groups:

Group 1 includes molecules with no lone electron pairs. The molecular geometry in this group is identical to the electron pair geometry.

Group 2 includes molecules containing one or more lone pairs. An extra step is required to translate from electron pair geometry to the final molecular geometry, because only the positions of bonded atoms are considered in molecular geometry.

Electron pairs	Geometrical arrangement		Predicted bond angles	Example
2	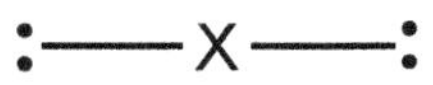 :—X—:	Linear	180°	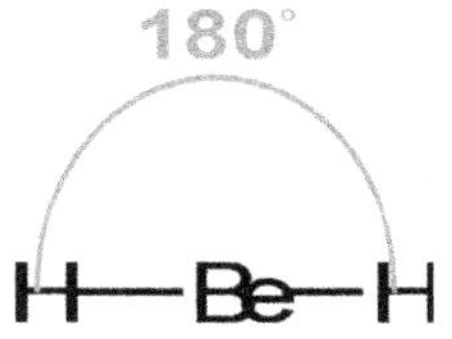 180° H—Be—H
3	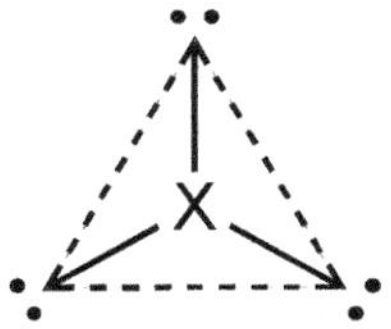 X	Trigonal planar	120°	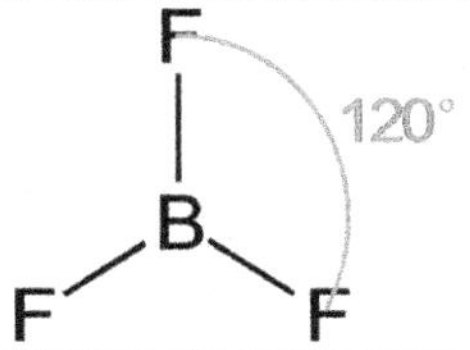 F, B, F, F, 120°
4	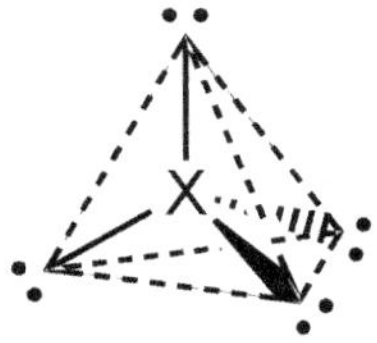 X	Tetrahedral	109.5°	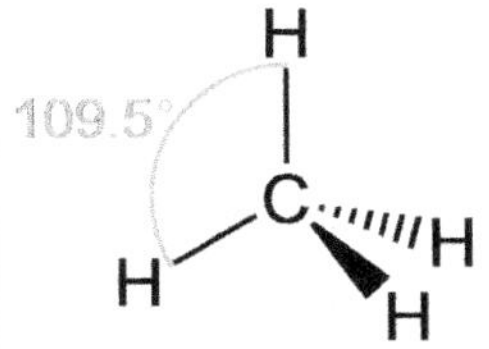 H, C, H, H, H, 109.5°
5	 X	Trigonal bipyramidal	120° and 90°	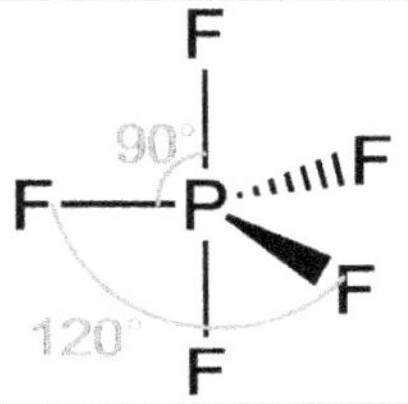 F, F, P, F, F, F, 90°, 120°
6	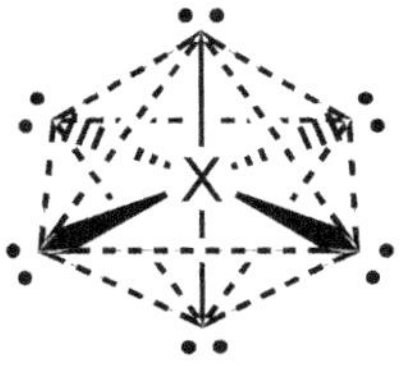 X	Octahedral	90°	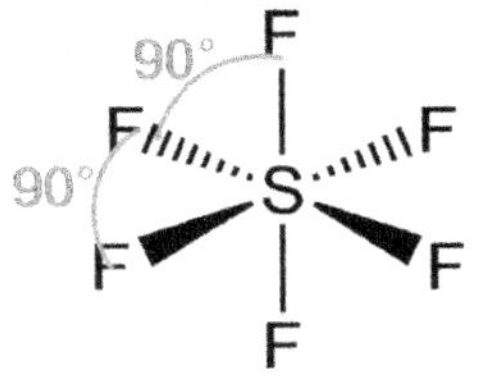 F, F, S, F, F, F, F, 90°, 90°

X represents a generic central atom. Lone pair electrons on F are not shown in the example molecules.

Unshared Electron Pairs

The shape of a molecule is given by the location of its atoms. These are connected to central atoms by shared electrons, but unshared electrons on a central atom also have an important impact on molecular shape. Unshared electrons also repel one another and so can alter the expected angles between atoms. Molecular shapes in the following table take into account total pairs and unshared electrons.

Electron pairs	Molecular shape				
	All shared pairs	1 unshared pair	2 unshared pairs	3 unshared pairs	4 unshared pairs
2	A—X—A Linear				
3	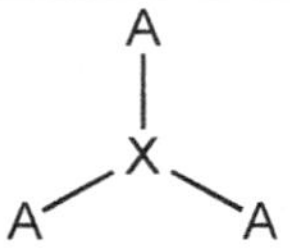 Trigonal planar	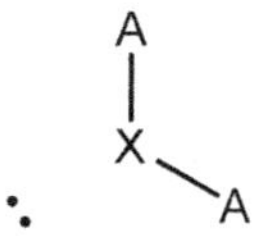 Bent			
4	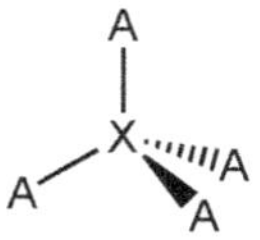 Tetrahedral	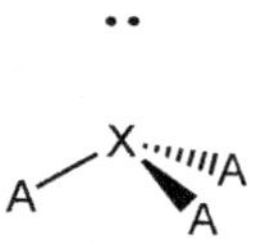 Trigonal pyramidal	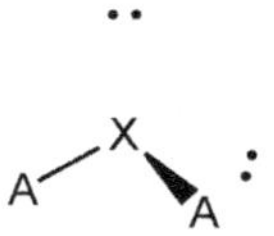 Bent		
5	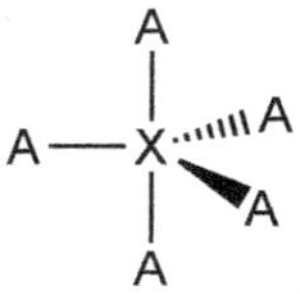 Trigonal bipyramidal	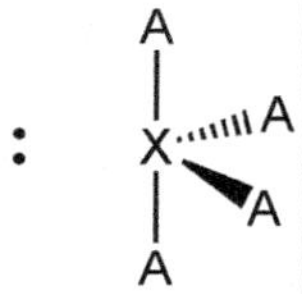 Seesaw or sawhorse	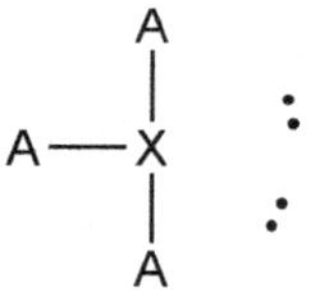 T-shaped	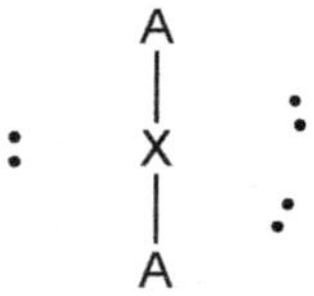 Linear	
6	 Octahedral	 Square pyramidal	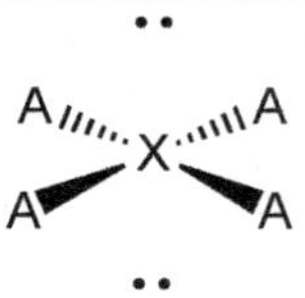 Square planar	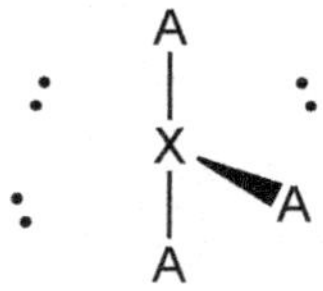 T-shaped	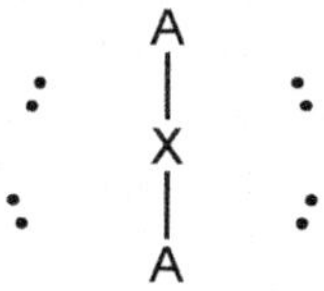 Linear

X represents a generic central atom bonded to atoms labeled A.

Altered Bond Angles

Unpaired electrons also have a less dramatic impact on molecular shape. Imagine you are an unshared electron pair around a molecule's central atom. The shared electron pairs are each attracted partially to the central atom and partially to the other atom in the bond, but you are different. You are attracted to the central atom, but there's nothing on your other side, so you are free to expand in that direction. That expansion means that you take up more room than the other electron pairs, and they are all squeezed a little closer together because of you. Multiple bonds have a similar effect because more space is required for more electrons. In general, unshared electron pairs and multiple bonds decrease the angles between the remaining bonds. A few examples are shown in the following tables.

Compound $CH_4NH_3H_2O$			
Unshared electrons	0	1	2
Shape	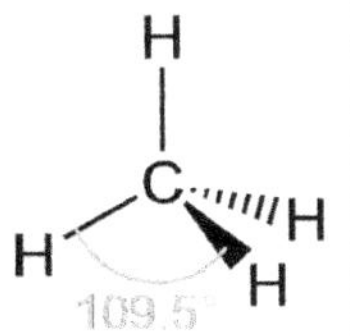 109.5° Tetrahedral	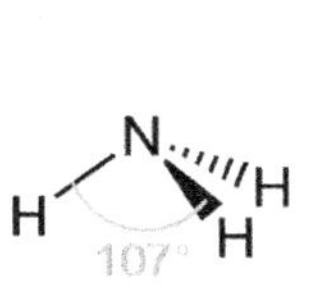 107° Trigonal pyramidal	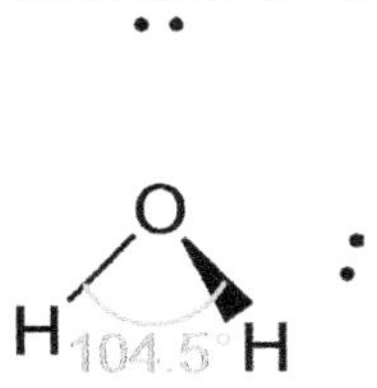 104.5° Bent

Compound $BF_3C_2H_4$		
Multiple bonds	0	1
Shape	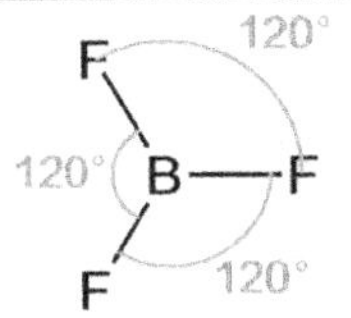 120° 120° 120° Trigonal planar	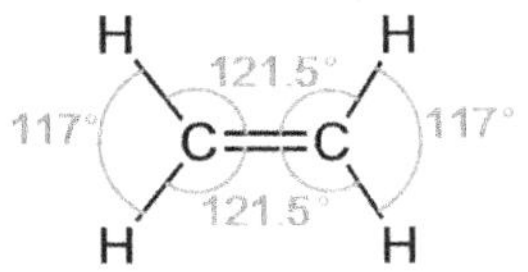 117° 121.5° 121.5° 117° Trigonal planar

Summary

In order to use VSEPR to predict molecular geometry, perform the following steps:

1. Write out Lewis dot structures.
2. Use the Lewis structure to determine the number of unshared electron pairs and bonds around each central atom counting multiple bonds as one (for now).
3. Use the information in the tables above to determine the arrangement of total and unshared electron pairs, accounting for electron repulsion around each central atom.
4. For multiple bonds or unshared electron pairs, decrease the angles slightly between the remaining bonds around the central atom.
5. Combine the results from the previous two steps to determine the shape of the entire molecule.

http://www.shef.ac.uk/chemistry/vsepr/ is a good site for explaining and visualizing molecular geometries using VSEPR.

2.4 Polarity

A polar molecule has positive and negative regions. Bond polarity is necessary but not sufficient for molecular polarity. A molecule containing polar bonds will still be nonpolar if the most negative and most positive location occurs at the same point. In other words, in a polar molecule, bond polarities must not cancel one another.

To determine if a molecule is polar, perform the following steps.

1. Draw the molecular structure.
2. Assign a polarity to each bond with an arrow (remember C-H is nonpolar). If none of the bonds are polar, the molecule is nonpolar.
3. Determine if the polarities cancel each other in space. If they do, the molecule is nonpolar. Otherwise the molecule is polar.

The polarity of molecules is critical for determining a good solvent for a given solute. Additional practice on the topic of polar bonds and molecules is available at http://cowtownproductions.com/cowtown/genchem/09_17M.htm.

Chapter 3 Bonding

Chemical compounds form when two or more atoms join together. A stable compound occurs when the total energy of the combination of the atoms together is lower than the atoms separately. The combined state suggests an attractive force exists between the atoms. This attractive force is called a chemical bond. There are various types of bonds depending on the nature of the elements involved.

3.1 Ionic bonds

Ionic Bonds

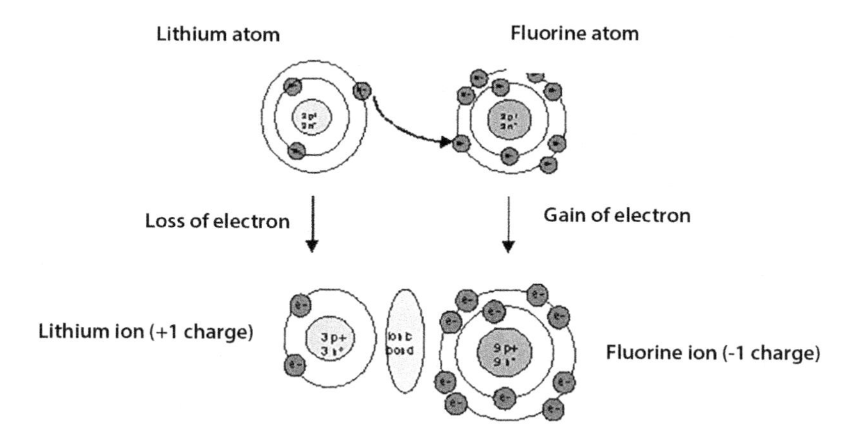

An **ionic bond** describes the electrostatic forces that exist between particles of opposite charge. Elements that form an ionic bond with each other have a large difference in their electronegativity.

Due to low ionization energies, metals have a tendency to lose valence electrons relatively easily, whereas non-metals, which have high ionization energies and high electronegativities, gain electrons easily. This produces cations (positively charged) and anions (negatively charged). Coulomb's Law says that opposites attract, and so do the oppositely charged ions. This *electrostatic interaction* between the anion and the cation results in an ionic bond. Ionic bonds will only occur when a metal is bonding to a non-metal, and is the result of the periodic trends (ionization energy and electronegativity). Elements that form an ionic bond with each other have a large difference in their electronegativity.

Anions and cations pack together into a crystal **lattice** as shown to the right for NaCl. Ionic compounds are also known as **salts**.

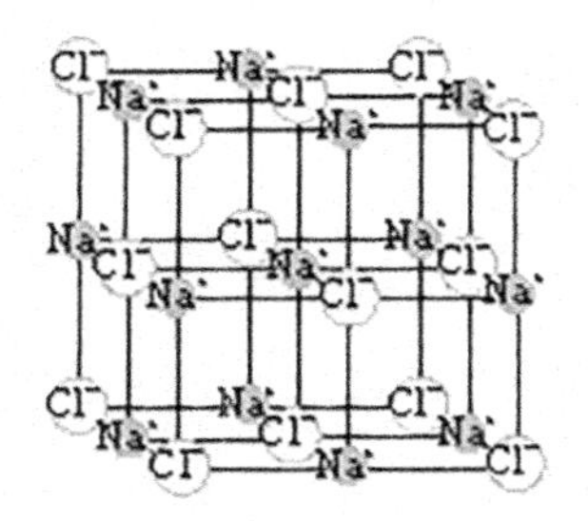

All common salts (compounds with ionic bonds) are **solids** at room temperature. Salts are **brittle**, have a **high melting point**, and **do not conduct electricity** because their ions are not free to move in the crystal lattice. Salts do conduct electricity in molten form. The formation of a salt is a highly exothermic reaction between a metal and a nonmetal.

Salts in solid form are generally stable compounds, but in molten form or in solution, their component ions often react to form a more stable salt. The reactivities of these ions vary with the electronegativity of the respective elements as described in Chapter 1.4.

Some salts decompose to form more stable salts, as in the decomposition of molten potassium chlorate to form potassium chloride and oxygen.

3.2 Covalent bonds

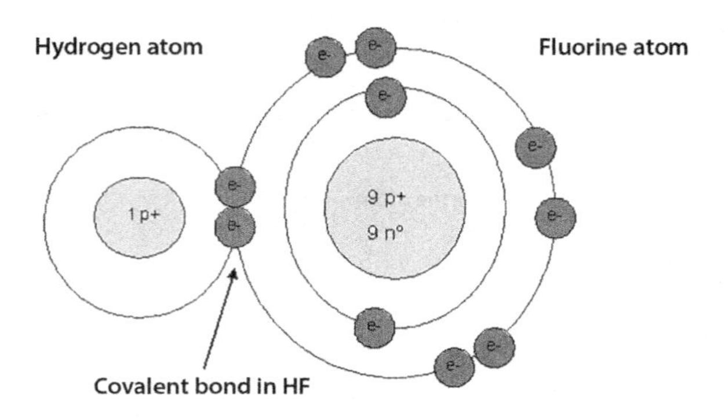

A **covalent bond** forms when at least one pair of electrons is shared by two atoms. The shared electrons are found in the outermost valence energy level and lead to a lower energy if they are shared in a way that creates a noble gas configuration (a full octet). Covalent, or molecular, bonds occur when a non-metal is bonding to another non-metal. This is due primarily to the fact that non-metals have high ionization energies and high electronegativities. Neither atom wants to give up electrons; both want to gain them. In order to fill both octets, the electrons can be shared between the two atoms.

Sharing of electrons can be equal or unequal, resulting in a separation of charge (polar) or an even distribution of charge (non-polar). The polarity of a bond can be determined through an examination of the electronegativities of the atoms involved in the bond. The more electronegative atom will have a stronger attraction to the electrons, thus possessing the electrons more of the time. This results in a partial negative charge (δ^-) on the more electronegative atom and a partial positive charge (δ^+) on the less electronegative atom.

Polar/nonpolar covalent bonds

Electron pairs shared between **two atoms of the same element are shared equally**. At the other extreme, **for ionic bonding there is no electron sharing** because the electron is transferred completely from one atom to the other. Most bonds fall somewhere between these two extremes, and the electrons are **shared unequally (a polar bond)**.

The polarity of a bond can be determined through an examination of the electronegativities of the atoms involved in the bond. The more electronegative atom will have a stronger attraction to the electrons, thus possessing the electrons more of the time. This results in a partial negative charge (δ^-) on the more electronegative atom and a partial positive charge (δ^+) on the less electronegative atom as shown below for gaseous HCl. Such bonds are referred to as **polar bonds**. A particle with a positive and a negative region is called a **dipole**. A lower-case delta (δ) is used to indicate partial charge or an arrow is draw from the partial positive to the partial negative atom.

$$\overset{\delta+}{\text{H}}\text{—}\overset{\delta-}{\text{Cl}} \qquad \overset{\longrightarrow}{\text{H—Cl}}$$

The simplest covalent bond is the nonpolar covalent bond between the two single electrons of hydrogen atoms. Covalent bonds may be represented by an electron pair (a pair of dots) or a line as shown below. The shared pair of electrons provides each H atom with two electrons in its valence shell (the 1s orbital), so both have the stable electron configuration of helium.

$$\text{H}\cdot + \cdot\text{H} \longrightarrow \text{H:H} \quad \text{H—H}$$

Chlorine molecules have seven electrons in their valence shell and share a pair of electrons so both have the stable electron configuration of argon.

$$:\ddot{\underset{..}{\text{Cl}}}\cdot + \cdot\ddot{\underset{..}{\text{Cl}}}: \longrightarrow :\ddot{\underset{..}{\text{Cl}}}:\ddot{\underset{..}{\text{Cl}}}: \quad :\ddot{\underset{..}{\text{Cl}}}\text{—}\ddot{\underset{..}{\text{Cl}}}:$$

In the previous two examples, a single pair of electrons was shared, and the resulting bond is referred to as a **single bond**. When two electron pairs are shared, two lines are drawn, representing a **double bond**, and three shared pairs of electrons represents a **triple bond** as shown below for CO_2 and N_2. The remaining electrons are in **unshared pairs**.

$$\ddot{\underset{..}{\text{O}}}::\text{C}::\ddot{\underset{..}{\text{O}}} \qquad \ddot{\underset{..}{\text{O}}}=\text{C}=\ddot{\underset{..}{\text{O}}} \qquad :\text{N}\vdots\vdots\text{N}: \qquad :\text{N}\equiv\text{N}:$$

3.3 Metallic bonds

Metallic bonds occur when the bonding is between two metals. Metallic properties, such as low ionization energies, conductivity, and malleability, suggest that metals possess strong forces of attraction between atoms, but still have electrons that are able to move freely in all directions throughout the metal. This creates a "sea of electrons" model where electrons are quickly and easily transferred between metal atoms. In this model, the outer shell electrons are free to move. The metallic bond is the force of attraction that results from the moving electrons and the positive nuclei left behind. The strength of metallic bonds usually results in regular structures and high melting and boiling points.

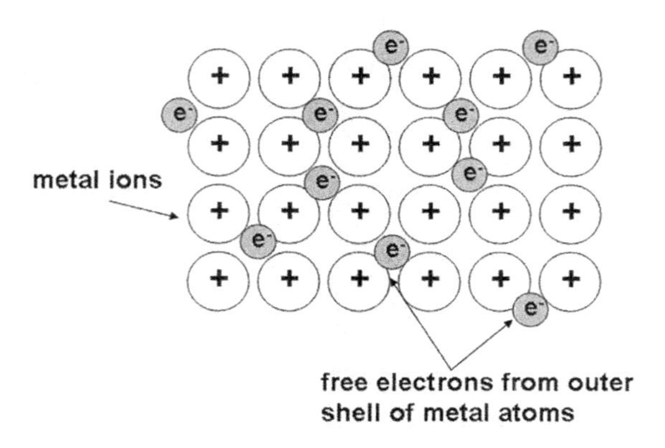

The physical properties of metals are attributed to the electron sea model of metallic bonds shown on the right. Metals **conduct heat and electricity** because their electrons are not incorporated into specific bonds between specific atoms. Rather, they are able to flow through the material. They are called **delocalized** electrons. Metals are **lustrous** because electrons at their surface reflect light at many different wavelengths.

Metals are **malleable** and **ductile** because the electrons are able to rearrange their positions to maintain the integrity of the solid when the metallic lattice is deformed, acting like glue between the cations. The strengths of different metallic bonds can be related to the relative amounts and positions of electrons present.

Alkali metals contain only one valence electron ("less glue"), and that electron is a considerable distance away from the nucleus ("weaker glue"). The result is a weak metallic bond and a low melting point. Heavier alkali metals contain a valence electron even further from the nucleus, resulting in a very weak metallic bond and a lower melting point. With two valence electrons and smaller atoms, alkaline earth metals form stronger metallic bonds than the alkali metals.

The metal with the weakest metallic bonds is mercury. Hg is a liquid at room temperature because Hg atoms hold on tightly to a stable valence configuration of full *s*, *f*, and *d* subshells. Fewer electrons are shared to create bonds than in other metals.

The reactivity of metals increases with lower electronegativity in reactions with nonmetals to form ionic bonds.

Polar/nonpolar covalent bonds

Electron pairs shared between **two atoms of the same element are shared equally**. At the other extreme, **for ionic bonding there is no electron sharing** because the electron is transferred completely from one atom to the other. Most bonds fall somewhere between these two extremes, and the electrons are **shared unequally (a polar bond)**.

The polarity of a bond can be determined through an examination of the electronegativities of the atoms involved in the bond. The more electronegative atom will have a stronger attraction to the electrons, thus possessing the electrons more of the time. This results in a partial negative charge (δ^-) on the more electronegative atom and a partial positive charge (δ^+) on the less electronegative atom as shown below for gaseous HCl. Such bonds are referred to as **polar bonds**. A particle with a positive and a negative region is called a **dipole**. A lower-case delta (δ) is used to indicate partial charge or an arrow is draw from the partial positive to the partial negative atom.

$$\overset{\delta+}{H}—\overset{\delta-}{Cl} \qquad \overrightarrow{H—Cl}$$

3.4 Intermolecular Bonds

The following bonds are typically weaker than metallic bonds, and are called intermolecular forces.

Hydrogen bonds

Hydrogen bonds are particularly strong dipole-dipole interactions that form between the H-atom of one molecule and an F, O, or N atom of an adjacent molecule. The partial positive charge on the hydrogen atom is attracted to the partial negative charge on the electron pair of the other atom. The hydrogen bond between two water molecules is shown as the dashed line below:

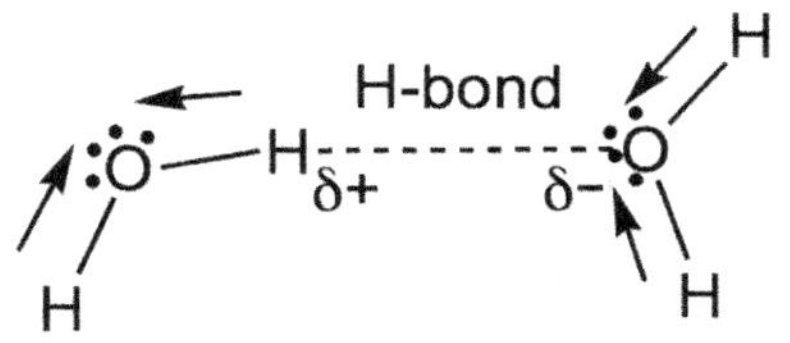

H-bond
H
:Ö—H δ+ - - - - - - - - δ- :O
H H

Dipole-dipole interactions

The intermolecular forces between polar molecules are known as dipole-dipole interactions. The partial positive charge of one molecule is attracted to the partial negative charge of its neighbor.

Ion-induced dipole

When a nonpolar molecule (or a noble gas atom) encounters an ion, its electron density is temporarily distorted resulting in an induced dipole that will be attracted to the ion. Intermolecular attractions due to induced dipoles in a nonpolar molecule are known as London forces or Van der Waals interactions. These are very weak intermolecular forces.

For example, carbon tetrachloride, CCl_4, has polar bonds but is a nonpolar molecule due to the symmetry of those bonds. An aluminum cation will draw the unbonded electrons of the chlorine atom towards it, distorting the molecule and creating an attractive force.

Dipole-induced dipole

The partial charge of **a permanent dipole may also induce a dipole in a nonpolar molecule**, resulting in an attraction similar to but weaker than that created by an ion.

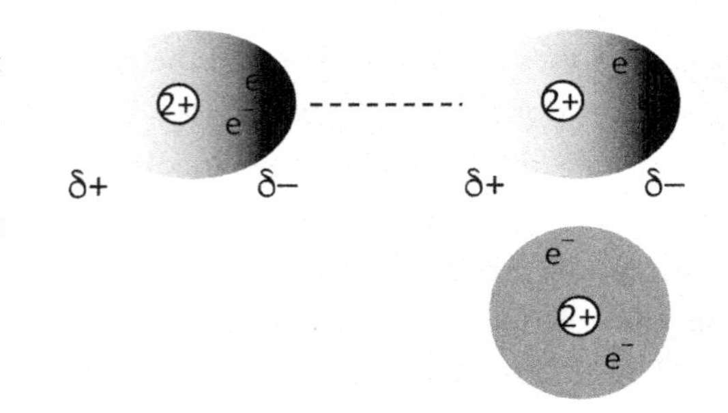

London dispersion force: induced dipole – induced dipole

The above two examples required a permanent charge to induce a dipole in a nonpolar molecule. A nonpolar molecule may also induce a temporary dipole on its identical neighbor in a pure substance. These forces occur because at any given moment, electrons are located within a certain region of the molecule, and the instantaneous location of electrons will **induce a temporary dipole** on neighboring molecules. For example, an isolated helium atom consists of a nucleus with a 2+ charge and two electrons in a spherical electron density cloud. An attraction between He atoms due to London dispersion forces (shown below by the dashed line) occurs when the electrons happen to be distributed unevenly on one atom, inducing a dipole on its neighbor. This dipole is due to intermolecular repulsion of electrons and the attraction of electrons to neighboring nuclei.

The strength of London dispersion forces **increases for larger molecules** because a larger electron cloud is more easily polarized. The strength of London dispersion forces also **increases for molecules with a larger surface area** because there is greater opportunity for electrons to influence neighboring molecules if there is more potential contact between the molecules. Paraffin in candles is an example of a solid held together by weak London forces between large molecules. These materials are soft.

3.5 Relationships of bonding to properties and structures: Intermolecular Forces

Relationships of bonding to properties and structures

In intramolecular bonding at least two factors must be considered and discussed: electron withdrawal and electron release. These result from the operation of two factors, the inductive effect and the resonance effect.

Inductive effect

The inductive effect is dependent upon the "intrinsic" tendency of an atom to release or withdraw electrons- by definition, its electronegativity – acting either through the molecular chain or through space. These effects steadily weaken with increasing distance from the atom. Most elements that are likely to be substituted for hydrogen in organic molecules are more electronegative than hydrogen. So, most substituents exert electron-withdrawing inductive effects: for example, -OH, -F, -Cl, - Br, -I, $-NO_2$, $-NH_2$.

The Resonance effect

The resonance effect involves delocalization of electrons- typically, those called pi electrons. It depends upon overlapping orbitals, and therefore can only operate when substituents are located in certain configurations relative to the charge center. Due to its nature, the resonance effect is a stabilizing effect, and so it amounts to electron withdrawal from a negatively charged center, and electron release to a positively charged center.

Resonance structures are used when one Lewis structure for a single molecule cannot fully describe the bonding that takes place between neighboring atoms relative to the empirical data for the actual bond lengths between those atoms. The net sum of valid resonance structures is defined as a resonance hybrid, which represents the overall delocalization of electrons within the molecule. A molecule that has several resonance structures is more stable than one with fewer. Some resonance structures are more favorable than others.

For example: Ozone resonance

$$:\ddot{O} = \ddot{O} - \ddot{\underset{\cdot\cdot}{O}}:$$

$$\updownarrow$$

$$:\ddot{\underset{\cdot\cdot}{O}} - \ddot{O} = \ddot{O}:$$

Delocalization and Resonance Structures Rules

In resonance structures, the electrons are able to move beyond one bond location to stabilize the molecule. This electron movement is called delocalization.

1. Resonance structures should have the same number of electrons; do not add or subtract any electrons. (Check the number of electrons by simply counting them.)
2. All resonance structures must follow the rules of writing Lewis Structures.
3. The hybridization of the structure should stay the same.
4. The skeleton of the structure cannot be changed and dislocated (only the electrons move).
5. Resonance structures must also all contain the same number of lone pairs.

Impact on physical properties

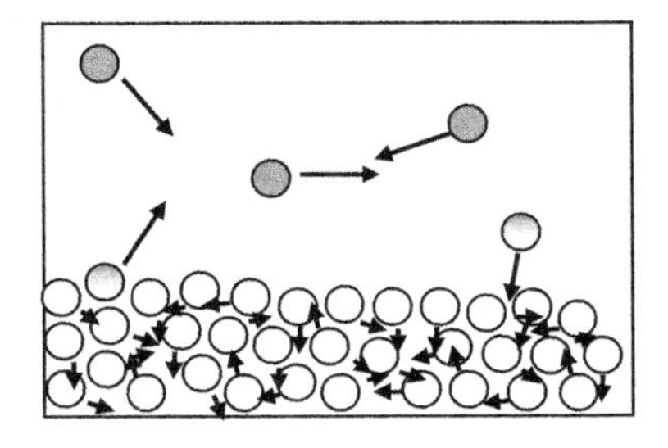

If two substances are being compared, the material with the **greater intermolecular attractive forces** (i.e. the stronger intermolecular bond) will require more energy to pull apart the molecules. Many properties affected by the strength of these forces are discussed in Chapter 5.

For example, H_2O and NH_3 are liquids at room temperature because they contain hydrogen bonds. These bonds are of intermediate strength, so the melting point of these compounds is lower than room temperature and their boiling point is higher than room temperature. H_2S contains weaker dipole-dipole interactions than H_2O because the sulfur atoms do not form hydrogen bonds. Therefore,

H_2S is a gas at room temperature due to its low boiling point. Small non-polar molecules such as CO_2, N_2, or atoms such as He are gases at room temperature due to very weak London forces, but larger non-polar molecules such as octane or CCl_4 may be liquids, and very large non-polar molecules such as paraffin will be soft solids.

Apply electronegativity to bond type.

Electronegativity is a measure of the ability of an atom to attract electrons in a chemical bond. Metallic elements have low electronegativities and nonmetallic elements have high electronegativities. Relative values of electronegativity are shown in the table below. Comparing the values allows us to determine the polarity of a bond, as described below.

H 2.2							
Li 1.0	Be 1.6	B 1.8	C 2.5	N 3.0	O 3.4	F 4.0
Na 0.9	Mg 1.3	Al 1.6	Si 1.9	P 2.2	S 2.6	Cl 3.2

Linus Pauling developed the concept of electronegativity and defined its relationship to different types of bonds in the 1930s.

A **large electronegativity difference** (greater than 1.7) results in an **ionic bond**. Any bond composed of two different atoms will be slightly polar, but for a **small electronegativity difference** (less than 0.4), the distribution of charge in the bond is so nearly equal that the result is called a **nonpolar covalent bond**. An **intermediate electronegativity difference** (from 0.4 to 1.7) results in a **polar covalent bond**. HCl is polar covalent because Cl has a very high electronegativity (it is near F in the periodic table) and H is a nonmetal (and so it will form a covalent bond with Cl), but H is near the dividing line between metals and nonmetals, so there is still a significant electronegativity difference between H and Cl. Using the numbers in the table above, the electronegativity for Cl is 3.2 and it is 2.2 for H. The difference of 3.2 : 2.2 = 1.0 places this bond in the middle of the range for polar covalent bonds.

Bond type is actually a continuum as shown in the following chart for common bonds. Note that the **C-H bond** is considered **nonpolar**.

Type of bonding	Electronegativity difference	Bond
Very ionic		Fr^+—F^-
		Na^+—F^-
	3.0	
Ionic		Na^+—Cl^-
	2.0	Na^+—Br^-
Mostly ionic		Na^+—I^-
Mostly polar covalent	1.5	$C^{\delta+}$—$F^{\delta-}$
		$H^{\delta+}$—$O^{\delta-}$
Polar covalent	1.0	$H^{\delta+}$—$Cl^{\delta-}$
		$C^{\delta+}$=$O^{\delta-}$
		$H^{\delta+}$—$N^{\delta-}$
		$C^{\delta+}$—$Cl^{\delta-}$
	0.5	$C^{\delta+}$≡$N^{\delta-}$
Mostly nonpolar covalent		C—H
Fully nonpolar covalent	0	H_2, N_2, O_2, F_2, Cl_2, Br_2, I_2, C—C, S—S

Increasing ionic character →

SECTION 1 PRACTICE QUESTIONS

CHAPTER 1

1. **Which of the following is the electron configuration of scandium?**

   (A) $1s^2$ $2s^2$ $2p^6$ $3s^2$ $3p^6$ $3d^1$ $4s^2$
   (B) $1s^2$ $2s^2$ $2p^6$ $3s^2$ $3p^6$ $3d^2$ $4s^1$
   (C) $1s^2$ $2s^2$ $2p^6$ $3s^2$ $3p^6$ $3d^3$
   (D) $1s^2$ $2s^2$ $2p^6$ $3s^2$ $3p^6$ $4s^2$ $3d^1$
   (E) $1s^2$ $2s^2$ $2p^6$ $3s^2$ $3p^6$ $3d^4$

   The correct answer is D.
   Recall that the 4s shell fills before the 3d.

Questions 2-4 refer to the following

(A) increase
(B) decrease
(C) stays the same

2. **The number of neutrons across the periodic table: The correct answer is A.**

3. **The electronegativity across the periodic table: The correct answer is A.**

4. **The atomic radius down a column of the periodic table: The correct answer is A**

5. **Cause and Effect**

   Statement I—True or False?
   Isotopes of the same element differ in atomic number
   This statement is False. Isotopes have the same number of protons and therefore the same atomic number.

   Statement II—True or False?
   The mass number reflects the number of isotopes of an element.
   This statement is True.

CHAPTER 2

1. Which of the following is a proper Lewis dot structure of CHClO?

(A)

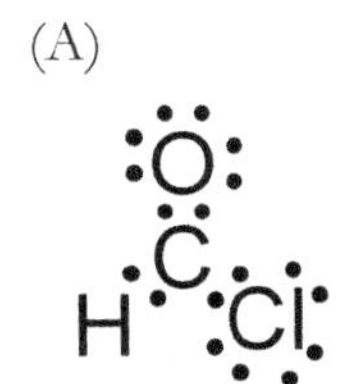

(B)

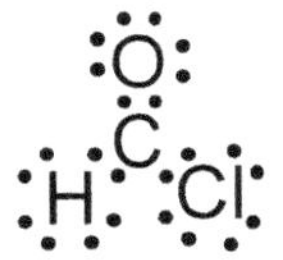

(C)

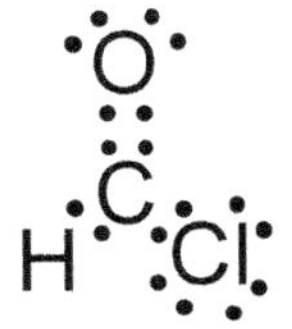

(D)

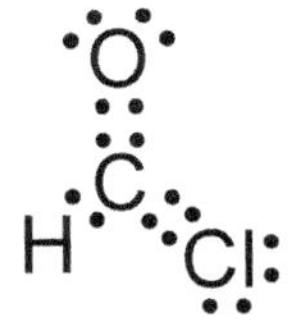

(E) None of the above

The correct answer is C.

C has 4 valence shell electrons, H has 1, Cl has 7, and O has 6. The molecule has a total of 18 valence shell electrons. This eliminates choice B which has 24. Choice B is also incorrect because it has an octet around a hydrogen atom instead of 2 electrons and because there are only six electrons surrounding the central carbon. A single bond connecting all atoms would give choice A. This is incorrect because there are only 6 electrons surrounding the central carbon. A double bond between C and O gives the correct answer, C. A double bond between C and O and also between C and Cl would give choice D. This is incorrect because there are 10 electrons surrounding the central carbon.

Questions 2-4 refer to the following:

(A) Benzene
(B) H_2O
(C) O_3
(D) CH_4
(E) None of the above

2. **Is aromatic: The correct answer is A.**

3. **Has 18 valence electrons: The correct answer is C.**

4. **Has a "bent" configuration: The correct answer is B.**

5. **Cause/Effect:**

   Statement I—True or False?
   CH_4 has a tetrahedral shape
   This statement is True.

   Statement II—True or False?
   Carbon has 4 unpaired electrons to use in bonding
   This statement is True.
   There is a cause and effect relationship.

CHAPTER 3

1. **Which of the following statements regarding ionic compounds are TRUE?**

   (A) They have high melting points
   (B) They do not conduct electricity
   (C) They are brittle
   (D) They are formed between elements with large differences in electronegativity
   (E) All of the Above

   The correct answer is E.
   All of the statements are true.

Questions 2-4 refer to the following:

   (A) Ionic Bonds
   (B) Covalent Bond
   (C) Hydrogen bonding
   (D) London Forces
   (E) Metallic Bonds

2. **NaCl: The correct answer is A.**

3. **CH_4: The correct answer is B.**

4. **H_2O: The correct answer is C.**

5. **Cause/Effect:**

   Statement I—True or False?
   For ionic bonding there is no electron sharing because the electron is transferred completely from one atom to the other.
   This statement is True.

   Statement II—True or False?
   When the electrons are shared unequally they form a polar bond.
   This statement is True.
   Both Statements are True but there is no Cause and Effect Relationship.

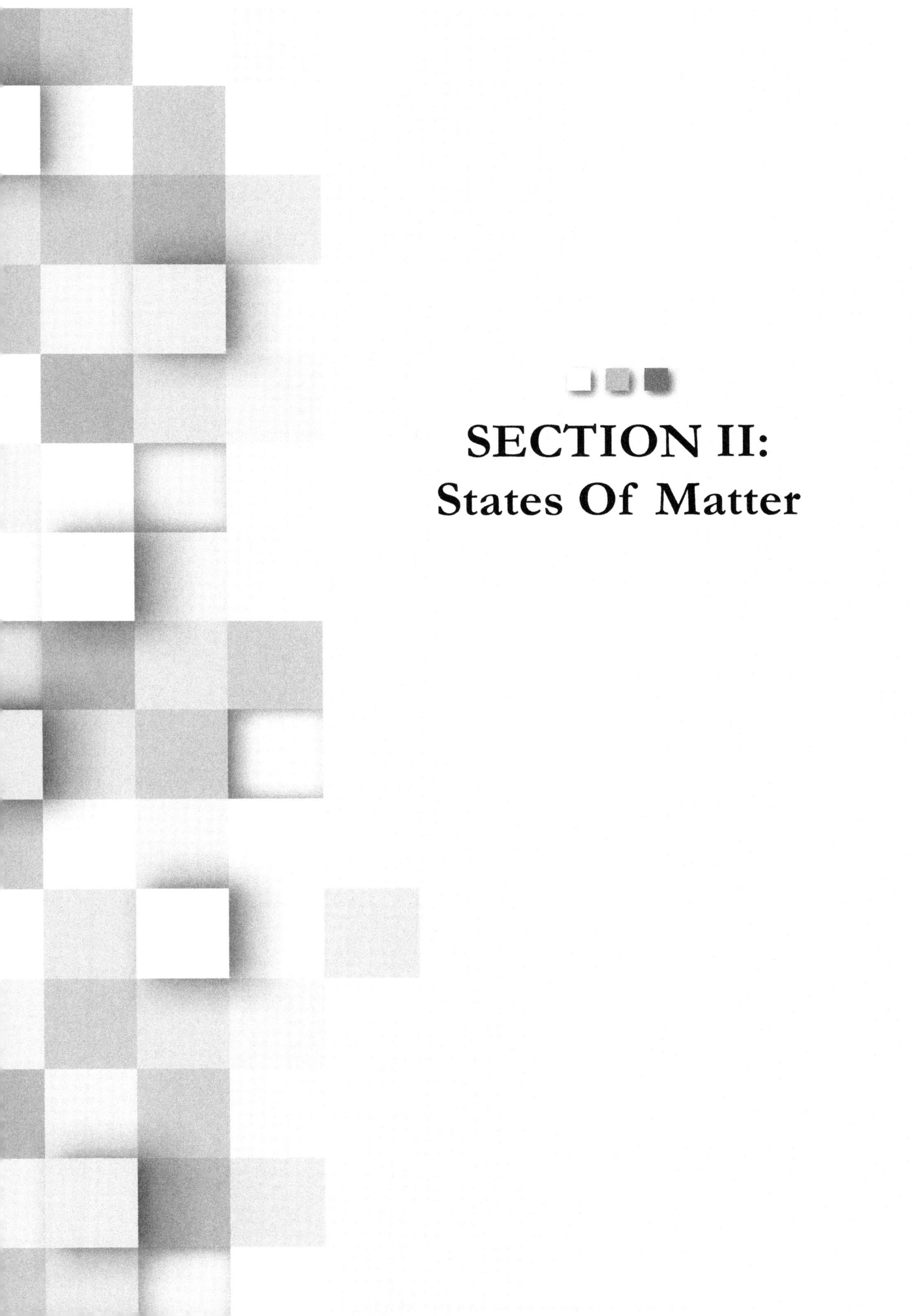

SECTION II: States Of Matter

Chapter 4: Gases

4.1 Kinetic Molecular Theory

Kinetic molecular theory says that the particles of a substance are in constant random motion. What causes this motion? Energy.

Heat energy is absorbed by atoms and molecules making up a substance and this causes the atoms and molecules to move more. In the solid state, the motion is little more than a vibration. During melting, this vibration separates the atoms from each other, moving them apart, decreasing the force of attraction between the atoms or molecules, causing a phase change to liquid. As a liquid, the molecules still have attractive forces acting between them. The addition of more heat forces the atoms or molecules farther apart, weakening the forces to the point where they can no longer keep the atoms or molecules together. At this point, the substance has entered the gas phase.

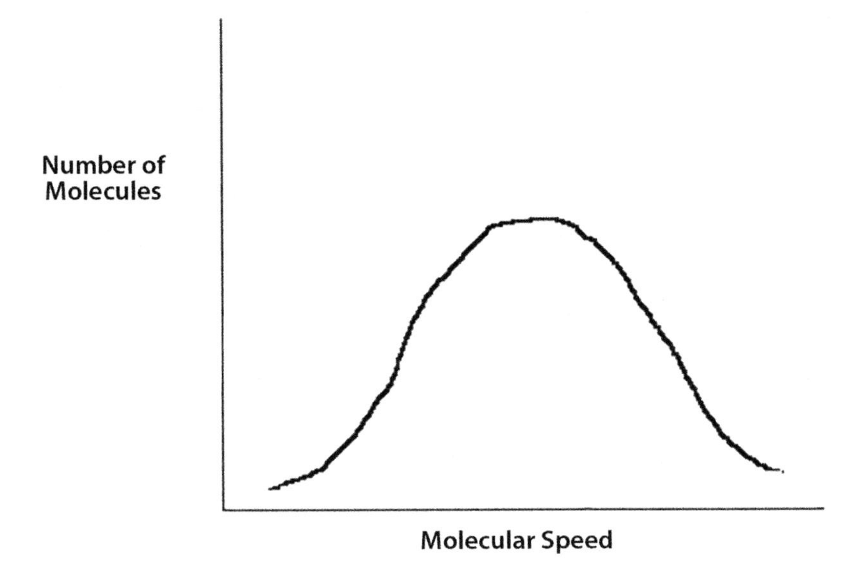

How is the heat energy of the molecule measured? No device will measure heat energy directly. However, a property called **temperature** can give us an idea of the heat energy of the atoms or molecules.

Temperature does not measure the individual heat of the molecules but rather how fast the molecules are moving (kinetic energy) which is tied to the heat possessed by the molecules.

A thermometer in a beaker of water will determine the temperature of the water. If you place the beaker on a hot plate and slowly add heat, the temperature will increase. If you place the beaker in ice water, the temperature will decrease. Why does the thermometer respond this way?

As heat energy is added to a system, the matter absorbs the heat. This increased heat causes the atoms or molecules to move more (increase in kinetic energy). Each water molecule has a kinetic energy equal to $1/2\ m \cdot v^2$.

As the molecules move about they will bump into one another, the walls of the beaker and the thermometer. Some of the energy they possess is transferred to the bulb of the thermometer. The energy is transferred to the glass and then to the liquid inside the thermometer. The liquid is made of molecules. The increased energy of the molecules causes them to move faster and increase the liquid's volume. With only one place to go, the liquid moves up through the space in the stem. The thermometer can't measure the individual energy of one molecule. It responds to all the molecules striking it. Some are moving faster than others. The observed temperature is the average kinetic energy of all the molecules striking the thermometer. The **temperature** of the water is a **measure of the average kinetic energy** of the water molecules involved in the process. How did the molecules get their kinetic energy? Heat.

The kinetic energy of molecules is unaltered during phase changes, but the freedom of molecules to move relative to one another increases dramatically. The following table summarizes the application of kinetic molecular theory to the addition of heat to ice, first changing it to liquid water and then to water vapor.

Effect at 1 atm of the addition of heat to:	**0** = no change, + = increase, ++ = strong increase			
	Temperature	Average speed of molecules	Average kinetic energy of molecules	Intermolecular freedom of motion
Ice at less than 0 °C	+	+	+	+
Ice at 0 °C	**0**	**0**	**0**	++ (melting)
Liquid water at 0 °C	+	+	+	+
Liquid water at 100 °C	**0**	**0**	**0**	++ (boiling)
Water vapor at 100 °C	+	+	+	**0** (complete freedom for an ideal gas)

Technically, heat is the measure of the internal energy gained or lost when two objects come in contact with one another and the temperature difference between the two objects is the measure of heat since it can not be directly determined.

The heat lost or gained is what is calculated.

Heat Lost = Heat Gained

$$m_{lost} * C_{lost} * \Delta T_{lost} = m_{gained} * C_{gained} * \Delta T_{gained}$$

where m= mass of substance undergoing temperature change,
c= heat capacity (specific heat) of the substance undergoing temperature change
and ΔT = change in temperature.

4.2 Gas law relationships

Gas **pressure** results from molecular collisions with container walls. The **number of molecules** striking an **area** on the walls and the **average kinetic energy** per molecule are the only factors that contribute to pressure. A higher **temperature** increases speed and kinetic energy. There are more collisions at higher temperatures, but the average distance between molecules does not change, and thus density does not change in a sealed container.

Kinetic molecular theory explains how pressure and temperature influence the behavior of gases, but in so doing makes a few assumptions, namely:

1. The energies of intermolecular attractive and repulsive forces may be neglected.
2. The average kinetic energy of the molecules is proportional to absolute temperature.
3. Energy can be transferred between molecules during collisions and the collisions are elastic, so the average kinetic energy of the molecules doesn't change due to collisions.
4. The volume of all molecules in a gas is negligible compared to the total volume of the container.

Strictly speaking, molecules also manifest some kinetic energy by rotating or vibrating. The motion of a molecule from one place to another is called **translation**. Translational kinetic energy is the form that is transferred by collisions, and kinetic molecular theory ignores other forms of kinetic energy because they are not proportional to temperature.

The following table summarizes the application of kinetic molecular theory to an increase in container volume, number of molecules, and temperature:

Effect of an *increase* in one variable with the other two constant	Impact on gas: – = decrease, 0 = no change, + = increase						
	Average distance between molecules	Density in a sealed container	Average speed of molecules	Average translational kinetic energy of molecules	Collisions with container walls per second	Collisions per unit area of wall per second	Pressure (*P*)
Volume of container (*V*)	+	–	0	0	–	–	–
Number of molecules	–	+	0	0	+	+	+
Temperature (*T*)	0	0	+	+	+	+	+

Additional details on the kinetic molecular theory may be found at http://hyperphysics.phy-astr.gsu.edu/hbase/kinetic/ktcon.html. An animation of gas particles colliding is located at http://comp.uark.edu/~jgeabana/mol_dyn/.

The kinetic molecular theory describes how an ideal gas behaves when conditions such as temperature, pressure, volume or quantity of gas are varied within a system. An **ideal gas** is an imaginary gas that obeys all of the assumptions of the kinetic molecular theory. While an ideal gas does not exist, most gases will behave like an ideal gas except when at very low temperatures or very high pressures.

Charles's law states that the volume of a fixed amount of gas at constant pressure is directly proportional to absolute temperature, or increasing the temperature causes a gas to expand, in a mathematically proportional manner:

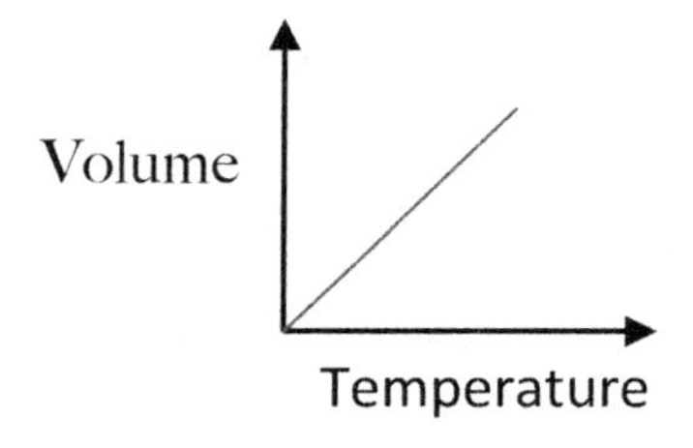

Or $V=kT$ where k is a constant.

This gives a mathematical equation:

$$\frac{V_1}{T_1} = \frac{V_2}{T_2}$$

Changes in temperature or volume can be found using Charles's law.

Example: What is the new volume of gas if 0.50 L of gas at 25°C is allowed to heat up to 35°C at constant pressure?

Solution: This is a volume-temperature change so use Charles's law. Temperature must be on the Kelvin scale. K= °C + 273.

T_1 = 298K (273 + 25)

V_1 = 0.50 L

T_2 = 308K (273 + 35)

V_2 = ?

Use the equation: $V_1 / T_1 = V_2 / T_2$ and rearrange for $V_2 = V_1 / T_1 * T_2$.

Substitute and solve V_2 = 0.52 L.

Boyle's law states that the volume of a fixed amount of gas at constant temperature is inversely proportional to the gas pressure, or increasing the pressure causes a gas to contract in a mathematically proportional manner:

$$V \propto 1 / P$$

Or $V = k/P$ where k is a constant. This gives a mathematical equation $P_1V_1 = P_2V_2$.

Pressure or volume changes (at a constant temperature) can be determined using Boyle's law.

Example: A 1.5 L gas has a pressure of 0.56 atm. What will be the volume of the gas if the pressure doubles to 1.12 atm at constant temperature?

Solution: This is a pressure-volume relationship at constant temperature so use Boyle's law.

P_1 = 0.56 atm

V_1 = 1.5 L

P_2 = 1.12 atm

V_2 = ?

Use the equation $P_1V_1 = P_2V_2$, rearrange to solve for $V_2 = P_1V_1/P_2$.

Substitute and solve. V_2 = 0.75 L

Gay-Lussac's law states that the pressure of a fixed amount of gas in a fixed volume is proportional to absolute temperature. In other words, increasing the temperature causes the pressure to increase in a mathematically proportional manner:

$$P \propto T$$

Or $P = kT$ where k is a constant. This gives a mathematical equation $P_1 / T_1 = P_2 / T_2$.

Changes in temperature or pressure (with a constant volume) can be found using Gay-Lussac's law.

Example: A 2.25 L container of gas at 25°C and 1.0 atm pressure is cooled to 15°C. How does the pressure change if the volume of gas remains constant?

Solution: This is a pressure-volume change so use Gay-Lussac's law.

P_1 = 1.0 atm

T_1 = 298 K (273 +25)

T_2 = 288 K (273 + 15)

Change the temperatures to the Kelvin scale. K = °C + 273.

Use the equation $P_1 / T_1 = P_2 / T_2$ to solve. Rearrange the equation to solve for P_2, substitute and solve.

$$P_2 = \frac{P_1 T_2}{T_1} = 0.97 \text{ atm}$$

The **combined gas law** uses the above laws to determine a proportionality expression that is used for a constant quantity of gas:

$$V \propto \frac{T}{P}.$$

The combined gas law is often expressed as an equality between identical amounts of an ideal gas at two different states ($n_1 = n_2$):

$$\frac{P_1 V_1}{T_1} = \frac{P_2 V_2}{T_2}$$

Example: 1.5 L of a gas at STP is allowed to expand to 2.0 L at a pressure of 2.5 atm. What is the temperature of the expanded gas?

Solution: Since pressure, temperature and volume are changing use the combined gas law to determine the new temperature of the gas.

P_1 = 1.0 atm

T_1 = 273K

V_1 = 1.5 L

V_2 = 2.0L

P_2 = 2.5 atm

T_2 = ?

Using this equation, rearrange to solve for T_2:

$T_2 = P_2 V_2 T_1 / P_1 V_1$

Substitute and solve T_2 = 910 K or, after subtracting 273: 637 °C

Avogadro's hypothesis states that equal volumes of different gases at the same temperature and pressure contain equal numbers of molecules. **Avogadro's law** states that the volume of a gas at constant temperature and pressure is directly proportional to the quantity of gas, or:

$V \propto n$ where *n* is the number of moles of gas.

Avogadro's law and the combined gas law yield:

$PV \propto nT$

The proportionality constant *R*--the **ideal gas constant**--is used to express this proportionality as the **ideal gas law**:

$PV = nRT$

The ideal gas law (PV = nRT) is useful because it contains all the information of Charles's, Avogadro's, Boyle's, and the combined gas laws in a single equality.

If pressure is given in atmospheres and volume is given in liters, a value for *R* of **0.08206 L-atm/(mol-K)** is used. If pressure is given in Pascal (newtons/m^2) and volume in cubic meters, then the SI value for *R* of **8.314 J/(mol-K)** may be used. This is because a joule is defined as a Newton-meter. A value for *R* of **8.314 m^3-Pa/(mol-K)** is identical to the ideal gas constant using joules.

Many problems are given at "**standard temperature and pressure**" or "**STP.**" Standard conditions are *exactly* **1 atm** (101.325 kPa) and **0 °C (273.15 K)**. At STP, one mole of an ideal gas has a volume of 22.4 L.

The value of 22.4 L is known as the **standard molar volume of any gas at STP**.

Solving gas law problems using these formulas is a straightforward process of algebraic manipulation. **Errors commonly arise from using improper units**, particularly for the ideal gas constant *R*. An absolute temperature scale must be used (never °C) and is usually reported using the Kelvin scale, but volume and pressure units often vary from problem to problem. Temperature in Kelvin is found from:

$K = °C + 273.15$

Example: What volume will 0.50 mole of an ideal gas occupy at 20.0 °C and 1.5 atm?

Solution: Since the problem deals with moles of gas with temperature and pressure, use the ideal gas law to find volume.

$PV = nRT$

$V = nRT/ P$

$V = 0.50 \text{ mol } (0.0821 \text{ atm L/mol K}) \; 293 \text{ K}/1.5 \text{ atm}$

$V = 8.0 \text{ L}$

Example: At STP, 0.250 L of an unknown gas has a mass of 0.491 g. Is the gas SO_2, NO_2, C_3H_8, or Ar? Support your answer.

Solution: Identify what is given and what is asked for.

Given: T = 273K

P =1.0 atm

V = 0.250 L

Mass = 0.419 g

Determine: Identity of the gas. In order to do this, must find molar mass of the gas. Find the number of moles of gas present using PV = nRT and then determine the MM to compare to choices given in the problem.

Solve for $n = \frac{PV}{RT}$ = 0.011 moles

MM = 38.1 g/mol

Compare to MM of SO_2 (96 g/mol), NO_2 (46 g/mol), C_3H_8 (44 g/mol) and Ar (39.9 g/mol). It is closest to Ar, so the gas is probably Argon.

4.3 Molar volumes

Molar volume

This is the volume occupied by one mole of a pure substance. Molar volume depends on the density of a substance and, like density, varies with temperature, owing to thermal expansion, and also with pressure. For solids and liquids, these variables ordinarily have little practical effect, so the values quoted for 1 atm pressure and 25°C are generally useful over a fairly wide range of conditions. This is definitely not the case with gases, whose molar volumes must be calculated for a specific temperature and pressure.

Problem Example 1: Molar volume of a liquid

Methanol, CH_3OH, is a liquid having a density of 0.79 g per milliliter. Calculate the molar volume of methanol.

Solution: The molar volume will be the volume occupied by one molar mass (32 g) of the liquid. Expressing the density in liters instead of mL, we have

$$V_M = (32\ g\ mol^{-1}) / (790\ g\ L^{-1}) = 0.0405\ L\ mol^{-1}$$

Problem Example 2: Molar volume of a liquid

We want to transfer 25 Kg of mercury (liquid metal with a density of 13.546 g cm^{-3}) to a far location and we have only one container of 1.5 liters. Can we transfer the whole amount of mercury? If not, how much of it will be left?

Solution:

The volume of mercury = M/d = (25×1000) g /13.546 = 1895 mL = 1.895 Liters of mercury

So, the amount of mercury that will be left is 1.895 – 1.5 = 0.395 Liters

Molar volume in gases

Molar volume in gases is, by definition, the volume occupied by one mole of a pure gas. From the results of many measurements, the average volume occupied by one mole of a gas at STP is 22.4 L, and this value is therefore taken to be the molar volume of an ideal gas at STP.

Many problems are given at "standard temperature and pressure" or "STP." Standard conditions are exactly 1 atm (101.325 kPa) and 0 °C (273.15 K). At STP, one mole of an ideal gas has a volume of:

$$V = \frac{nRT}{P}$$

$$= \frac{(1\ \text{mole})\left(0.08206\ \frac{\text{L-atm}}{\text{mol-K}}\right)(273\ \text{K})}{1\ \text{atm}} = 22.4\ \text{L}.$$

The value of 22.4 L is known as the standard molar volume of any gas at STP. Some typical values of molar volumes of gases are shown in the Table 1:

Table 1. Molar volumes of several gases at STP

Gas	Molar Volume (liters)
Oxygen, O_2	22.397
Nitrogen, N_2	22.402
Hydrogen, H_2	22.433
Helium, He	22.434
Argon, Ar	22.397
Carbon dioxide, CO_2	22.260
Ammonia, NH_3	22.079

4.4 Density

Solve problems involving an intensive property (e.g., density, specific heat) of matter.

An **intensive property** is one that is independent of the amount of matter present in a sample. Some examples of intensive physical properties include density, specific heat, viscosity, malleability, conductivity, melting point temperature, and boiling point temperature. Luster is a shine, as seen in steel or silver. Conductivity is a measure of how well heat and electricity pass through the matter. A substance that is ductile can be drawn into wire while one that is malleable can be hammered into thin sheets.

Density

Density is a physical property that can be used to help identify a substance. If the density of an object is less than the density of a liquid, the object will float in the liquid. If the object is denser than the liquid, then the object will sink.

Density is defined as mass per unit volume. Density is calculated by dividing the mass of an object by its volume. This is shown in equation form, as follows:

Density = mass / volume

The unit for density is grams/cubic centimeters for solids and grams/ milliliters for liquids and gases.

To discover an object's density, first use a balance to find its mass. Then calculate its volume. If the object is a regular shape, you can find the volume by multiplying the length, width, and height together. However, if it is an irregular shape, you can find the volume by seeing how much water it displaces. Measure the water in the container before and after the object is submerged. The difference will be the volume of the object.

Example: A sample of kerosene has a mass of 36.4 g. Its volume is measured to be 45.6 mL. What is the density of the kerosene?

Solution: density = M/V = 36.4 g/45.6 mL = 0.80 g/mL

Example: What is the volume in cubic meters of a femur (thigh bone) with a mass of 1.18 kg? The density of bone is 1.8 kg/m^3.

Solution: Volume = mass/density or 1.18 kg / 1.8 kg/m^3 = 0.66 m^3

Viscosity

Viscosity is a measure of the resistance of a liquid to flow. Liquids with high viscosity flow less easily because they have strong intermolecular forces relative to kinetic energy. The viscosity of liquids decreases with temperature because it is easier for rapidly moving molecules to flow. For most liquids (water is an exception), viscosity increases with pressure because the molecules are squeezed together forcing a greater interaction, but this dependence is not as strong as the dependence on temperature.

Specific Heat will be covered in detail in Chapter 14.

Chapter 5: Liquids and Solids

5.1 Types of solids

A **solid** has a definite volume and definite shape. The kinetic model for a solid is a collection of molecules attracted to each other with sufficient strength to essentially lock them in place. Each molecule may vibrate, but it has an average position relative to its neighbors. If these positions form an ordered pattern, the solid is called crystalline. Otherwise, it is called amorphous. Solids have a high density and are almost incompressible because the molecules are close together. Diffusion occurs extremely slowly because the molecules almost never alter their position.

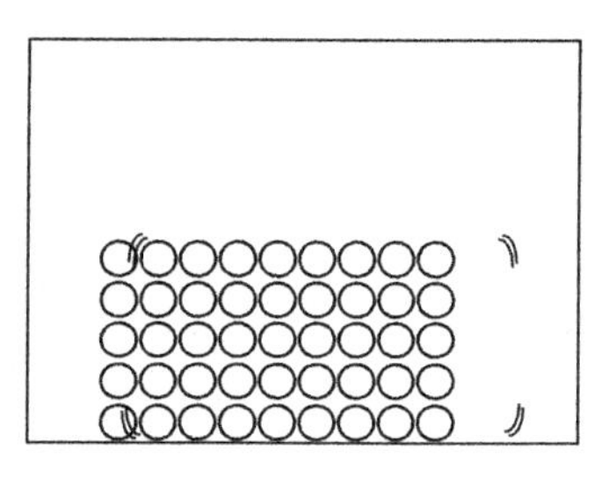

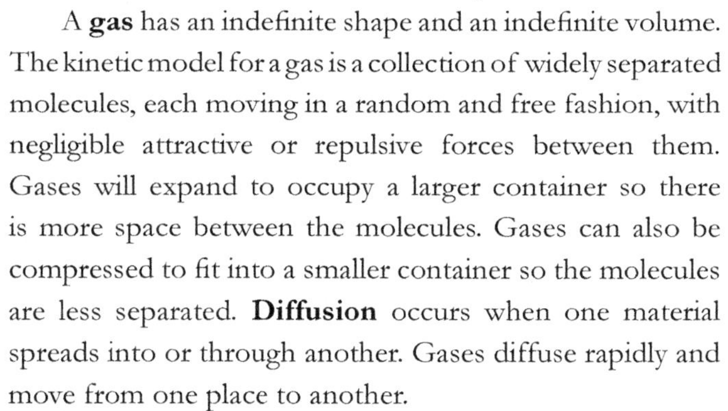

A **gas** has an indefinite shape and an indefinite volume. The kinetic model for a gas is a collection of widely separated molecules, each moving in a random and free fashion, with negligible attractive or repulsive forces between them. Gases will expand to occupy a larger container so there is more space between the molecules. Gases can also be compressed to fit into a smaller container so the molecules are less separated. **Diffusion** occurs when one material spreads into or through another. Gases diffuse rapidly and move from one place to another.

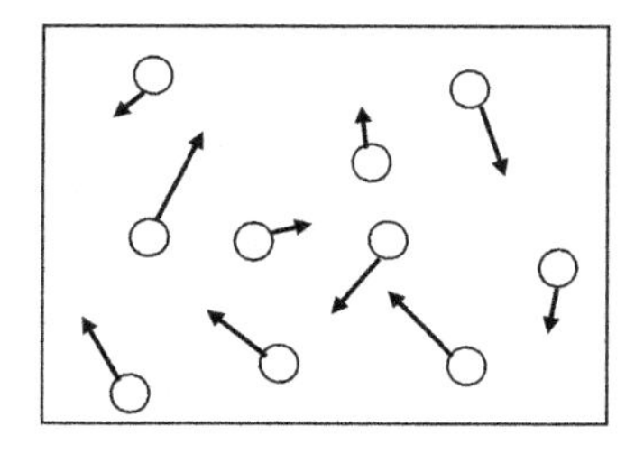

A **liquid** assumes the shape of the portion of any container that it occupies and has a specific volume. The kinetic model for a liquid is a collection of molecules attracted to each other with sufficient strength to keep them close to each other but with insufficient strength to prevent them from moving around randomly. Liquids have a higher density and are much less compressible than gases because the molecules in a liquid are closer together. Diffusion occurs more slowly in liquids than in gases because the molecules in a liquid stick to each other and are not completely free to move.

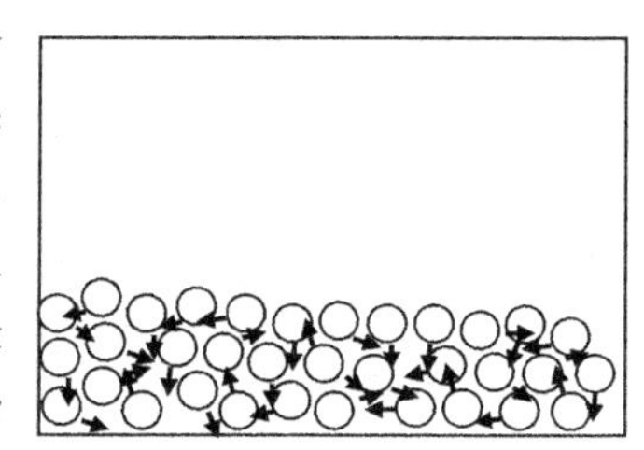

When thinking of these properties, it is useful to imagine analogous groups of people. Whereas a gas is like strangers moving at random, getting close only if they randomly move near each other or if they are compressed in a tight space, a liquid is like a group of friends walking through a park-- they move around, with the shape of the group changing as they do, but they maintain contact with other members of the group so that the group moves as a whole even as individuals are constantly changing who they are talking to or walking next to. Whereas gases and liquids are like random strangers in a space or a group of close friends, respectively, solids are like people seated in a classroom—they can move their bodies, but they don't switch places. If you don't know the answer to a question involving physical states of matter, think of the molecules as tiny people who are either strangers, close friends, or a formal seated group, and you will be in a good position to make an educated guess.

Types of crystalline solids and their properties

There is only a limited number of ways of arranging particles in a crystalline solid. The particular arrangements, as well as the physical properties of the solid, are determined by the types of particles present at the lattice point and the nature of attractive forces between them. As a result, we can divide crystals into distinct types that are each characterizes by certain kinds of properties.

Molecular Crystals

In molecular crystals, either molecules or individual atoms occupy lattice sites. The attractive forces between them are much weaker than the covalent bonds that exist within individual molecules. London forces are present in crystals of nonpolar substances such as Ar, O_2, naphthalene (moth crystals), and CO_2 (dry ice). In crystals of polar molecules like SO_2, there are also dipole-dipole attractions, and in solids such as ice (H_2O), NH_3, and HF, the molecules are held in place primarily by hydrogen bonding. Since these are relatively weak forces (compared to ionic or covalent attractions), molecular crystals tend to have smaller lattice energies and are easily deformed; we say that they are soft. Also, relatively little thermal energy is required to overcome these attractions, and molecular solids tend to have low melting points.

Molecular crystals are poor conductors of electricity because the electrons are bound to individual molecules and are not able to move freely through the solid.

Ionic Crystals

In an ionic crystal, such as NaCl, ions are located at lattice sites and binding between them is mainly electrostatic (which is essentially nondirectional). As a result, the kind of lattice that is formed is determined mostly by the relative sizes of the ions and their charges. When the crystal forms, the ions arrange themselves to maximize attractions and minimize repulsions.

Because electrostatic forces are strong, ionic forces have large lattice energies. They are often hard and are characterized by relatively high melting points. They are also very brittle. When struck, they tend to shatter, because as planes of ions slip by one another, they pass from a condition of mutual attraction to one of mutual repulsion.

In the solid state, ionic compounds are poor conductors of electricity because the ions are held rigidly in place. When melted, however, the ions are free to move about and ionic substances become good conductors.

Covalent Crystals

In a covalent crystal there is a network of covalent bonds between the atoms that extends throughout the entire solid. An example of such a substance is diamond. Diamond is a form of elemental carbon in which each atom is covalently bonded to four neighbors. Other common examples are carborundum (silicon carbide, SiC) and quartz (silicon dioxide, SiO_2, commonly recognized as the major constituent of many sands).

Because of the interlocking framework of covalent bonds, covalent crystals have very high melting points and are usually extremely hard. Diamond, of course, is the hardest substance known and is used in grinding and cutting tools. Silicon carbide is like diamond, except that half of the carbon atoms in the structure have been replaced by silicon atoms. It too is very hard and is used as an abrasive in sandpaper, as well as in other grinding and cutting applications.

Covalent crystals are poor conductors of electricity because the electrons in the solid are localized in the covalent bonds and are not free to move through the crystal.

Metallic crystals

The simplest picture of a metallic crystal has positive ions (nuclei plus core electrons) situated at lattice points, with the valence electrons belonging to the crystal as a whole instead of to any single atom. The solid is held together by the electrostatic attraction between the lattice of positive ions and this sort of "sea of electrons." These electrons can move freely, so we find metals to be good conductors of electricity. Since the melting point and hardness of metals vary over wide ranges, there must also be at least some degrees of covalent bonding between atoms in the solid.

Amorphous solids

Some substances, such as glass, rubber, and many plastics, never do achieve a crystalline state when their liquids solidify on cooling. These compounds consist of long chainlike molecules that intertwine. As they are cooled, their molecules move so slowly that they never do find the proper orientation to form a crystalline solid and an amorphous solid will result instead. The term amorphous comes from the Greek word meaning "without shape".

5.2 Intermolecular Forces

The impact of intermolecular forces on substances is best understood by imagining ourselves shrinking down to the size of molecules and picturing what happens when we stick more strongly to molecules nearby. It will take more energy (higher temperatures) to pull us away from our neighbors.

If two substances are being compared, the material with the **greater intermolecular attractive forces** (i.e. the stronger intermolecular bond) will have the following properties relative to the other substance:

For solids:
Higher melting point
Higher enthalpy of fusion
Greater hardness
Lower vapor pressure

For liquids :
Higher boiling point
Higher critical temperature
Higher critical pressure
Higher enthalpy of vaporization
Higher viscosity
Higher surface tension
Lower vapor pressure

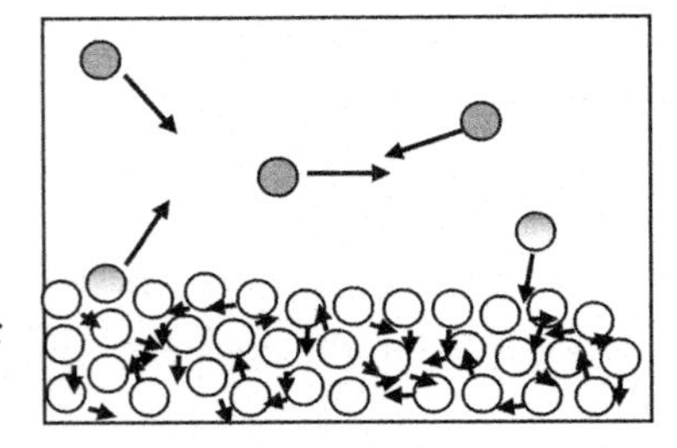

For gases:
Intermolecular attractive forces are neglected for ideal gases.

Vapor Pressure

When a liquid is placed in an open container or a closed container that is not entirely filled, there are always some molecules at the surface of the liquid (e.g., the half-shaded molecules to the left of the diagram) which have enough kinetic energy to overcome the attraction of their neighbors and escape into the gas above the liquid. This process is known as **evaporation**. In a closed container, these gas molecules develop a pressure until a dynamic equilibrium is achieved, at which point the rate of their return to the liquid phase by **condensation** (e.g., the half-shaded molecule on the right in the diagram) equals the rate of their escape by evaporation:

$$\text{Liquid} \underset{\text{Condensation}}{\overset{\text{Evaporation}}{\longleftrightarrow}} \text{Vapor}$$

At equilibrium, the partial pressure of the substance in the gas phase is at its **saturated vapor pressure**. Solids are also in equilibrium with vapor and have a saturated vapor pressure. There is no real difference between the terms *gas* and *vapor*, but *gas* is often used to describe a substance that appears in the gaseous state under standard temperature and pressure and *vapor* is normally used to describe the gaseous state of a substance that is ordinarily a liquid or solid at standard temperature and pressure.

A dynamic equilibrium consists of two **opposing reversible processes** that both occur at the **same rate**. *Balance* is a synonym for equilibrium. A system at equilibrium is stable; it does not change with time. Equilibria are drawn with a double arrow.

The saturated vapor pressure of a liquid is often simply called its **vapor pressure**. This term can sometimes lead to confusion when equilibrium is not present, but equilibrium is usually assumed.

An increase in temperature raises vapor pressure (making the liquid more **volatile**) because kinetic energy opposes intermolecular attractions and permits more molecules to escape from the liquid phase. See section 5.3 for information on phase changes.

More information on vapor pressure may be found at:

http://hyperphysics.phy-astr.gsu.edu/hbase/kinetic/vappre.html. A flash animation of liquid/vapor equilibrium showing how vapor pressure is measured is located at: http://www.mhhe.com/physsci/chemistry/essentialchemistry/flash/vaporv3.swf.

Surface Tension

In a situation like that described above, there are very few molecules in the gas phase and so the intermolecular attractive force pulling a surface molecule (labeled in the diagram) away from the liquid is very weak, but there are many molecules with intermolecular attractive forces within the liquid phase. This leads to an imbalance of forces and it makes the surface molecule stick to the liquid molecules nearby more strongly. There is a net inwards pull away from the interface to minimize its surface area. This is not the case for molecules in the interior of the liquid where the forces are balanced. **Surface tension** is the energy required to increase the surface area of a liquid by a unit amount. Because of surface tension, friction exists at the liquid-gas interface that makes it more difficult to move a solid object through the surface than to move it when it's completely submerged. It is as if there is a "film" at the interface that makes it more difficult for a solid object to "break through" the surface than to move around when it's already completely submerged.

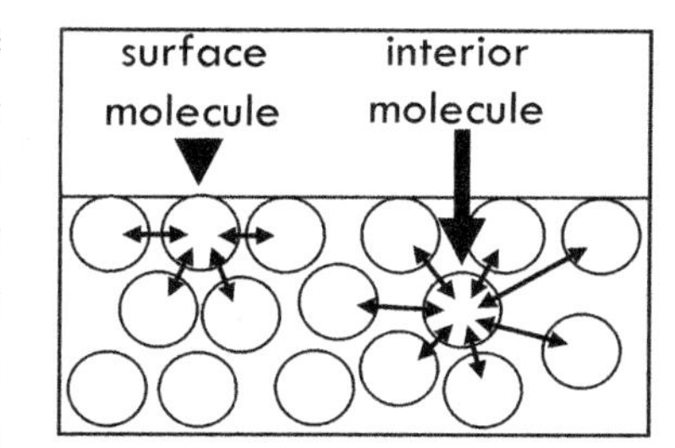

An increase in temperature decreases the surface tension because the kinetic energy acts in opposition to the intermolecular attractive forces. Chemicals with strong intermolecular attractive forces have a high surface tension. Surface tension can also be altered by adding other substances. For example, NaOH added to water will raise its surface tension, and soap added to water lowers surface tension (because the soap acts as a surfactant).

Boiling Point as a Function of Pressure

For a liquid in an open container, vapor pressure increases with temperature until the vapor pressure is equal to the external pressure, and the boiling point occurs at that temperature. Boiling is defined as the process of vapor bubbles forming and escaping from the liquid by breaking the intermolecular attractive forces within the liquid. Substances with stronger intermolecular attractive forces have a higher boiling point. An increase in the surrounding pressure forces molecules closer together and increases their intermolecular attractive forces. More kinetic energy is then required to break these bonds, so the boiling point of a liquid increases with increases in pressure.

Critical Point

See Chapter 5.3 for phase diagrams and a complete discussion on phase changes and the critical point. Rising temperature at a gas-liquid interface increases vapor pressure and decreases surface tension. All the liquid will become a gas if left at the boiling point long enough, but if the external pressure is increased above the vapor pressure, some material will remain in the liquid phase and the boiling point will increase. A pressure cooker is a good example of this. Finally, however, a temperature is reached at which no amount of pressure will keep the material in a liquid state. The highest temperature at which a substance can exist as a liquid is its **critical temperature**. **Critical pressure** is the vapor pressure of a liquid at its critical temperature. Surface tension shrinks to zero and there is no longer a gas-liquid interface when critical conditions are reached.

Above its critical temperature and pressure, a substance takes the shape and fills the volume of its container like a gas, but its density and intermolecular attractive forces are similar to those of a liquid. This phase is called a **supercritical fluid**. Like liquids and gases, they are able to flow from one place to another.

Summary

The following table summarizes the properties of a liquid as temperature and pressure are altered. The speed and kinetic energy of molecules are only dependent on temperature.

Effect on a liquid of an increase in one variable with the other constant	– = decrease, 0 = no change, + = increase, NA=not applicable					
	Average speed of molecules	Average translational kinetic energy of molecules	Viscosity	Vapor pressure	Surface tension	Boiling point
Temperature	+	+	–	+	–	NA
External pressure	0	0	+/–[1]	NA[2]	NA[2]	+

[1]A slight increase for most materials but a slight decrease for water at some temperatures.

[2]Not applicable. For a pure substance in a closed container at equilibrium, external pressure forces more vapor into the liquid phase. The volume of each phase is altered but conditions at the interface remain unchanged.

5.3 Phase changes

Phase changes occur when the relative importance of kinetic energy and intermolecular forces is altered sufficiently for a substance to change its state. The kinetic energy within the molecule changes as heat is added or removed.

In a solid, the energy of the intermolecular attractive forces (such as ionic or network covalent bonds) is much stronger than the kinetic energy of the molecules (the vibrational energy within the molecules themselves). As temperature increases in a solid, the vibrations of individual molecules grow more intense and the molecules spread slightly further apart, decreasing the density of the solid.

In a liquid, the energy of the intermolecular attractive forces (such as dipole-dipole and London dispersion forces) is about as strong as the kinetic energy of the molecules. Therefore, both play a role in the properties of liquids. Liquids will be discussed in detail later.

In a gas, the energy of intermolecular forces is much weaker than the kinetic energy of the molecules. Kinetic molecular theory is usually applied to gases.

Substances with strong intermolecular attractive forces will remain solid even when kinetic energy is high, whereas those with weak intermolecular attractive forces will be gases even when kinetic energy is low, and those with moderate intermolecular attractive forces will fall in between. For example, at 100°C, water is boiling into a gas, but the steel pot containing the water is nowhere near melting, let alone evaporating. At 0°C, water is transitioning into ice (a solid), but oxygen will still be a gas at 100 degrees below this.

The transition from gas to liquid is called **condensation** and from liquid to gas is called **vaporization**. The transition from liquid to solid is called **freezing** and from solid to liquid is called **melting.** The transition from gas to solid is called **deposition** and from solid to gas is called **sublimation**.

Interpreting a heating/cooling curve of a substance.

Kinetic models are used to describe gases, liquids, and solids.

Heat removed from a substance during condensation, freezing, or deposition permits new intermolecular bonds to form, and heat added to a substance during vaporization, melting, or sublimation breaks intermolecular bonds. During these phase transitions, this **latent heat** is removed or added with **no change in the temperature** of the substance because the heat is not being used to alter the speed of the molecules or the kinetic energy when they strike each other or the container walls. Latent heat alters intermolecular bonds.

If a time graph was made of a pure substance being heated or cooled, it would look something like the following graph for the heating of water and be called a heating curve for water. Different changes are taking place during each interval on the graph.

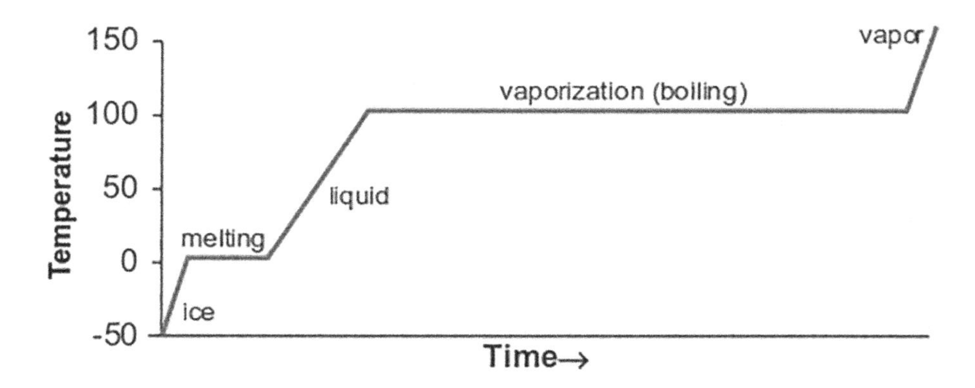

When the system is heated, energy is transferred into it. In response to the energy it receives, the system changes, for example by increasing its temperature.

During the first interval on the graph, energy is being absorbed by the water molecules to increase the temperature to water's melting point, 0°C. The slope of the line for this interval shows the increase in temperature and is related to the heat capacity of the substance. During the next interval on the graph, energy is still being added to the water but the temperature remains the same, at 0°C or water's melting point temperature. The additional energy is being used to overcome the intermolecular forces holding the water molecules in their solid pattern. This energy is moving the particles apart, breaking or weakening the forces of attraction trying to keep the water molecules aligned. The solid water (ice) is being converted to liquid water; a phase change is occurring. The temperature will not increase until every solid particle has melted and the entire sample is liquid.

Temperature again increases as energy is being absorbed by the liquid water molecules. Notice that the slope of the line during this interval is different than the slope of the line during the first interval. Again, this is due to differences in the heat capacity of ice and liquid water.

The flat line during the next interval indicates that a phase change is occurring here. The additional energy is being used to overcome the attractive forces holding the liquid water molecules together. The water molecules increase their kinetic energies and move farther apart, changing to water vapor molecules. This occurs at the boiling point temperature, or 100°C in the case of water. The temperature stays at the boiling point temperature until all liquid water molecules are converted to vapor molecules.

Once this conversion occurs, the temperature increases as energy is added, reflective of the heat capacity of the substance as a vapor.

5.4 Phase diagrams

A **phase diagram** is a graphical way to summarize the conditions under which the different states of a substance are stable. The diagram is divided into three areas representing each state of the substance. The curves separating each area represent the boundaries of phase changes

Below is a typical phase diagram. It consists of three curves that divide the diagram into regions labeled "solid, liquid, and gas".

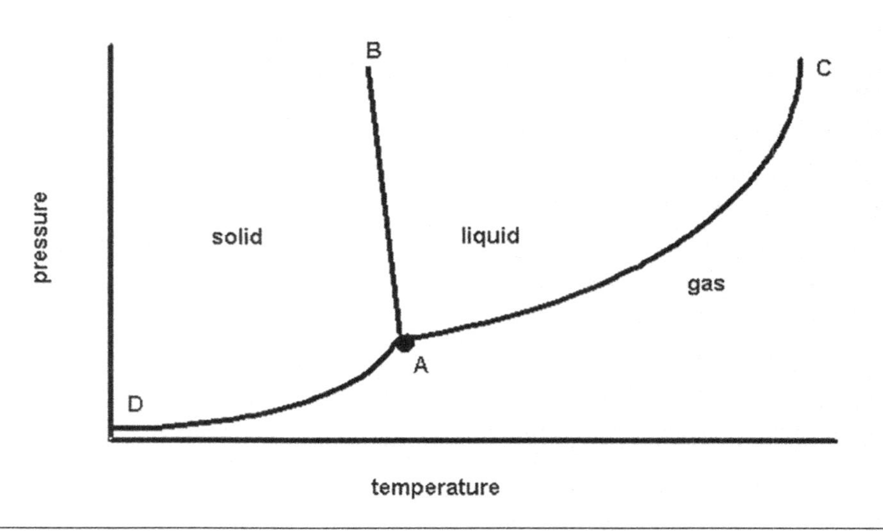

Curve **AB**, dividing the solid region from the liquid region, represents the conditions under which the solid and liquid are in equilibrium.

Usually, the melting point is only slightly affected by pressure. For this reason, the melting point curve, AB, is nearly vertical.

Curve **AC**, which divides the liquid region from the gaseous region, represents the boiling points of the liquid for various pressures.

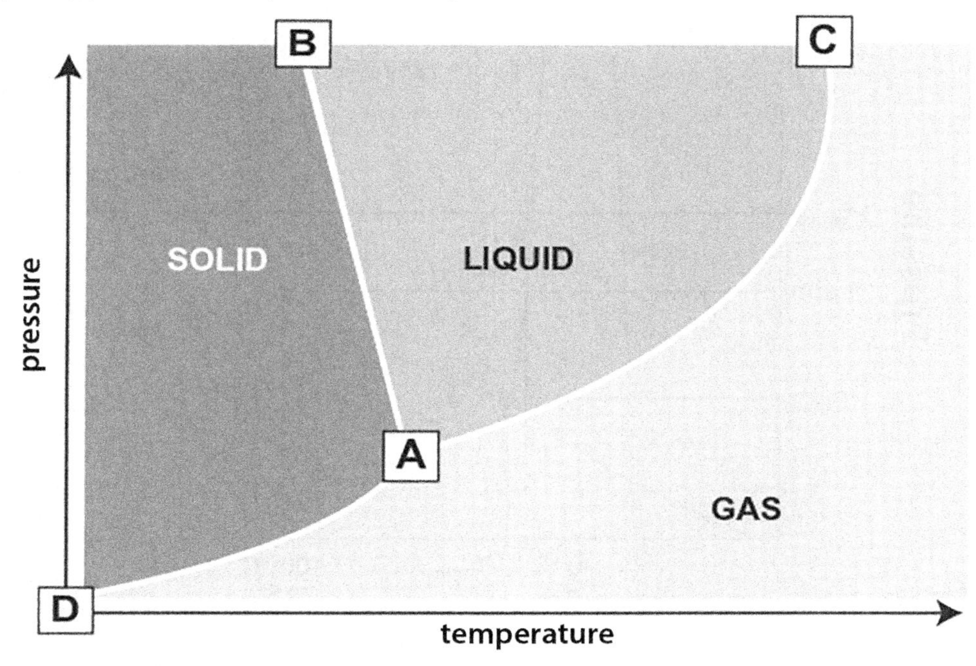

Curve **AD**, which divides the solid region from the gaseous region, represents the vapor pressures of the solid at various temperatures.

The curves intersect at **A**, the **triple point**, which is the temperature and pressure where three phases of a substance exist in equilibrium.

The temperature above which the liquid state of a substance no longer exists regardless of pressure is called the **critical temperature**.

The vapor pressure at the critical temperature is called the **critical pressure**. Note that curve AC ends at the **critical point, C.**

The phase diagram for water (shown on this page) is unusual. The solid/liquid phase boundary slopes to the left with increasing pressure because the melting point of water decreases with increasing pressure. Note that the normal melting point of water is lower than its triple point. The diagram is not drawn to a uniform scale. Many anomalous properties of water are discussed here: http://www.lsbu.ac.uk/water/anmlies.html.

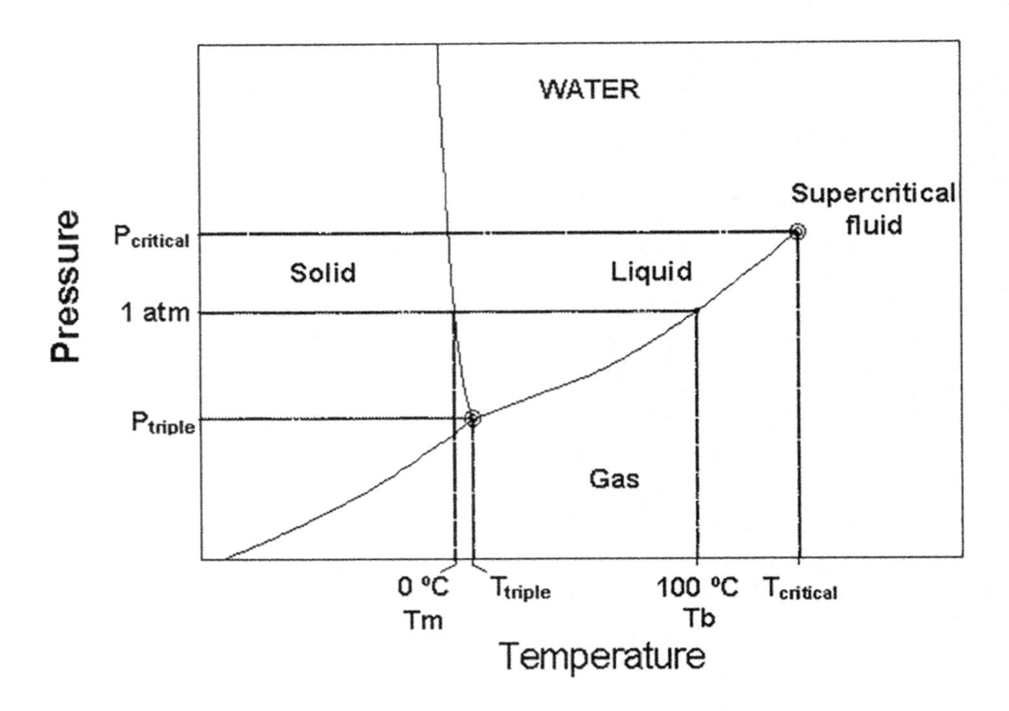

Normal melting point (T_m) and **normal boiling point** (T_b) are defined at 1 atm. Note that freezing point and melting point refer to an identical temperature approached from different directions, but they represent the same concept.

Solid lines dividing these regions are located at conditions under which two phases may exist at equilibrium and a phase change may occur. All three phases may coexist at the **triple point** of a substance. **Vapor pressure** at a given temperature is the pressure of the phase transition line to a gas at that temperature. **Normal melting point** (T_m) and **normal boiling point** (T_b) are defined at 1 atm. Note that freezing point and melting point refer to an identical temperature approached from different directions, but they represent the same concept. At temperatures and pressures above the **critical point**, the substance becomes too dense with too much kinetic energy for a gas-liquid interface to form. Matter under these conditions forms a **supercritical fluid** with properties of gases and of liquids.

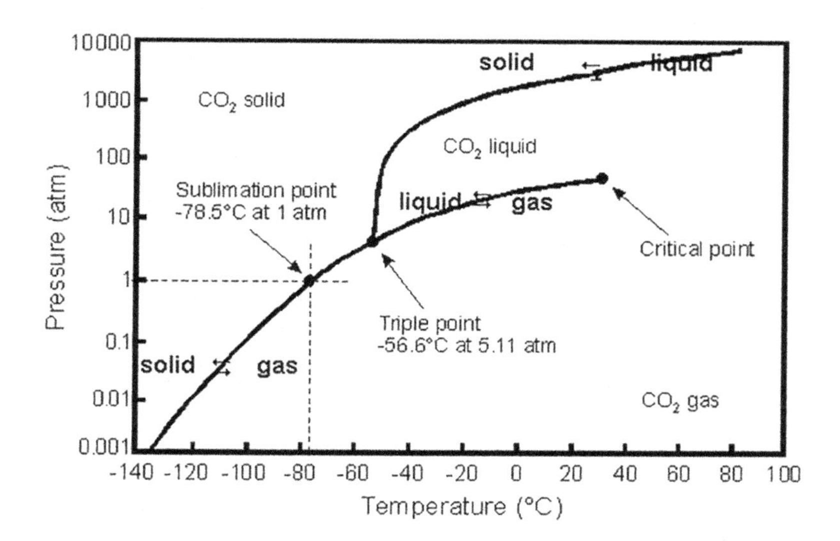

This is the phase diagram for carbon dioxide, CO_2. It shows the same features, only at different temperatures and pressures.

The triple point pressure of CO_2 is greater than 1 atm, so dry ice sublimates at atmospheric pressure with no liquid phase.

Chapter 6: Solutions

6.1 Molarity and percent by mass concentrations

Concentration

In a solution, the amount of solute dissolved in a given amount of solvent is called concentration. Solutions that cannot dissolve any additional solute are referred to as saturated solutions. Qualitative terms such as concentrated and dilute are used to describe solutions with relativity high or low solute concentrations, respectively. However, these expressions give very little information about the actual quantities present in the solution.

Expressions of percent by mass or volume, molarity, molality, as well as part per million give more information about the quantity of solute and solvent present in a solution.

Percent by mass or volume expresses the amount of solute present as a percentage of the total solution present, either by mass or by volume.

$$\% \text{ by mass} = \frac{\text{mass of solute}}{\text{mass of solute} + \text{mass of solvent}} \times 100\%$$

$$\% \text{ by volume} = \frac{\text{volume of solute}}{\text{volume of solute} + \text{volume of solvent (or total volume of solution)}} \times 100\%$$

Example: What is the percent by mass of a solution prepared by dissolving 4.0 g of CH_3COOH is 35.0 g of water?

Solution: $\% \text{ by mass} = \frac{\text{mass of solute}}{\text{mass of solute} + \text{mass of solvent}} \times 100\%$

$= 4\text{ g} / (4\text{ g} + 35\text{ g}) = 10\%$

$$\% \text{ by volume} = \frac{\text{volume of solute}}{\text{volume of solute} + \text{volume of solvent}} \times 100\%$$

Note: remember that water volume and mass can be easily interchanged in the metric system, since one milliliter of water weighs one gram. Also, one cubic centimeter (cc) is the same volume as one milliliter.

This concept will be discussed in detail in Chapter 10.1

6.2 Solution preparation

When two or more pure materials mix in a homogeneous way (with their molecules intermixing on a molecular level), the mixture is called a **solution**. Dispersions of small particles that are larger than molecules are called **colloids**. Liquid solutions are the most common, but any two phases may form a solution. When a pure liquid and a gas or solid form a liquid solution, the pure liquid is called the **solvent** and the non-liquids are called **solutes**. When all components in the solution were originally liquids, then the one present in the greatest amount is called the solvent and the others are called solutes. Solutions with water as the solvent are called **aqueous** solutions. The amount of solute in relation to the amount of solvent is called its **concentration**. Qualitative terms such as **concentrated** and **dilute** are used to describe the relative amounts of solute that could be dissolved. A solution with a small concentration of solute is called **dilute**, and a solution with a large concentration of solute is called **concentrated**. However, these expressions give very little information about the actual quantities present in the solution.

Electrolytes

Water is the solvent in an aqueous solution. The solute is the substance or compound being dissolved in the solvent. A solute has fewer particles than a solvent, and its particles are in random motion. It is interesting that aqueous solutions with ions conduct electricity to some degree. Pure water, having a very low ion concentration cannot conduct electricity. When a solute dissociates in water to form ions, it is called an electrolyte, because the solution becomes a good electrical conductor. When no ions are produced, or the ion content is low, the solute is a non-electrolyte. Non-electrolytes do not conduct electricity or conduct it to a very small degree. In an aqueous solution a strong electrolyte is completely ionized, or dissociated, in water, meaning it is soluble. Strong acids and bases are usually strong electrolytes. Electrolytes are considered weak if they do not completely dissociate, resulting in whole compounds and ions in the solution. Weak acids and bases are usually weak electrolytes. In other words, strong electrolytes have a better tendency to supply ions to the aqueous solution than weak electrolytes, and therefore strong electrolytes create an aqueous solution that is a better conductor of electricity.

Dilution

Adding more solvent to a solution to decrease the concentration is known as dilution. Starting with a known volume of a solution of known molarity, we would be able to prepare a more dilute solution of any desired concentration.

Ion Concentrations

In an aqueous solution the amount of ions of a compound, that is, the number of moles of that compound, per volume of solution is the concentration of the substance. Molarity is the number of moles of a solute divided by the total volume (V) of the solution.

Precipitation Reactions

Precipitation occurs when the product of a reaction is insoluble; in other words, when a solid is produced. This solid is called the precipitate. The precipitate is a combination of cation and anions forming an ionic bond. Precipitates are used in chemical manufacturing, such as isolating certain ions by forming precipitates with them.

These will all be discussed in more detail in the following sections.

6.3 Factors affecting solubility of solids, liquids, and gases

There are a number of factors that effect the solubility of a solute in a given solvent.

As more solid solute particles dissolve in a liquid solvent, the concentration of solute increases, and the chance that dissolved solute will collide with the remaining undissolved solid also increases. A collision may result in the solute particle either bouncing off the solid or reattaching itself to the solid. If it reattaches, the process is called **crystallization**, and is the opposite of the solution process. An animation of the solution process may be found here: http://www.mhhe.com/physsci/chemistry/essentialchemistry/flash/molvie1.swf.

Equilibrium occurs when no additional solute will dissolve because the rates of crystallization and solution are equal.

$$\textbf{Solute + Solvent} \underset{\text{Crystallize}}{\overset{\text{Dissolve}}{\longleftrightarrow}} \textbf{Solution}$$

A solution at equilibrium with non-dissolved solute is a **saturated** solution. The amount of solute required to form a saturated solution in a given amount of solvent is called the **solubility** of that solute. If less solute is present, the solution is called **unsaturated**. It is also possible under certain special conditions to have more solute than the equilibrium amount, resulting in a solution that is termed **supersaturated**.

Pairs of liquids that mix in all proportions are called **miscible**. Liquids that don't mix are called **immiscible**.

Gas solubility

Pressure does not dramatically alter the solubility of solids or liquids, but kinetic molecular theory predicts that **increasing the partial pressure of a gas will increase the solubility of the gas** in a liquid. If a substance is distributed between gas and solution phases and pressure is exerted, more gas molecules will impact the gas/liquid interface per second, so more will dissolve into the solution until a new equilibrium is reached at a higher concentration. **Henry's law** describes this relationship as a direct proportionality:

$$\text{Solubility} \propto \text{Pressure}$$

Carbonated drinks are bottled under high CO_2 pressure, permitting the gas to dissolve into aqueous solution. When the bottle is opened, the partial pressure of CO_2 in the gas phase rapidly decreases to the value in the atmosphere, and the gas bubbles out of solution. When the bottle is closed again, CO_2 gas pressure builds until a saturated solution at equilibrium is again obtained. The solubility of gas in a liquid also increases in the bloodstream of deep-sea divers when they experience high pressures. If they return to atmospheric pressure too rapidly, large bubbles of nitrogen gas will form in their blood and cause a potentially lethal condition known as **the bends** or **decompression sickness**. The diver must enter a hyperbaric (high pressure) chamber to redissolve the nitrogen back into the blood.

Increasing temperature decreases the solubility of a gas in a liquid because kinetic energy opposes intermolecular attractions and permits more molecules to escape from the liquid phase. The vapor pressure of a pure liquid increases with temperature for the same reason. Greater kinetic energy will favor material in the gas phase.

Liquid and solid solubility

For solid and liquid solutes, the effect of temperature is dependent on whether the reaction absorbs heat (endothermic) or releases heat (exothermic). If a compound is dissolved in water and the solution becomes hot, the solution process is exothermic. If the solution becomes cold, the solution process is endothermic. For more on endothermic and exothermic, see Chapter 14 Thermodynamics. The graphs show how the temperature of water may change upon the addition of an inorganic compound.

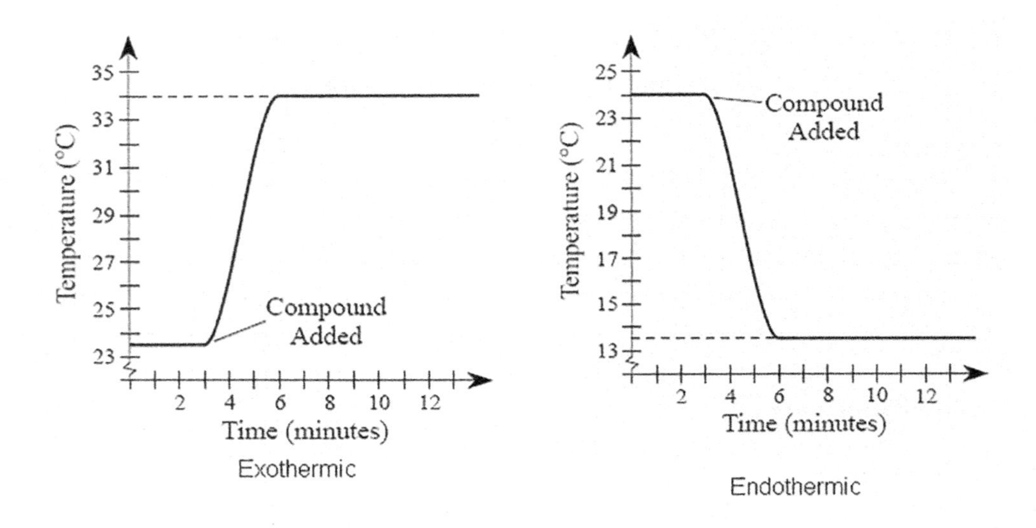

Intermolecular forces in the solution process

Solutions tend to form when the intermolecular attractive forces between solute and solvent molecules are about as strong as those that exist in the solute alone or in solvent alone. NaCl dissolves in water because the water molecules interact with the Na^+ and Cl^- ions with sufficient strength to overcome the attraction between them in the crystal.

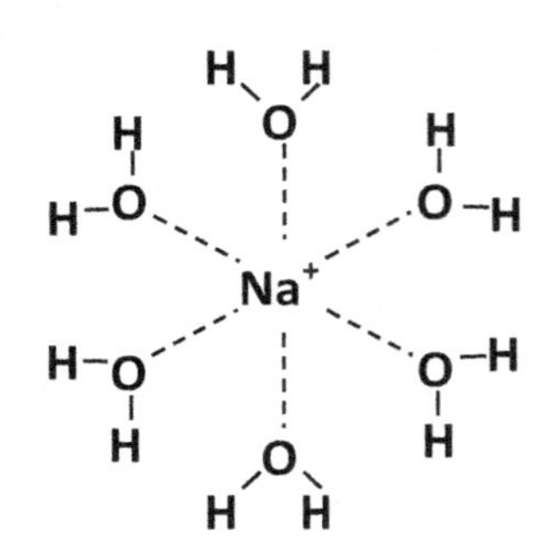

The intermolecular attraction between solute and solvent molecules is known as **solvation**. When the solvent is water, it is known as **hydration**. The figure to the left shows a hydrated Na^+ ion.

Polar and nonpolar solutes and solvents

A nonpolar liquid like heptane (C_7H_{16}) has intermolecular bonds with relatively weak London dispersion forces. Heptane is immiscible in water because the attraction that water molecules have for each other via hydrogen bonding is strong in comparison. Unlike Na^+ and Cl^- ions, heptane molecules cannot break these bonds. Because bonds of similar strength must be broken and formed for solvation to occur, nonpolar substances tend to be soluble in nonpolar solvents, and ionic and polar substances are soluble in polar solvents like water. Polar molecules are often called **hydrophilic** and non-polar molecules are called **hydrophobic**. This observation is often stated as "**like dissolves like**." Network solids (e.g., diamond) are soluble in neither polar nor nonpolar solvents because the covalent bonds within the solid are too strong for these solvents to break.

Electrolytes and precipitates

Compounds that are completely ionized in water are called **strong electrolytes** because these solutions easily conduct electricity. Most salts are strong electrolytes. For example, all NaCl is present in solution as ions. Other compounds (including many acids and bases) may dissolve in water without completely ionizing. These compounds are referred to as **weak electrolytes** and their state of ionization is in equilibrium with the larger molecule. Those compounds that dissolve with no ionization (e.g., glucose, $C_6H_{12}O_6$) are called **nonelectrolytes**.

Particles in solution are free to move about and collide with each other, vastly increasing the likelihood that a reaction will occur compared with particles in a solid phase. Aqueous solutions may react to produce an insoluble substance that will fall out of solution as a solid or gas precipitate in a precipitation reaction. Aqueous solutions may also react to form additional water, or a different chemical that remains in solution.

Solubility rules for ionic compounds

Given a cation and anion in aqueous solution, we can determine if a precipitate will form according to some common solubility rules.

1. Salts with NH_4^+ or with a cation from group 1 of the periodic table are *soluble* in water.
2. Nitrates (NO_3^-), acetates ($C_2H_3O_2^-$), chlorates (ClO_3^-), and perchlorates (ClO_4^-) are *soluble* in water.
3. Cl^-, Br^-, and I^- salts are water soluble except in the presence of Ag^+, Hg_2^{2+}, or Pb^{2+}, with which they will form precipitates.
4. Sulfates (SO_4^{2-}) are water soluble except in the presence of Ca^{2+}, Ba^{2+}, Ag^+, Hg_2^{2+}, or Pb^{2+}.
5. Hydroxides (OH^-) are *insoluble* in water unless they have a cation from group 1 of the periodic table or Ca^{2+}, Sr^{2+}, or Ba^{2+} are present.
6. Sulfides (S^{2-}), sulfites (SO_3^{2-}), phosphates (PO_4^{3-}), and carbonates (CO_3^{2-}) are *insoluble* in water unless they have a cation from group 1 of the periodic table.

Whether precipitation occurs among a group of ions—and which compound will form the precipitate—may be determined by the solubility rules on the following page.

Summary

The following table summarizes the impact of temperature and pressure on solubility:

Ion	Solubility	Exception
All compounds containing alkali metals	All are soluble	
All compounds containing ammonium, NH_4^+	All are soluble	
All compounds containing nitrates, NO_3^-	All are soluble	
All compounds containing chlorates, ClO_3^- and perchlorates, ClO_4^-	All are soluble	
All compounds containing acetates, $C_2H_3O_2^-$	All are soluble	
Compounds containing Cl^-, Br^-, and I^-	Most are soluble	Except halides of Ag^+, Hg_2^{2+}, and Pb^{2+}
Compounds containing F^-	Most are soluble	Except fluorides of Mg^{2+}, Ca^{2+}, Sr^{2+}, Ba^{2+}, and Pb^2
Compounds containing sulfates, SO_4^{2-}	Most are soluble	Except sulfates of Mg^{2+}, Ca^{2+}, Sr^{2+}, Ba^{2+}, and Pb^2
All compounds containing carbonates, CO_3^{2-}	Most are insoluble	Except those containing alkali metals or NH_4^+
All compounds containing phosphates, PO_4^{3-}	Most are insoluble	Except those containing alkali metals or NH_4^+
All compounds containing oxalates, $C_2O_4^{2-}$	Most are insoluble	Except those containing alkali metals or NH_4^+
All compounds containing chromates, CrO_4^{2-}	Most are insoluble	Except those containing alkali metals or NH_4^+
All compounds containing oxides, O^{2-}	Most are insoluble	Except those containing alkali metals or NH_4^+
All compounds containing sulfides, S^{2-}	Most are insoluble	Except those containing alkali metals or NH_4^+
All compounds containing sulfites, SO_3^{2-}	Most are insoluble	Except those containing alkali metals or NH_4^+
All compounds containing silicates, SiO_4^{2-}	Most are insoluble	Except those containing alkali metals or NH_4^+
All compounds containing hydroxides	Most are insoluble	Except those containing alkali metals or NH_4^+, or Ba^{2+}

Effect on solution of an increase in one variable with the other constant	− = decrease, 0 = no/small change, + = increase, ++ = strong increase				
	Gas solute in liquid solvent			Solid and liquid solutes	
	Average kinetic energy of molecules	Collisions of gas with liquid interface	Solubility	Solubility for an endothermic heat of solution	Solubility for an exothermic heat of solution
Pressure	0	++	+	0	0
Temperature	+	+	−	+	−

6.4 Qualitative aspects of colligative properties

A **colligative** property is a physical property of a solution that **depends on the number of solute particles present in solution** and usually not on the identity of the solutes involved. Colligative properties may be predicted by imagining ourselves shrinking down to the size of molecules in a solution and visualizing the impact of an increasing number of generic solute particles around us.

Vapor pressure lowering, boiling point elevation, freezing point lowering

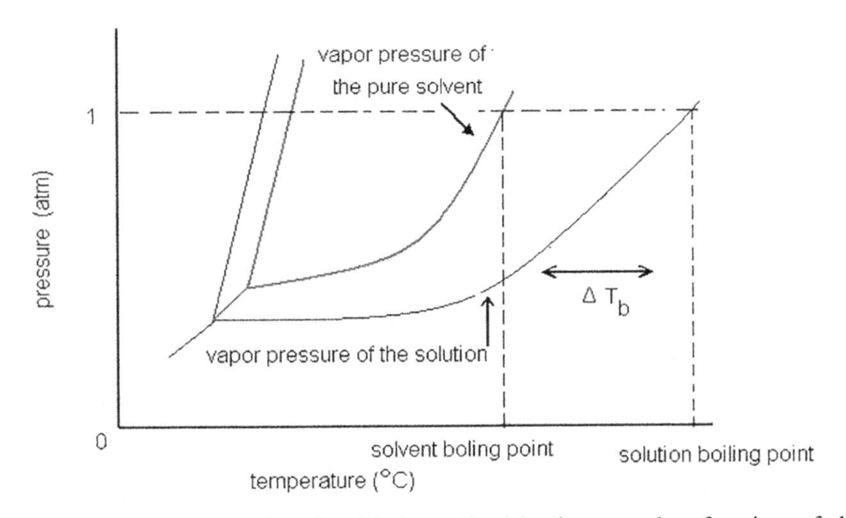

After a nonvolatile solute is added to a liquid solvent, only a fraction of the molecules at a liquid-gas interface are now volatile and capable of escaping into the gas phase. The vapor consists of essentially pure solvent that is able to condense freely.

This imbalance drives equilibrium away from the vapor phase and into the liquid phase and **lowers the vapor pressure** by an amount proportional to the solute particles present.

It follows from a lowered vapor pressure that a higher temperature is required for the vapor pressure to become equal to the external pressure over the liquid. Thus, **the boiling point is raised** by an amount proportional to the solute particles present.

Solute particles in a liquid solvent are not normally soluble in the solid phase of that solvent. When solvent crystals freeze, they typically align themselves with each other to keep the solute out. This means that only a fraction of the molecules in the liquid at the liquid-solid interface are capable of freezing while the solid phase consists of essentially pure solvent that is able to melt freely. This imbalance drives equilibrium away from the solid phase and into the liquid phase and **lowers the freezing point** by an amount proportional to the solute particles present.

Boiling point elevation and freezing point depression are both caused by a higher fraction of solvent molecules in the liquid phase than in the other phase. For pure water at 1 atm there is equilibrium at the normal boiling and freezing points; for water with a high solute concentration, equilibrium is not present at the normal boiling and freezing points.

Osmotic pressure

A **semipermeable membrane** is a material that permits some particles to pass through it but not others. The diagram below shows a membrane that permits solvent but not solute to pass through it.

When a semipermeable membrane separates a dilute solution from a concentrated solution, the solvent flows from the dilute to the concentrated solution (i.e., from higher solvent to lower solvent concentration) in a process called **osmosis** until equilibrium is achieved.

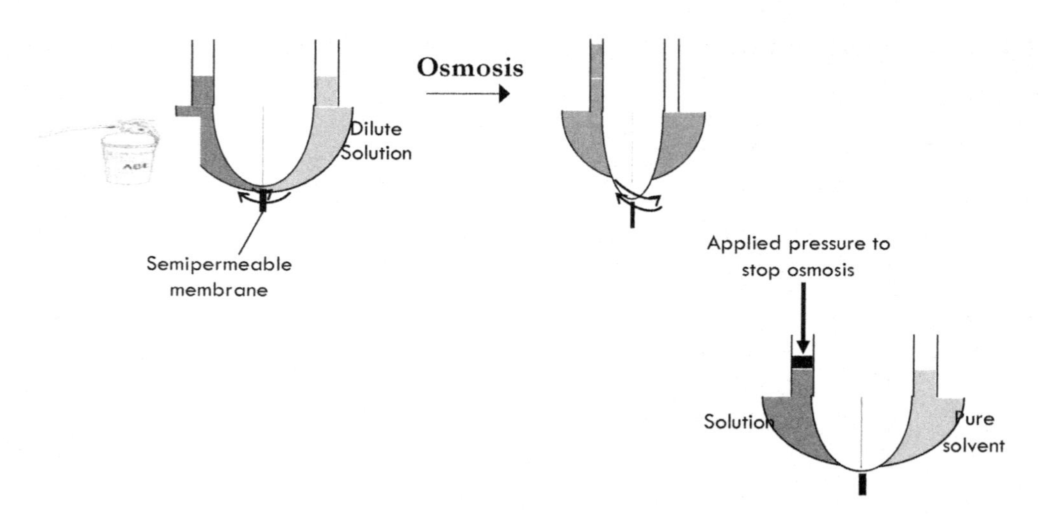

The pressure required to prevent osmosis from a pure solvent into a solution is called **osmotic pressure**. Osmotic pressure is proportional to the molarity of the solution and thus it is a colligative property of solution. The osmotic pressure of a pure solvent is zero.

Colligative Property Problems

Quantitative colligative property problems typically involve a change related to a solute concentration by a direct proportionality.

Raoult's law states that the vapor pressure of a solution with nonvolatile solutes is the mole fraction of solvent multiplied by the pure solvent vapor pressure:

$$P = XP^{\circ}$$

where P = the vapor pressure of the solution, X = the mole fraction of the solvent, and P° = the vapor pressure of the pure solvent.

Different concentration units are used for other colligative properties to express a **change from pure solvent**. The following table summarizes these expressions for nonelectrolytes:

Colligative property	Equation for property	Proportionality constant
Vapor pressure lowering	$P_{solvent} = X_{solvent} P°_{solvent}$	Pure solvent vapor pressure
Boiling point elevation	$\Delta_B = K_B$ (moles of solute/kg of solution)	Solvent-dependant constant K_b
Freezing point lowering	$\Delta_F = K_F$ (moles of solute/kg of solution)	Solvent-dependant constant K_f
Osmotic pressure	$\pi = mRT$	Gas constant (R)

For solutions that contain electrolytes, the change from the pure solvent to solution is different from what is predicted by the above equations. Due to their ionic nature, these substances will dissociate to put many more ions in solution than their molal concentration would predict. The total number of ions affects the colligative properties just as the number of molecules would for a nonpolar solute.

The **van't Hoff factor** is an important factor in the change in boiling point or freezing point of a solution after solute has been added. The van't Hoff factor is symbolized by the lower-case letter *i*. It is a unitless constant directly associated with the degree of dissociation of the solute in the solvent.

This pattern continues for any number of particles into which a solute can dissociate:

Substances which do not ionize in solution, like sugar, have i = 1. Substances which ionize into two ions, like NaCl, have i = 2. Substances which ionize into three ions, like $MgCl_2$, have i = 3.

What this implies is exactly what van't Hoff observed as he was compiling and examining boiling point and freezing point data. At the time, he did not understand what i meant. His use of i was strictly to try and make the data fit together.

He took a 1.0 molal aqueous solution of sugar, NaCl, and $MgCl_2$ and checked the boiling point temperature of each solution. He found that the NaCl had a boiling point elevation twice as high as that of the sugar and the $MgCl_2$ elevated the boiling point temperature three times higher than the sugar. Upon testing other substances, he found groups but he had no idea why until Svante Arrhenius' theory of electrolytic dissociation was published. The explanation seemed clear then.

Many colligative property problems compare one solution to another and may be solved without the use of the above expressions. All that is required for these comparison problems is knowledge of what the colligative properties are, how they are altered, and which solution contains the greater concentration of dissolved particles.

Example: One mole of each of the following compounds is added to water in separate flasks to make 1.0 L of solution.

Potassium phosphate
Silver chloride
Sodium chloride
Sugar (sucrose)

(A) Which solution will exhibit the greatest change in the freezing point temperature?
(B) Which solution will exhibit the least change in the boiling point temperature?

Be sure to explain you choices.

Solution:
Determine the molecular formulas and analyze the choices for solubility and dissociation:

Potassium phosphate	soluble in water →	4 ions
Silver chloride	soluble in water →	2 ions
Sodium chloride	soluble in water →	2 ions
Sugar (sucrose)	soluble in water →	1 molecule (nonelectrolyte)

(A) Of the choices, potassium phosphate forms the most ions so given that the molar amount added is the same for all of the choices, K_3PO_4 will effect the boiling point temperature and the freezing point temperature the most. For every 1 mole of K_3PO_4 that dissolves, 4 moles of ions will be present in solution.
(B) Sugar, a nonelectrolyte, will have the least effect on the freezing point and boiling point temperatures due to the fact that it is a molecular substance and does not dissociate into ions. For every 1 mole of sugar added, only 1 mole of molecules will be present in solution.

Changes to boiling point and freezing point may be determined by looking at the molal concentration of the solute, according to the equations in the table above.

Remember that a colligative property is one that depends only on the number of particles present, not their nature.

For **freezing point temperature changes,**

$$\Delta T_f = m k_f i$$

where m is the molal concentration of the solute (in moles/kg of solvent), K_f is a constant specific to each solvent, and i is the number of particles or ions in solution. For water, K_f = -1.86 °C/m

For **boiling point temperature changes,**

$$\Delta T_b = mk_b i$$

where m is the molal concentration of the solute (in moles/kg of solvent), K_b is a constant specific for each solvent and *i* is the number of particles or ions in solution. For water, K_b = 0.52 °C/m

Example: How much will boiling point increase when 31.5 grams of potassium chloride are added to 225 g of water?

Solution:

KCl is an electrolyte with *i* = 2
Mass of water = 225 g = 0.225 kg
Mass KCl = 31.5 g, moles KCl = 0.423 mole
m = 0.423/0.225 kg =1.88 m
K_b = 0.52 °C/m

$\Delta T_b = mk_b i$
= 1.88 m (0.56 °C/m) 2 = 1.96 °C

Example: How many grams of benzoic acid, $C_7H_6O_2$, a nonelectrolyte, must be added to 178 g of water to increase the boiling point temperature 4°C?

Solution: First convert the units and identify the constants. Benzoic acid is a weak enough electrolyte to be considered a nonelectrolyte, so *i* = 1.

Mass of water = 178 g = 0.178 kg
Molecular weight of benzoic acid = 122 g/mol
ΔT = 4 °C
K_b = 0.52 °C/m

$\Delta T_b = mk_b i$

Rearranging the equation above and first solving for molality of benzoic acid:

$m = \Delta T_b / K_b i$ = 4/0.52 °C/m (1) = 7.69 m

7.69 m= moles benzoic acid/0.178 kg water = 1.37 mole benzoic acid

Next, solve for grams of benzoic acid:

1.37 moles benzoic acid = x grams benzoic acid / 122 g/mol
167 g benzoic acid are needed.

The most common errors in solving all types of colligative property problems arise from considering some value other than **the number of particles in solution**. Remember that one mole of glucose(*aq*) forms one mole of hydrated particles, but one mole of NaCl(*aq*) forms two moles of hydrated particles, and one mole of K_3PO_4 forms four moles of them. We would expect a 0.5 M solution of glucose to have roughly the same colligative properties as a 0.25 M solution of sodium hydroxide and a 0.1 M solution of aluminum sulfate. Also remember that undissolved solids do not contribute to colligative properties.

SECTION II PRACTICE QUESTIONS

CHAPTER 4

1. **Heat is added to a pure solid at its melting point until it all becomes liquid at its freezing point. Which of the following occur(s)?**

   (A) Intermolecular attractions are increased.
   (B) The kinetic energy of the molecules increases.
   (C) The freedom of the molecules to move about decreases.
   (D) The temperature of the system remains constant.
   (E) All of the above.

   The correct answer is D.
   There is no temperature change during a phase change.

Questions 2-4 refer to the following:

(A) Increases
(B) Decreases
(C) Remains the same
(D) Can not be determined/varies

2. **As the pressure is increased at constant T, the volume: The correct answer is B.**

3. **As the number of moles in the system increases at constant pressure, the volume: The correct answer is A.**

4. **If the temperature decreases, at constant V, the pressure: The correct answer is B.**

   (You can use PV = nRT to answer all of these; just ignore R since these are qualitative/ proportional questions rather than asking for an exact value.)

5. **Cause and Effect:**

   Statement I
   Equal volumes of different gases at the same temperature and pressure contain equal numbers of molecules.
   This statement is True.

   Statement II
   The volume occupied by one mole of a pure substance, molar volume, depends on the density of a substance and, like density, varies with temperature.
   This statement is True.
   Both Statements are TRUE but there is no cause/effect relationship

CHAPTER 5

1. **During a phase change**

   (A) Heat is added to a system
   (B) The molecules move faster
   (C) The temperature does not change
   (D) The volume increases
   (E) All of the above.

   The correct answer is C.
   The others may be true if you are heating the system, however the cooling of a system also may result in a phase change during which heat is removed, the volume decreases and the molecules slow down. Only the constant temperature is true for a phase change no matter which way the change is occurring.

Questions 2-4 refer to the following:

(A) Solid
(B) Liquid
(C) Gas
(D) Ionic Crystal
(E) Metallic Crystal

2. **Held together by a "sea of electrons": The correct answer is E.**

3. **Have a defined shape and volume: The correct answer is A, D, and E.**

4. **Have high melting points: The correct answer is D and E.**

5. **Cause and Effect**

   Statement I
   Diamond is a form of elemental carbon in which each atom is covalently bonded to four nearest neighbors
   This statement is True.

   Statement II
   Covalent crystals are poor conductors of electricity
   This statement is True.
   Both Statements are TRUE but there is no cause/effect relationship

CHAPTER 6

1. **Which of the following is an example of a colligative property?**

    (A) Boiling Point
    (B) Melting Point Depression
    (C) Molarity
    (D) Concentration
    (E) All of the above

    The correct answer is B.
    Colligative properties of solutions are **properties** that depend upon the **concentration** of solute molecules or ions, but not upon the identity of the solute. **Colligative properties** include vapor pressure lowering, boiling point elevation, freezing point depression, and osmotic pressure.

Questions 2-4 refer to the following:

(A) unsaturated
(B) saturated
(C) concentrated
(D) dilute
(E) supersaturated

2. **A solution at equilibrium with non-dissolved solute is: The correct answer is B.**

3. **A solution that has more solute than the equilibrium amount is: The correct answer is E.**

4. **A solution with less solute than the equilibrium amount is: The correct answer is A.**

5. **Cause and Effect**

    Statement I
    Increasing the partial pressure of a gas will increase the solubility of the gas in a liquid
    This statement is True.

    Statement II
    More gas molecules will impact the gas/liquid interface per second, so more will dissolve into the solution
    This statement is True.
    Both statements are TRUE and there IS a cause/effect relationship.

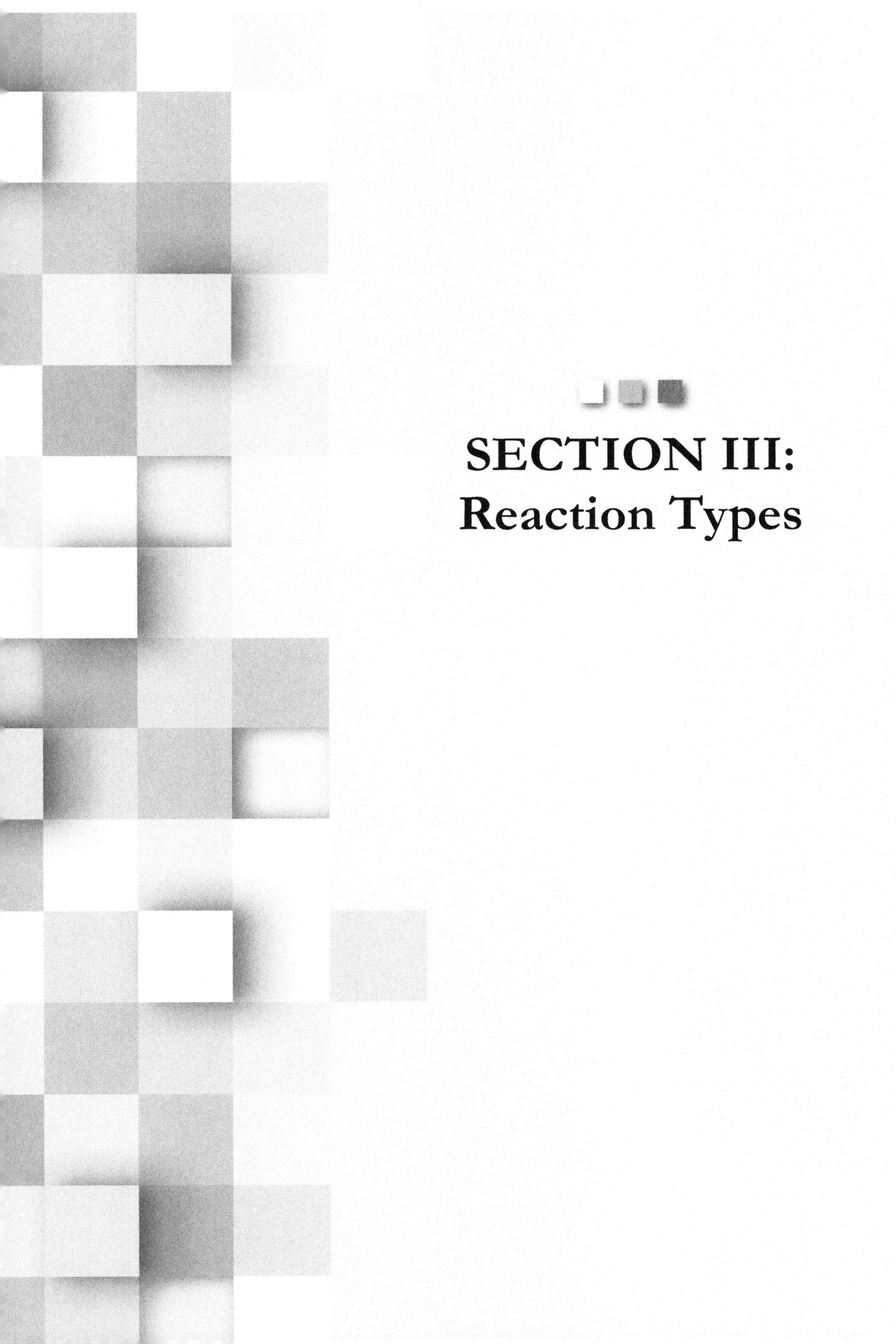

SECTION III: Reaction Types

Chapter 7: Acids and Bases

7.1 Brønsted-Lowry theory

Brønsted-Lowry definition of acids and bases

In the 1920s, Johannes **Brønsted** and Thomas **Lowry** recognized that **acids can transfer a proton to bases** regardless of whether an OH^- ion accepts the proton. In an equilibrium reaction, the direction of proton transfer depends on whether the reaction is read left to right or right to left, so **Brønsted acids and bases exist in conjugate pairs with and without a proton**. Acids that are able to transfer more than one proton are called **polyprotic acids**.

Examples:

1. In the reaction:

   $$HF\ (aq) + H_2O\ (l) \longrightarrow F^-\ (aq) + H_3O^+\ (aq)$$

   HF transfers a proton to water. Therefore HF is the Brønsted acid and H_2O is the Brønsted base. But in the reverse direction, hydronium ions transfer a proton to fluoride ions. H_3O^+ is the conjugate acid of H_2O because it has an additional proton, and F^- is the conjugate base of HF because it lacks a proton.

2. In the reaction:

   $$NH_3\ (aq) + H_2\ (l) \longrightarrow NH_4^+\ (aq) + OH^-\ (aq)$$

   water transfers a proton to ammonia. H_2O is the Brønsted acid and OH^- is its conjugate base. NH_3 is the Brønsted base and NH_4^+ is its conjugate acid.

3. In the reaction:

   $$H_3PO_4\ HS^- \longrightarrow H_2PO_4^- + H_2S$$

   $H_3PO_4/H_2PO_4^-$ is one conjugate acid-base pair and H_2S/HS^- is the other.

4. H_3PO_4 is a polyprotic acid. It may further dissociate to transfer more than one proton:

$$H_3PO_4 \longrightarrow H_2PO_4^- + H^+$$

$$H_2PO_4^- \longrightarrow HPO_4^{2-} + H^+$$

$$HO_4^{2-} \longrightarrow PO_4^{3-} + H^+$$

7.2 Strong and weak acids and bases

Strong acids and bases completely dissociate in water, but weak acids and bases do not.

> **Example**: $HCl \longrightarrow H^+ + Cl^-$ goes to completin because HCl is a strong acid. The acids in the previous examples were all weak.

The aqueous dissociation constants K_a and K_b quantify acid and base strength. Another way of looking at acid dissociation is that strong acids transfer protons more readily than H_3O^+ transfers protons, so they protonate water, the conjugate base of H_3O^+. In general, **if two acid/base conjugate pairs are present, the stronger acid will transfer a proton to the conjugate base of the weaker acid**.

Acid and base **strength is not necessarilty related to safety**. Weak acids like HF may be extremely corrosive and dangerous.

The most **common strong acids and bases** are listed here:

Strong acid		Strong base	
HCl	Hydrochloric acid	LiOH	Lithium hydroxide
HBr	Hydrobromic acid	NaOH	Sodium hydroxide
HI	Hydroiodic acid	KOH	Potassium hydroxide
HNO_3	Nitric acid	$Ca(OH)_2$	Calcium hydroxide
H_2SO_4	Sulfuric acid	$Sr(OH)_2$	Strontium hydroxide
$HClO_4$	Perchloric acid	$Ba(OH)_2$	Barium hydroxide

A flash animation tutorial demonstrating the difference between strong and weak acids is located at: http://www.mhhe.com/physsci/chemistry/essentialchemistry/flash/acid13.swf.

Trends in acid and base strength

The strongest acid in a polyprotic series is always **the acid with the most protons** (e.g. H_2SO_4 is a stronger acid than HSO_4^-). The strongest acid in a series with the same central atom is always **the acid with the central atom at the highest oxidation number** (e.g. $HClO_4 > HClO_3 > HClO_2 > HClO$). The strongest acid in a series with different central atoms at the same oxidation number is usually **the acid with the central atom at the**

highest electronegativity (e.g. the K_a of HClO > HBrO > HIO). This electronegativity trend stretches across the periodic table for oxides.

Lewis definition of acids and bases

The transfer of a proton from a Brønsted acid to a Brønsted base requires that the base accept the proton. When Lewis diagrams are used to draw the proton donation of Brønsted acid-base reactions, it is always clear that the base must contain an unshared electron pair to form a bond with the proton. For example, ammonia contains an unshared electron pair in the following reaction:

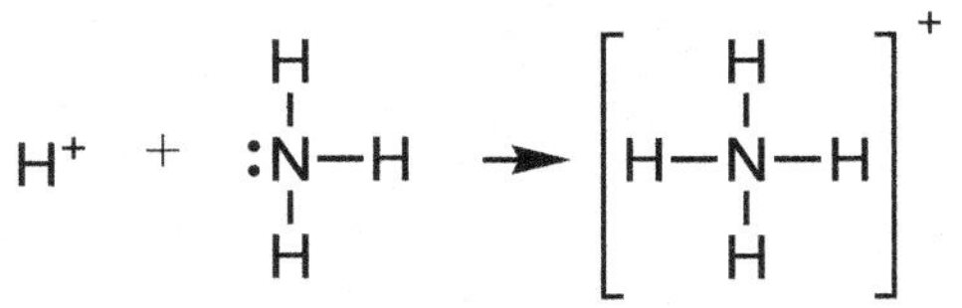

$$H^+ + :NH_3 \rightarrow [NH_4]^+$$

In the 1920s, Gilbert N. **Lewis** proposed that **bases donate unshared electron pairs to acids**, regardless of whether the donation is made to a proton or to another atom. Boron trifluoride is an example of a Lewis acid that is not a Brønsted acid because it is a chemical that accepts an electron pair without involving an H^+ ion:

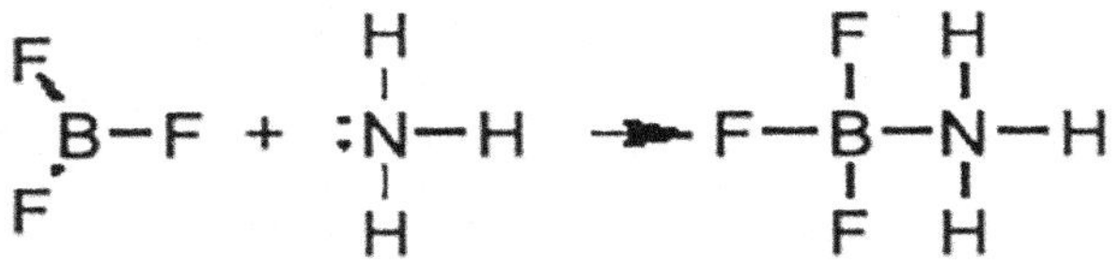

$$BF_3 + :NH_3 \rightarrow F_3B{-}NH_3$$

The Lewis theory of acids and bases is more general than the Brønsted-Lowry theory, but Brønsted-Lowry's definition is used more frequently. The terms "acid" and "base" most often refer to Brønsted acids and bases, and the term "Lewis acid" is usually reserved for chemicals like BF_3 that are not also Brønsted acids.

Summary of definitions

A Lewis base transfers an electron pair to a Lewis acid. A Brønsted acid transfers a proton to a Brønsted base. These exist in conjugate pairs at equilibrium. In an Arrhenius base, the proton acceptor (electron pair donor) is OH^-. All Arrhenius acids/bases are Brønsted acids/bases and all Brønsted acids/bases are Lewis acids/bases. Each definition contains a subset of the one that comes after it.

7.3 pH, titrations

Interpret graphical and numerical titration data.

In a typical acid-base **titration, an acid-base indicator** (such as phenolphthalein) or a **pH meter** is used to monitor the course of a **neutralization reaction**. The usual goal of titration is to determine an unknown concentration of an acid (or base) by neutralizing it with a known concentration of base (or acid).

The reagent of known concentration is usually used as the **titrant**. The titrant is poured into a **buret** (also spelled burette) until it is nearly full, and an initial buret reading is taken. Buret numbering is close to zero when nearly full. A known volume of the solution of unknown concentration is added to a flask and placed under the buret. The indicator is added (or the pH meter probe is inserted). The initial state of a titration experiment is shown to the right above.

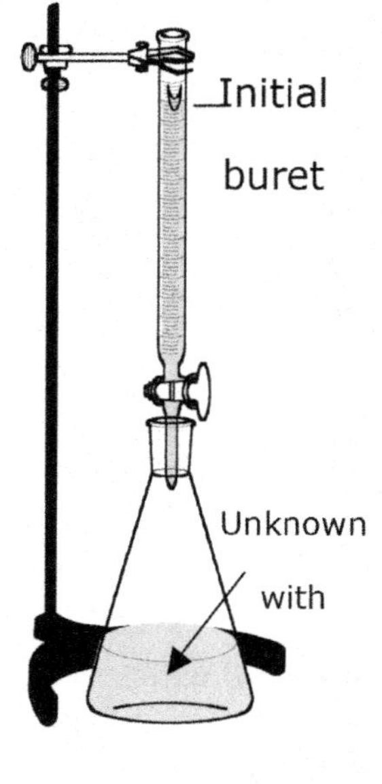

The buret stopcock is opened, and titrant is slowly added until the solution permanently changes color, which indicates a rapid change in pH. This is the titration **endpoint**, and a final buret reading is made. The final state of a titration experiment is shown below, to the right. The endpoint occurs when the number of **acid and base equivalents in the flask are identical.**

$$N_{acid} = N_{base}. \text{ Therefore, } C_{acid}V_{acid} = C_{base}V_{base}.$$

The endpoint is also known as the titration **equivalence point**.

Titration data typically consist of:

$V_{initial}$ = Initial buret volume

V_{final} = Final buret volume

C_{known} = Concentration of known solution

V_{known} = Volume of unknown solution

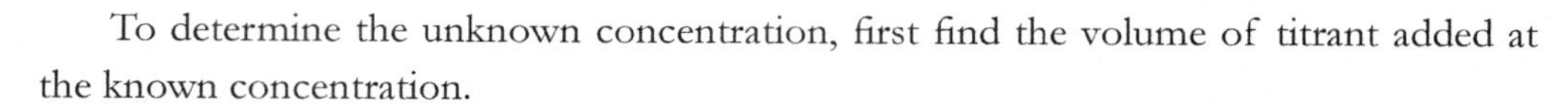

To determine the unknown concentration, first find the volume of titrant added at the known concentration.

$$V_{unknown} = V_{final} - V_{initial}$$

At the equivalence point, $N_{unknown} = N_{known}$

$$\text{Therefore, } C_{unknown} = \frac{C_{known}V_{known}}{V_{unknown}} = \frac{C_{known}\left(V_{final} - V_{initial}\right)}{V_{unknown}}$$

Units of molarity may be used for concentration in the previous expressions **unless a mole of either solution yields more than one acid or base equivalent.** In that case, concentration must be expressed using **normality**.

> **Example**: A 20.0 mL sample of an HCl solution is titrated with 0.200 M NaOH. The initial buret volume is 1.8 mL and the final buret volume at the titration endpoint is 29.1 mL. What is the molarity of the HCl sample?
>
> **Solution**: Two solution methods will be used. HCl contains one acid equivalent and NaOH contains one base equivalent, so we may use molarity in all our calculations.

1) Calculate the moles of the known substance added to the flask:

$$0.200\frac{\text{mol}}{\text{L}}\times\frac{1\text{ L}}{1000\text{ mL}}\times(29.1\text{ mL} - 1.8\text{ mL})=0.00546\text{ mol NaOH.}$$

At the endpoint, this base will neutralize 0.00546 mol HCl. Therefore, this amount of HCl must have been present in the sample before the titration.

$$\frac{0.00546\text{ mol HCl}}{0.0200\text{ L}}=0.273\text{ M HCl.}$$

2) Utilize the formula: $C_{unknown}=\frac{C_{known}(V_{final}-V_{initial})}{V_{unknown}}$

$$C_{HCl}=\frac{C_{NaOH}(V_{final}-V_{initial})}{V_{HCl}}=\frac{0.200\text{ M }(29.1\text{ mL} - 1.8\text{ mL})}{20.0\text{ mL}}=0.273\text{ M HCl.}$$

Titrating with the unknown

In a common variation of standard titration, an unknown is added to the buret as a titrant, and the reagent of known concentration is placed in the flask. The chemistry involved is the same as in the standard case, and the mathematics is also identical except for the identity of the two volumes. For this variation, V_{known} will be the volume added to the flask before titration begins and $V_{unknown}=V_{final}-V_{initial}$ Therefore:

$$C_{unknown}=\frac{C_{known}V_{known}}{V_{unknown}}=\frac{C_{known}V_{known}}{V_{final}-V_{initial}}$$

Example: 30.0 mL of a 0.150 M HNO_3 solution is titrated with $Ca(OH)_2$. The initial buret volume is 0.6 mL and the final buret volume at the equivalent point is 22.2 mL. What is the molarity of $Ca(OH)_2$ used for the titration?

Solution: The same two solution methods will be used as in the previous example. 1 mol $Ca(OH)_2$ contains 2 base equivalents because it reacts with 2 moles of H^+ via this reaction

$$Ca(OH)_2 + 2HNO_3 \rightarrow Ca(NO_3)_2 + 2H_2O.$$

Therefore, normality must be used in the formula for solution method 2.

1. First calculate the moles of the substance in the flask:

$$0.150\frac{\text{mol}}{\text{L}}\text{ X }0.0300\text{ L} = 0.00450\text{ mol }HNO_3.$$

This acid must be titrated with 0.00450 base equivalents for neutralization to occur at the end point. We calculate moles $Ca(OH)_2$ used in the titration from stoichiometry:

$$0.00450 \text{ base equivalents} \times \frac{1 \text{ mol } Ca(OH)_2}{2 \text{ base equivalents}} = 0.00225 \text{ mol } Ca(OH)_2.$$

The molarity of $Ca(OH)_2$ is found from the volume used in the titration:

$$\frac{0.00225 \text{ mol } Ca(OH)_3}{0.0222 \text{ L} - 0.0006 \text{ L}} = 0.104 \text{ M } Ca(OH)_3.$$

2. Utilize the formula: $C_{unknown} = \frac{C_{known}V_{known}}{V_{final} - V_{initial}}$ using units of normality. For HNO_3, molarity=normality because 1 mol contains 1 acid equivalent.

3. $$C_{Ca(OH)_2} = \frac{C_{HNO_3}V_{HNO_3}}{V_{final} - V_{initial}} = \frac{\left(0.150 \text{ M} \times \frac{1 \text{ N}}{1 \text{ M}}\right)(30.0 \text{ mL})}{22.2 \text{ mL} - 0.6 \text{ mL}} = 0.208 \text{ N } Ca(OH)_2$$

This value is converted to molarity. For $Ca(OH)_2$, normality is twice molarity because 1 mol contains 2 base equivalents.

7.4 Indicators

Acid-Base indicators

This section describes how simple acid-base indicators work, and how to choose the right one for a particular titration.

How simple indicators work

Acid - Base indicators (also known as pH indicators) are substances which change color with pH. They are usually weak acids or bases, which when dissolved in water dissociate slightly and form ions. The acid and its conjugate base have different colors.

Consider an indicator which is a weak acid, with the formula HIn. At equilibrium, the following equilibrium equation is established with its conjugate base:

$$HIn\ (aq) + H_2O\ (l) \rightleftharpoons H_3O^+\ (aq) + In^-\ (aq)$$

acid (colour A) — conjugate base (colour B)

At low pH values the concentration of H_3O^+ is high and so the equilibrium position lies to the left and the solution has the color A. At high pH values, the concentration of H_3O^+ is low - the equilibrium position lies to the right and the solution has color B.

Litmus

Litmus is a weak acid. It has a complicated molecular strucure which we will simplify to HLit. In this molecule, H is a proton which can be donated to another molecule and the "Lit" is the rest of the molecule.

When this acid is dissolved in water, equilibrium will be established whose simplified version can be written as follows:

$$HLit_{(aq)} \rightleftharpoons H^+_{(aq)} + Lit^-_{(aq)}$$

The un-ionized litmus is red, whereas the ion is blue.

By adding hydroxide ion to the solution having this equilibrium the hydroxide ion will react with the hydrogen ion and now according to the Le Châtelier's principle the equilibrium position moves to replace the lost hydrogen ions and so the litmus turns blue.

On the contrry, if we add extra hydrogen ions to this equilibrium, the equilibrium position moves to remove the extra hydrogen ions and therefore the equilibrium shifts to the left and litmus turns red.

If the concentrations of HLit and Lit^- are equal:

At some point during the movement of the position of equilibrium, the concentrations of the two colors will become equal. The color you see will be a mixture of the two.

Methyl orange

Methyl orange is another indicator commonly used in titrations. In an alkaline solution, methyl orange is yellow and the structure is:

$$^-O-S(=O)_2-C_6H_4-N=N-C_6H_4-N(CH_3)_2$$

You have the same sort of equilibrium between the two forms of methyl orange as in the litmus case - but the colors are different.

$$H\text{-}Meor_{(aq)} \rightleftharpoons H^+_{(aq)} + Meor^-_{(aq)}$$

Now, when you add an acid, the hydrogen ion attaches to one of the nitrogens in the nitrogen-nitrogen double bond in yellow (base) version of the molecule, turning it into the acid (red) version of the molecule. According to Le Châtelier's principle, the equilibrium position moves to remove the added hydrogens. Hence, the equilibrium shifts to the left and the methyl orange turns from yellow to red.

On the other hand, if you add an alkali, the hydroxide ion (OH^-) will react with hydrogen ions and, again based on the Le Châtelier's principle, the equilibrium position moves to replace the reacted hydrogen ions. Therefore the equilibrium shifts to the right. The methyl orange turns yellow.

the red form of methyl orange

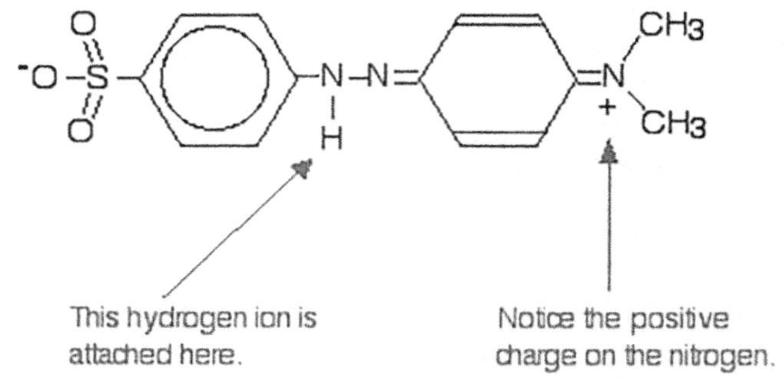

This hydrogen ion is attached here.

Notice the positive charge on the nitrogen.

In the methyl orange case, the half-way stage where the mixture of red and yellow produces an orange color happens at pH 3.7 - nowhere near neutral. This will be explored further down this page.

Phenolphthalein

Phenolphthallso is another commonly used indicator for titrations, and is another weak acid.

$$\text{H-phph}_{(aq)} \rightleftharpoons H^+_{(aq)} + \text{phph}^-_{(aq)}$$

In this case, the acid form of phenolphthalein, H-phph(aq), is colorless, and its ion, phph^- (aq), is bright pink. If we add hydrogen ions to the bright pink basic form, it shifts the position of equilibrium to the left, and turns the indicator colorless. Adding hydroxide ions removes the hydrogen ions from the equilibrium and therefore tips the equilibrium to the right to replace them — turning the indicator pink.

The pH range of indicators

Indicators don't change color sharply at one particular pH (given by their pK_{ind}). Instead, they change over a narrow range of pH.

Suppose the equilibrium is firmly to one side, but now you add something in order to shift it. As the equilibrium shifts, you will start to see more and more of the second color formed, and at some point the eye will start to detect it.

For example, if you had methyl orange in an alkaline solution you will have the dominant color yellow. Now start to add acid so that the equilibrium begins to shift.

At some point there will be enough of the red form of the methyl orange present in the solution that the solution will begin to take on an orange tint. As you continue adding more acid, the red will eventually become so dominant that you can no longer see any yellow color.

There is a gradual change from one color to the other, taking place over a range of pH. As a rough "rule of thumb", the visible change takes place about 1 pH unit either side of the pK_{ind} value.

The exact values for the three indicators we have discussed are:

indicator	pK_{ind}	pH range
litmus	6.5	5 - 8
methyl orange	3.7	3.1 - 4.4
phenolphthalein	9.3	8.3 - 10.0

The litmus color change takes place over an unusually wide range, but it is useful for detecting acids and bases in the lab because it changes color around pH 7. Phenolphthalein or methyl orange would be less useful.

Indicators chosen for titrations

Since the equivalence point of a titration is the point at which you have mixed the two substances in exactly equal proportions, you have to choose an indicator that changes color as close as possible to that point.

Strong acid versus strong base

If you are using phenolphthalein, you have to titrate until it just becomes colorless (at pH 8.3) because that is the closest point to the equivalence point. On the other hand, for methyl orange, you have to titrate until the very first trace of orange appears in the solution. If the solution becomes red it means you are getting away from the equivalence point.

Strong acid versus weak base

In this case, phenolphthalein would be completely useless, but methyl orange starts to change from yellow towards orange close to the equivalence point.

Weak acid versus strong base

Methyl orange is useless. However, the phenolphthalein changes color exactly where you want it to.

Weak acid versus weak base

Neither indicator is any use. Phenolphthalein will have finished changing well before the equivalence point, and methyl orange falls off the graph altogether.

As a whole, you would never titrate a weak acid and a weak base in the presence of an indicator.

Chapter 8: Oxidation-Reduction

8.1 Recognition of oxidation-reduction reactions

Identification of oxidation-reduction processes

An oxidation-reduction reaction is a type of chemical reaction that involves a transfer of electrons between two species. **Redox** is shorthand for *reduction* and *oxidation*. **Reduction** is the **gain of an electron** by a molecule, atom, or ion. **Oxidation** is the **loss of an electron** by a molecule, atom, or ion. These two processes always occur together. Electrons lost by one substance are gained by another. In a redox process, the **oxidation numbers** of atoms are altered. Reduction decreases, or reduces, the oxidation number of an atom. Oxidation increases the oxidation number.

The easiest redox processes to identify are those involving monatomic ions with altered charges. For example, the reaction

$$Zn(s) + Cu^{2+}(aq) \rightarrow Zn^{2+}(aq) + Cu(s)$$

is a redox process because electrons are transferred from Zn to Cu.

However, many redox reactions involve the transfer of electrons from one molecular compound to another. In these cases, **oxidation numbers must be determined** as follows:

Oxidation numbers, sometimes called oxidation states, are positive or negative (+ or –) numbers assigned to atoms in molecules and ions. They allow us to keep track of the electrons associated with each atom. Oxidation numbers are frequently used to write chemical formulas, to help predict properties of compounds, and to help balance equations in which electrons are transferred. Knowledge of the oxidation state of an atom gives us an idea about its positive or negative character. In themselves, oxidation numbers have no physical meaning; they are used to simplify tasks that are more difficult to accomplish without them.

Rules for determining oxidation state

1. Free elements are assigned an oxidation state of 0.
   Example: Al, Na, Fe, H_2, O_2, N_2, Cl_2 etc. have zero oxidation states

2. The oxidation state for any simple one-atom ion is equal to its charge.
   Example: The oxidation state of Na^+ is +1, Be^{2+}, +2, and of F^-, -1

3. The alkali metals (Li, Na, K, Rb, Cs and Fr) in compounds are always assigned an oxidation state of +1.
   Example: In LiOH, Li = +1; in Na_2SO_4, Na = +1

4. Fluorine in compounds is always assigned an oxidation state of -1.
   Example: In HF_2^- and BF_2^-, F=-1

5. The alkaline earth metals (Be, Mg, Ca, Sr, Ba, and Ra) and also Zn and Cd in compounds are always assigned an oxidation state of +2. Similarly, Al & Ga are always +3.
   Example: In $CaSO_4$, Ca = +2; in $AlCl_3$, Al = +3

6. Hydrogen in compounds is assigned an oxidation state of +1, except in hydrides, where it is assigned -1.
   Example: In H_2SO_4, H = +1; in LiH, H = -1

7. Oxygen in compounds is assigned an oxidation state of -2. An exception is peroxide, where in H_2O_2, O = -1.
   Example: In H_3PO_4, O = -2

8. The oxidation states of many other atoms may vary from compound to compound. However, their oxidation states can frequently be determined based on the following rule: the sum of the oxidation states of all the atoms in a species must be equal to the net charge on the species.
   Example: The net charge of $HClO_4$ = 0. Applying the above rules, H = +1 and 4O = -2 x 4 = -8. Therefore, the oxidation state of Cl must be +7 so that the overall charge = 1 + -8 + 7 = 0.

   Example: The net charge of CrO_4^{2-} = -2. Applying the above rules, 4O = -2 x 4 = -8. Therefore, the oxidation state of Cr must be +6 so that the overall charge = -8 + 6 = -2.

For example, the reaction:

$$H_2 + F_2 \rightarrow 2HF$$

is a redox process because the oxidation numbers of atoms are altered. The oxidation numbers of elements are always zero, and oxidation numbers in a compound are never zero. Fluorine is the more electronegative element, so in HF it has an oxidation number of –1 and hydrogen has an oxidation number of +1. This is a redox process where electrons are transferred from H_2 to F_2 to create HF.

In the reaction:

$$NaOH \rightarrow H_2O + NaCl$$

the H-atoms on both sides of the reaction have an oxidation number of +1, the Cl atom has an oxidation number of –1, the Na atom has an oxidation number of +1, and the O atom has an oxidation number of –2. **This is not a redox process because oxidation**

numbers remain unchanged by the reaction. **See section 8.3** for more details on balancing incomplete redox equations in acidic and basic solutions.

8.2 Combustion

Characteristics of biochemical and fossil fuel combustion reactions.

Most combustion reactions are the oxidation of a fuel material with oxygen gas. Complete burning produces carbon dioxide from all the carbon in the fuel, and water from the hydrogen in the fuel. These reactions are used mainly for the production of heat energy. The fuel value is the energy released when 1 g of a material is combusted or burned. This number is a positive number since energy is released, and it is measured by calorimetry.

Biochemical combustion involves the use of enzymes to divide the overall combustion reaction into several smaller, more manageable reactions from which usable energy can be harvested. In our bodies, most of the energy we use comes from the combustion of carbohydrates and fats. Starch, a carbohydrate, is decomposed in the intestines into glucose, $C_6H_{12}O_6$. Glucose is soluble in blood and is transported to cells where it reacts with oxygen, producing carbon dioxide and water while releasing 2803 kJ/mol of energy.

$$C_6H_{12}O_6\ (s) + 6\ O_2 g) \longrightarrow 6CO_2\ (g) + 6H_2O\ (l) \quad \Delta H = -2803\ kJ$$

Carbohydrates supply energy quickly due to their rapid breakdown. Storage of carbohydrates is limited, however.

Fats also undergo combustion in the cells to produce carbon dioxide and water.

$$2C_{57}H_{110}O_6\ (s) + 163\ O_2(g) \longrightarrow 114\ CO_2\ (g) + 110\ H_2O\ (l) \quad \Delta H = -75.520\ kJ$$

(tristearin)

Fuel Values of Carbohydrates, Fats and Proteins	
	Fuel Value (kJ/g)
Carbohydrate	17
Fat	38
Protein	17

This energy is used to regulate body temperature, contract muscles, and build and repair tissues. Any excess energy is stored as fat. Fats are insoluble in water, and they produce more energy per gram than carbohydrates and proteins, so they are prefect for storing excess energy.

However, combustion of protein in the body releases less energy than combustion in a calorimeter, since the products are slightly different. In the body, protein, which contains nitrogen, forms urea with the nitrogen, $(NH_2)_2CO$. But in the calorimeter, proteins form N_2 when combusted.

The combustion of fossil fuels releases energy as well. This energy is in the form of heat. Carbon dioxide and water are formed as the bonds in the fossil fuel rearrange themselves during the combustion reaction. Coal, petroleum and natural gas are known as fossil fuels and are formed from the decay of living matter. Natural gas is a gaseous type of hydrocarbon, primarily methane with small amounts of ethane, propane and butane, and it has a fuel value of 49 kJ/g. Petroleum, with a fuel value of around 48 kJ/g for gasoline, is a liquid that contains mostly hydrocarbons, but also has some compounds containing sulfur, nitrogen or oxygen. These impurities result in air pollution beyond the greenhouse gas CO_2 and soot, carbon-based particles resulting from incomplete combustion. Coal is a solid that consists of large hydrocarbons, as well as some impurities of sulfur, nitrogen or oxygen. Anthracite coal has a fuel value of 31 kJ/g while bituminous coal has a fuel value of 32 kJ/g.

$$CH_4\,(g) + O_2\,(g) \longrightarrow CO_2\,(g) + 2H_2O\,(g) \quad \Delta H = -802\text{ kJ}$$

The combustion of methane

Heat of combustion ΔH_c (also called enthalpy of combustion) is the heat of reaction when a chemical **burns in O_2** to form completely oxidized products such as **CO_2 and H_2O**. It is also the heat of reaction for **nutritional molecules that are metabolized** in the body. The standard heat of combustion **ΔH_c°** takes place at 25 °C and 100 kPa. **Combustion is always exothermic**, so the negative sign for values of ΔHc is often omitted. If a combustion reaction is used in Hess's Law, the value must be negative.

8.3 Oxidation numbers

Balance incomplete redox equations in acidic and basic solutions.

An **oxidizing agent** (also called an oxidant or oxidizer) has the ability to oxidize other substances by removing electrons from them. The **oxidizing agent is reduced** in the process. A **reducing agent** (also called a reductive agent, reductant or reducer) is a substance that has the ability to reduce other substances by transferring electrons to them. The **reducing agent is oxidized** in the process.

Redox reactions may always be written as **two half-reactions**, a **reduction half-reaction** with **electrons as a reactant** and an **oxidation half-reaction** with **electrons as a product**.

Example: The redox reaction:

$$Zn(s) + Cu^{2+}(aq) \rightarrow Zn^{2+}(aq) + Cu(s) \quad \text{and} \quad H_2 + F_2 \rightarrow 2HF$$

may be written in terms of the half-reactions:

$$2e^- + Cu^{2+}(aq) \rightarrow Cu(s) \qquad 2e^- + F_2 \rightarrow 2F^-$$

$$Zn(s) \rightarrow Zn^{2+}(aq) + 2e^-. \quad \text{and} \quad H_2 \rightarrow 2H^+ + 2e^-.$$

An additional (non-redox) reaction, $2F^- + 2H^+ \rightarrow 2HF$ achieves the final products for the second reaction.

Determining whether a chemical equation is balanced (see 11.1 **Balancing of equations**) requires an additional step for redox reactions because there must be a **charge balance.** For example, the equation:

$$Sn^{2+} + Fe^{3+} \rightarrow Sn^{4+} + Fe^{2+}$$

contains one Sn and one Fe on each side, but it is not balanced because the sum of charges on the left side of the equation is +5 and the sum on the right side is +6. A charge balance is obtained by considering each half-reaction separately before multiplying by an appropriate factor so that the number of electrons gained by one half-reaction is the same as the number lost in the other. The two half-reactions may then be combined again into one reaction.

From the example above, one electron is gained in the reduction half-reaction ($Fe^{3+} + e^- \rightarrow Fe^{2+}$), but two are lost in the oxidation half-reaction ($Sn^{2+} \rightarrow Sn^{4+} + 2e^-$). A charge balance is obtained by multiplying the reduction half-reaction by two to obtain the half-reactions:

$$2Fe^{3+} + 2e^- \rightarrow 2Fe^{2+}$$

$$Sn^{2+} \rightarrow Sn^{4+} + 2e^-.$$

The equation: $Sn^{2+} + 2Fe^{3+} \rightarrow Sn^{4+} + 2Fe^{2+}$
is properly balanced because both sides contain the same sum of charges (+8) and electrons cancel from the half-reactions.

Oxidation Number Method:

Redox reactions must be balanced to observe the **Law of Conservation of Mass.** This process is a little more complicated than balancing other reactions because the number of electrons lost must equal the number of electrons gained. Balancing redox reactions, then, conserves not only matter but also charge or electrons. It can be accomplished by slightly varying our balancing process.

Example: $Cr_2O_3(s) + Al\ (s) \longrightarrow Cr\ (s) + Al_2O_{3\ (s)}$

Solution: First, assign oxidation numbers to each atom in order to identify which are losing or gaining electrons.

$Cr_2O_3(s) + Al\ (s) \longrightarrow Cr\ (s) + Al_2O_{3\ (s)}$
3+ 2- 0 0 3+ 2-

Identify those atoms gaining and losing electrons

$Cr^{3+} \longrightarrow Cr^0$ gained 3 electrons (reduction)

$Al^0 \longrightarrow Al^{3+}$ lost 3 electrons (oxidation)

Then, balance the atoms and electrons:

$Cr_2O_3(s) \longrightarrow 2Cr\ (s)$ + 6 electrons

$2Al\ (s)$ + 6 electrons $\longrightarrow Al_2O_3\ (s)$

Half reactions are balanced by adding missing elements. Ignore elements whose oxidation number does not change, and add an H_2O for oxygen and H^+ for hydrogen, if the reaction is to take place in an acidic solution.

$$Cr_2O_3(s) \longrightarrow 2Cr\ (s) + 6\ \text{electrons} + \mathbf{3\ H_2O}$$

In the first half-reaction, there are 3 O atoms needed on the product side. This requires 6 H^+ ions on the reactant side.

$$Cr_2O_3(s) + \mathbf{6\ H^+} \longrightarrow 2Cr\ (s) + 6\ \text{electrons} + \mathbf{3\ H_2O}$$

$$2Al\ (s) + 6\ \text{electrons} + \mathbf{3\ H_2O} \longrightarrow Al_2O_{3\ (s)} + \mathbf{6\ H^+}$$

In the second half-reaction, there are 3 O atoms required on the reactant side. This requires 6 H^+ ions on the product side.

Put the two half reactions together and add the species. Then, cancel out the species that occur in both the reactant and product sides of the equation.

$$Cr_2O_3(s) + 6\ H^+ + 2Al\ (s) + 6\ \text{electrons} + 3\ H_2O \longrightarrow$$

$$2Cr\ (s) + 6\ \text{electrons} + 3\ H_2O + Al_2O_{3\ (s)} + 6\ H^+$$

Therefore, the balanced equation is:

$$Cr_2O_3(s) + 2Al\ (s) \longrightarrow 2Cr\ (s) + Al_2O_{3\ (s)}$$

Example: Balance the following redox reaction:

$$AgNO_3 + Cu \longrightarrow Cu(NO_3)_2 + Ag$$

Solution: First, assign oxidation numbers to each atom in order to identify which are losing or gaining electrons.

$$AgNO_3 + Cu \longrightarrow Cu(NO_3)_2 + Ag$$
$$1+\ 5+\ 2^-\quad 0 \qquad\qquad 2+\ 5+\ 2^-\quad 0$$

Then identify the atoms that are gaining and losing electrons:

$Ag^{1+} + 1\ e^- \longrightarrow Ag^0$ 1 electron gained (reduction)

$Cu^0 \longrightarrow Cu^{2+} + 2\ e^-$ 2 electrons lost (oxidation)

Balance each of the atoms and electrons:

$$AgNO_3 + 1\ e^- \longrightarrow Ag^0$$

$$Cu^0 \longrightarrow Cu(NO_3)_2 + 2\ e^-$$

$$Cu^0 \longrightarrow Cu^{2+} + 2\ e^-$$

There is one electron gained and two electrons lost. This must be balanced, so 2 electrons must be gained.

$$2\ [Ag + 1\ e^- \longrightarrow Ag^0] = \mathbf{2\ AgNO_3 + 2\ e^- \text{—>}\ 2\ Ag^{0+}}$$

Reduction: $2\ AgNO_3 + 2\ e^- \longrightarrow 2\ Ag^{0+}$

Oxidation: $Cu^0 \longrightarrow Cu(NO_3)_2 + 2\ e^-$

Balance the half reactions by adding missing elements. Ignore elements whose oxidation number does not change. Add H_2O for oxygen and H^+ for hydrogen

$2\ AgNO_3 + 2\ e^- \longrightarrow 2\ Ag^{0+}$ **+ 2 NO_3-**

$2NO_3$ + $Cu^0 \longrightarrow Cu(NO_3)_2 + 2\ e^-$

Put the two half reactions together and add the species. Cancel out the species that occur in both the reactants and products.

Reduction: $2\ AgNO_3 + 2\ e^- \longrightarrow 2\ Ag^{0+} + 2\ NO_3$

Oxidation: $2NO_3 + Cu^0 \longrightarrow Cu(NO_3)_2 + 2\ e^-$

The balanced reaction is then:

$2\ AgNO_3 + Cu \longrightarrow Cu(NO_3)_2 + 2\ Ag$

Example: Balance the following redox reaction:

$Ag_2S + HNO_3 \longrightarrow AgNO_3 + NO + S + H_2O$

Solution: First, assign oxidation numbers to each atom:

$Ag_2S + HNO_3 \longrightarrow AgNO_3 + NO + S + H_2O$
$1 + 2^-\ \ 1 + 5 + 2^- \qquad 1 + 5 + 2^-\ \ 2 + 2^-\ \ 0\ \ 1 + 2^-$

Identify those atoms gaining and losing electrons:

Oxidation: $S^{2-} \longrightarrow S^0 + 2\ e^-$

Reduction: $N^{5+} + 3\ e^- \longrightarrow N^{2+}$

Balance the atoms in the equation:

$2NO_3$ - $+ Ag_2S \longrightarrow S + 2\ e^- + 2\ AgNO_3$ (oxidation)

$3\ H^+ + HNO_3 + 3\ e^- \longrightarrow NO + 2\ H_2O$ (reduction)

Balance the electrons that are lost or gained:
There are 2 electrons that are lost and 3 electrons that are gained. This must balance, so find the lowest multiple of the two numbers, which is 6 in this case.

Oxidation:

$3[\ 2NO_3 + Ag_2S \longrightarrow S + 2\ e^- + 2\ AgNO_3]$

$6NO_3 + 3\ Ag_2S \longrightarrow 3\ S + 6\ e^- + 6\ AgNO_3$

Reduction:

$$2[\,3\,H^+ + HNO_3 + 3\,e^- \longrightarrow NO + 2\,H_2O]$$

$$6H^+ + 2\,HNO_3 + 6\,e^- \longrightarrow 2\,NO + 4\,H_2O$$

Finally, put the two half reactions together and add the species. Cancel out the species that occur on both the reactant and product sides of the equation.

$$6\,NO_3^- + 3\,Ag_2S + 6\,H^+ + 2\,HNO_3 + 6\,e^- \longrightarrow$$

$$3\,S + 6\,e^- + 6\,AgNO_3 + 2\,NO + 4\,H_2O$$

Therefore, the balanced equation is:

$$3\,Ag_2S + 8\,HNO_3 \longrightarrow 6\,AgNO_3 + 2\,NO + 3\,S + 4\,H_2O$$

Balancing a redox reaction which occurs in a basic solution is very similar to balancing a redox reaction which occurs in an acidic solution. First, balance the reaction as you would for an acidic solution and then make appropriate adjustments for the basic solution.

Example: Solid chromium(III) hydroxide, $Cr(OH)_3$, reacts with aqueous chlorate ions, ClO_3^-, in basic conditions to form chromate ions, CrO_4^{2-}, and chloride ions, Cl^-.

$$Cr(OH)_3(s) + ClO_3^-(aq) \longrightarrow CrO_4^{2-}(aq) + Cl^-(aq) \quad \text{(basic)}$$

Solution:

1. Write each of the half-reactions:

   $$Cr(OH)_3(s) \longrightarrow CrO_4^{2-}(aq) \text{ and}$$

   $$ClO_3^-(aq) \longrightarrow Cl^-(aq)$$

2. Balance the atoms in each half-reaction. Use H_2O to add oxygen atoms and H^+ to add hydrogen atoms.

   $$H_2O\,(l) + Cr(OH)_3(s) \longrightarrow CrO_4^{2-}(aq) + 5\,H^+\,(aq)$$

   $$6H^+\,(aq) + ClO_3^-(aq) \longrightarrow Cl^-(aq) + 3\,H_2O\,(l)$$

3. Balance the charges of both half-reactions by adding electrons.

   $$H_2O\,(l) + Cr(OH)_3(s) \longrightarrow CrO_4^{2-}(aq) + 5\,H^+\,(aq)$$

   has a charge of +3 on the right and 0 on the left. Adding 3 electrons to the right side will give that side a 0 charge as well.

   $$H_2O\,(l) + Cr(OH)_3(s) \longrightarrow CrO_4^{2-}(aq) + 5\,H^+\,(aq) + 3e^-$$

   The other half-reaction,

   $$6H^+\,(aq) + ClO_3^-\,(aq) \longrightarrow Cl^-(aq) + 3\,H_2O\,(l)$$

   has a charge of -1 on the right side and +5 on the left. Six electrons need to be added to the left side to equal the -1 charge on the right side.

$$6\,e^- + 6H^+\,(aq) + ClO_3^-(aq) \longrightarrow Cl^-(aq) + 3\,H_2O\,(l)$$

4. The number of electrons lost must equal the number of electrons gained, so multiply each of the half-reactions by a number that will give an equal number of electrons lost and gained.

$$H_2O\,(l) + Cr(OH)_3(s) \longrightarrow CrO_4^{2-}(aq) + 5\,H^+\,(aq) + 3e^-$$

$$6\,e^- + 6H^+\,(aq) + ClO_3^-(aq) \longrightarrow Cl^-(aq) + 3\,H_2O\,(l)$$

The first half reaction needs to be multiplied by 2 to equal the 6 electrons gained in the second half-reaction.

$$2[H_2O\,(l) + Cr(OH)_3(s) \longrightarrow CrO_4^{2-}(aq) + 5\,H^+\,(aq) + 3e^-] =$$

$$2H_2O\,(l) + 2\,Cr(OH)_3(s) \longrightarrow 2\,CrO_4^{2-}(aq) + 10\,H^+\,(aq) + 6e^-$$

5. Add the two half-reactions together; canceling out species that appear on both sides of the reaction.

$$\cancel{2H_2O} + 2\,Cr(OH)_3(s) + \cancel{6e^-} + \cancel{6H^+}\,(aq) + ClO_3^-(aq) \longrightarrow$$

$$2\,CrO_4^{2-}(aq) + \underset{\downarrow\;4\,H+}{\cancel{10\,H^+}}\,(aq) + \cancel{6e^-} + Cl^-(aq) + \underset{\downarrow\;4\,H+}{\cancel{3H_2O}}$$

Which simplifies to:

$$2\,Cr(OH)_3(s) + ClO_3^-(aq) \longrightarrow 2\,CrO_4^{2-}(aq) + 4\,H^+ + Cl^-(aq) + H_2O(l)$$

6. Since the reaction occurs in a basic solution and there are 4 H^+ ions on the right side, 4 OH^- need to be added to both sides.

$$4\,OH^-\,(aq) + 2\,Cr(OH)_3(s) + ClO_3^-(aq) \longrightarrow$$

$$2\,CrO_4^{2-}(aq) + 4\,H^+\,(aq) + Cl^-(aq) + 1H_2O\,(l) + 4\,OH^-\,(aq)$$

Combine the H^+ and OH^- where appropriate to make water molecules:

$$4\,OH^-\,(aq) + 2\,Cr(OH)_3(s) + ClO_3^-(aq) \longrightarrow 2\,CrO_4^{2-}(aq) + Cl^-(aq) + 1H_2O(l) + 4\,H_2O(l)$$

Which simplifies to:

$$4\,OH^-\,(aq) + 2\,Cr(OH)_3(s) + ClO_3^-(aq) \longrightarrow 2\,CrO_4^{2-}(aq) + Cl^-(aq) + 5\,H_2O(l)$$

Write the final balanced equation:

$$4\,OH^-\,(aq) + 2\,Cr(OH)_3(s) + ClO_3^-(aq) \longrightarrow 2\,CrO_4^{2-}(aq) + Cl^-(aq) + 5\,H_2O(l)$$

Determine the spontaneity of a chemical reaction using standard reduction potentials.

A **standard cell potential,** E°_{cell}, is the voltage generated by an electrochemical cell at **100 kPa and 25°C,** when all components of the reaction are pure materials (or solutes) at a **concentration of 1 M**. Older textbooks may use 1 atm instead of 100 kPa. Standard solute concentrations may differ from 1 M for solutions that behave in a non-ideal way, but this difference is beyond the scope of general high school chemistry.

Standard cell potentials are calculated from the **sum of the two half-reaction potentials** for the reduction and oxidation reactions occurring in the cell:

$$E^{\circ}_{cell} = E^{\circ}_{red}(\text{cathode}) + E^{\circ}_{ox}(\text{anode}).$$

All half-reaction potentials are relative to the reduction of H^+ to form H_2. This potential is assigned a value of zero:

$$\text{For } 2H^+(aq \text{ at } 1\text{ M}) + 2e^- \rightarrow H_2(g \text{ at } 100\text{ kPa}), \quad E^{\circ}_{red} = 0\text{ V}.$$

The standard potential of an oxidation half-reaction, E°_{ox}, **is equal in magnitude but has the opposite sign to the potential of the reverse reduction reaction.** Standard half-cell potentials are calculated as reduction potentials. These are sometimes referred to as **standard electrode potentials** $E°$. Therefore,

$$E^{\circ}_{cell} = E^{\circ}(\text{cathode}) - E^{\circ}(\text{anode})$$

Example: Given

$E°=0.34$ V for $Cu^{2+}(aq)+e^- \longrightarrow Cu(s)$ and

$E°= -0.76$ V for $Zn^{2+}(aq)+e^- \longrightarrow Zn(s)$,

what is the standard cell potential of the system $Zn(s) + Cu^{2+}(aq) \rightarrow Zn^{2+}(aq) + Cu(s)$

Solution:

$$E^{\circ}_{cell} = E^{\circ}(\text{cathode}) - E^{\circ}(\text{anode})$$
$$= E^{\circ}\left(Cu^{2+}(aq) + 2e^- \rightarrow Cu(s)\right) - E^{\circ}\left(Zn^{2+}(aq) + 2e^- \rightarrow Zn(s)\right)$$
$$= 0.34\text{ V} - (-0.76\text{ V}) = 1.10\text{ V}.$$

Spontaneity

When the value of E° is positive, the reaction is spontaneous. If the E° value is negative, an outside energy source is necessary for the reaction to occur. In the above example, the E° is a positive 1.10 V, therefore this reaction is spontaneous.

Spontaneity will be discussed in more detail in the chapters on thermodynamics.

8.4 Use of Activity Series

Valence and oxidation numbers

The term **valence** is often used to describe the number of atoms that may react to form a compound with another atom by sharing, removing, or losing **valence electrons**. A more useful term is **oxidation number**. The **oxidation number of an ion is its charge**. The oxidation number of an atom sharing its electrons is **the charge it would have if the bonding were ionic**. There are four rules for determining oxidation number:

1. The oxidation number of an element (i.e., a Cl atom in Cl_2) is zero because the electrons in the bond are shared equally.
2. In a compound, the more electronegative atoms are assigned negative oxidation numbers, and the less electronegative atoms are assigned positive oxidation numbers equal to the number of shared electron-pair bonds. For example, hydrogen may only have an oxidation number of –1 when bonded to a less electronegative element or +1 when bonded to a more electronegative element. Oxygen almost always has an oxidation number of –2. Fluorine always has an oxidation number of –1 (except in F_2).
3. The oxidation numbers in an uncharged compound must add up to zero, and the sum of oxidation numbers in a polyatomic ion must equal the overall charge of the ion.
4. The charge on a polyatomic ion is equal to the sum of the oxidation numbers for the species present in the ion. For example, the sulfate ion, SO_4^{2-}, has a total charge of –2. This comes from adding the –2 oxidation number for 4 oxygen (total -8) and the +6 oxidation number for sulfur.

Example: What is the oxidation number of nitrogen in the nitrate ion, NO_3^-?

Solution: Oxygen has the oxidation number of –2 (rule 2), and the sum of the oxidation numbers must be –1 (rule 3). Since the 3 oxygens give a total oxidation number of -–6, and the total charge of the ion must equal -1, the oxidation number of N is +5.

There is a **periodicity in oxidation numbers** as shown in the table below for examples of oxides with the maximum oxidation number. Remember that an element may occur in different compounds in several different oxidation states.

Group	1	2	13	14	15	16	17	18
Oxide with maximum oxidation number	Li_2O	BeO	B_2O_3	CO_2	N_2O_5		Cl_2O_7	XeO_4
	Na_2O	MgO	Al_2O_3	SiO_2	P_2O_5	SO_3	Br_2O_7	
Oxidation number	+1	+2	+3	+4	+5	+6	+7	+8

They are called "oxidation numbers" because oxygen was the element of choice for reacting with materials when modern chemistry began, and as a result, Mendeleev arranged his first table to look similar to this one.

Chapter 9: Precipitation

9.1 Basic solubility rules

Electrolytes and precipitates

Compounds that are completely ionized in water are called **strong electrolytes** because these solutions easily conduct electricity. Most salts are strong electrolytes. For example, all NaCl is present in solution as ions. Other compounds (including many acids and bases) may dissolve in water without completely ionizing. These compounds are referred to as **weak electrolytes** and their ionized state is in equilibrium with the larger un-ionized molecule. Those compounds that dissolve with no ionization (e.g., glucose, $C_6H_{12}O_6$) are called **nonelectrolytes**.

Solubility rules for ionic compounds

Given a cation and anion in aqueous solution, we can determine if a precipitate will form according to some common solubility rules.

1. Particles in solution are free to move about and collide with each other, vastly increasing the likelihood that a reaction will occur compared with particles in a solid phase. Aqueous solutions may react to produce an insoluble substance that will fall out of solution as a solid or gas precipitate in a precipitation reaction. Aqueous solutions may also react to form additional water, or a different chemical in aqueous solution.
2. Salts with NH_4^+ or with a cation from group 1 of the periodic table are *soluble* in water.
3. Nitrates (NO_3^-), acetates ($C_2H_3O_2^-$), chlorates (ClO_3^-), and perchlorates (ClO_4^-) are *soluble*.
4. Cl^-, Br^-, and I^- salts are soluble except in the presence of Ag^+, Hg_2^{2+}, or Pb^{2+} with which they will form precipitates.
5. Sulfates (SO_4^{2-}) are soluble except in the presence of Ca^{2+}, Ba^{2+}, Ag^+, Hg_2^{2+}, or Pb^{2+}.
6. Hydroxides (OH^-) are *insoluble* except with cations from group 1 of the periodic table or in the presence of Ca^{2+}, Sr^{2+}, or Ba^{2+}.
7. Sulfides (S^{2-}), sulfites (SO_3^{2-}), phosphates (PO_4^{3-}), and carbonates (CO_3^{2-}) are *insoluble* except with cations from group 1 of the periodic table.

Ion	Solubility	Exception
All compounds containing alkali metals	All are soluble	
All compounds containing ammonium, NH_4^+	All are soluble	
All compounds containing nitrates, NO_3^-	All are soluble	
All compounds containing chlorates, ClO_3^- and perchlorates, ClO_4^-	All are soluble	
All compounds containing acetates, $C_2H_3O_2^-$	All are soluble	
Compounds containing Cl^-, Br^-, and I^-	Most are soluble	Except halides of Ag^+, Hg_2^{2+}, and Pb^{2+}
Compounds containing F^-	Most are soluble	Except fluorides of Mg^{2+}, Ca^{2+}, Sr^{2+}, Ba^{2+}, and Pb^2
Compounds containing sulfates, SO_4^{2-}	Most are soluble	Except sulfates of Mg^{2+}, Ca^{2+}, Sr^{2+}, Ba^{2+}, and Pb^2
All compounds containing carbonates, CO_3^{2-}	Most are insoluble	Except those containing alkali metals or NH_4^+
All compounds containing phosphates, PO_4^{3-}	Most are insoluble	Except those containing alkali metals or NH_4^+
All compounds containing oxalates, $C_2O_4^{2-}$	Most are insoluble	Except those containing alkali metals or NH_4^+
All compounds containing chromates, CrO_4^{2-}	Most are insoluble	Except those containing alkali metals or NH_4^+
All compounds containing oxides, O^{2-}	Most are insoluble	Except those containing alkali metals or NH_4^+
All compounds containing sulfides, S^{2-}	Most are insoluble	Except those containing alkali metals or NH_4^+
All compounds containing sulfites, SO_3^{2-}	Most are insoluble	Except those containing alkali metals or NH_4^+
All compounds containing silicates, SiO_4^{2-}	Most are insoluble	Except those containing alkali metals or NH_4^+
All compounds containing hydroxides	Most are insoluble	Except those containing alkali metals or NH_4^+, or Ba^{2+}

Summary

The following table summarizes the impact of temperature and pressure on solubility:

Effect on solution of an increase in one variable with the other constant	– = decrease, 0 = no/small change, + = increase, ++ = strong increase				
	Gas solute in liquid solvent			Solid and liquid solutes	
	Average kinetic energy of molecules	Collisions of gas with liquid interface	Solubility	Solubility for an endothermic heat of solution	Solubility for an exothermic heat of solution
Pressure	0	++	+	0	0
Temperature	+	+	–	+	–

SECTION III PRACTICE QUESTIONS

CHAPTER 7

1. **What is the pH of a 0.1 M solution of NaOH ?**

   (A) 1
   (B) 2
   (C) 13
   (D) 14
   (E) 7

   The correct answer is C.
   The concentration of OH is 0.1 so the pOH is 1. The pH + pOH = 14; therefore, the pH is 14 minus the pOH, so pH = 13.

Questions 2-4 refer to the following

(A) NH_4^+
(B) NH_3
(C) NaOH
(D) All of the above
(E) B and C

2. **Would turn litmus paper blue: The correct answer is E.**

3. **Is the conjugate base of NH4+: The correct answer is B.**

4. **Is a Brønsted Acid: The correct answer is A.**

5. **Cause /Effect**

   Statement I
   The equivalence point is when the **acid and base equivalents in the flask are identical**.
   This statement is True.

   Statement II
   The usual goal of titration is to determine an unknown concentration of an acid (or base) by neutralizing it with a known concentration of base (or acid).
   This statement is True.
   Both statements are TRUE – but there is no cause/effect relationship.

Practice Questions

CHAPTER 8

1. **For the reaction: $AgNO_3$ + Cu —> $Cu(NO_3)_2$ + Ag**

   (A) The oxidation reaction is Cu^0 —> $Cu(NO_3)_2 + 2\ e^-$
   (B) The reduction reaction i Cu^0 —> $Cu(NO_3)_2 + 2\ e^-$
   (C) There is one electron gained and two electrons lost.
   (D) There is only one half-reaction
   (E) This is not a oxidation-reduction reaction

   The correct answer is A.

Questions 2-4 refer to the following

(A) 0
(B) +1
(C) +2
(D) -1
(E) cannot be determined

2. **The oxidation number of a free element : The correct answer is A.**

3. **The oxidation number of alkali metals (Li, Na, K, Rb, Cs and Fr): The correct answer is B.**

4. **The net change in electrons in an oxidation-reduction reaction: The correct answer is A.**

5. **Cause and Effect**

   Statement I
   Standard cell potentials are calculated from the sum of the two half-reaction potentials for the reduction and oxidation reactions occurring in the cell.
   This statement is True.

   Statement II
   Standard half-cell potentials are calculated as oxidation potentials.
   This statement is False.

Practice Questions

CHAPTER 9

1. **Which of the following is a good solubility rule?**

   (A) Salts with NH_4^+ or with a cation from group 1 of the periodic table are soluble in water
   (B) Organic compounds tend to be soluble in water
   (C) Acids and bases are only slightly soluble
   (D) All compounds containing NH_4^+ are insoluble
   (E) None of the above

   The correct answer is A.

Questions 2-4 refer to the following

(A) Increases
(B) Decreases
(C) Stays the same
(D) Cannot be predicted

2. **Increasing the temperature ________ the solubility: The correct answer is D.**

3. **Increasing the pressure ____________the solubility of a gas: The correct answer is A.**

4. **Increasing the pressure ____________ the solubility of a solid: The correct answer is C.**

5. **Cause and Effect**

   Statement I
   Compounds that are completely ionized in water are called strong electrolytes.
   This statement is True.

   Statement II
   Compounds that are completely ionized in water easily conduct electricity.
   This statement is True.
   Both statements are TRUE and there IS a cause and effect relationship.

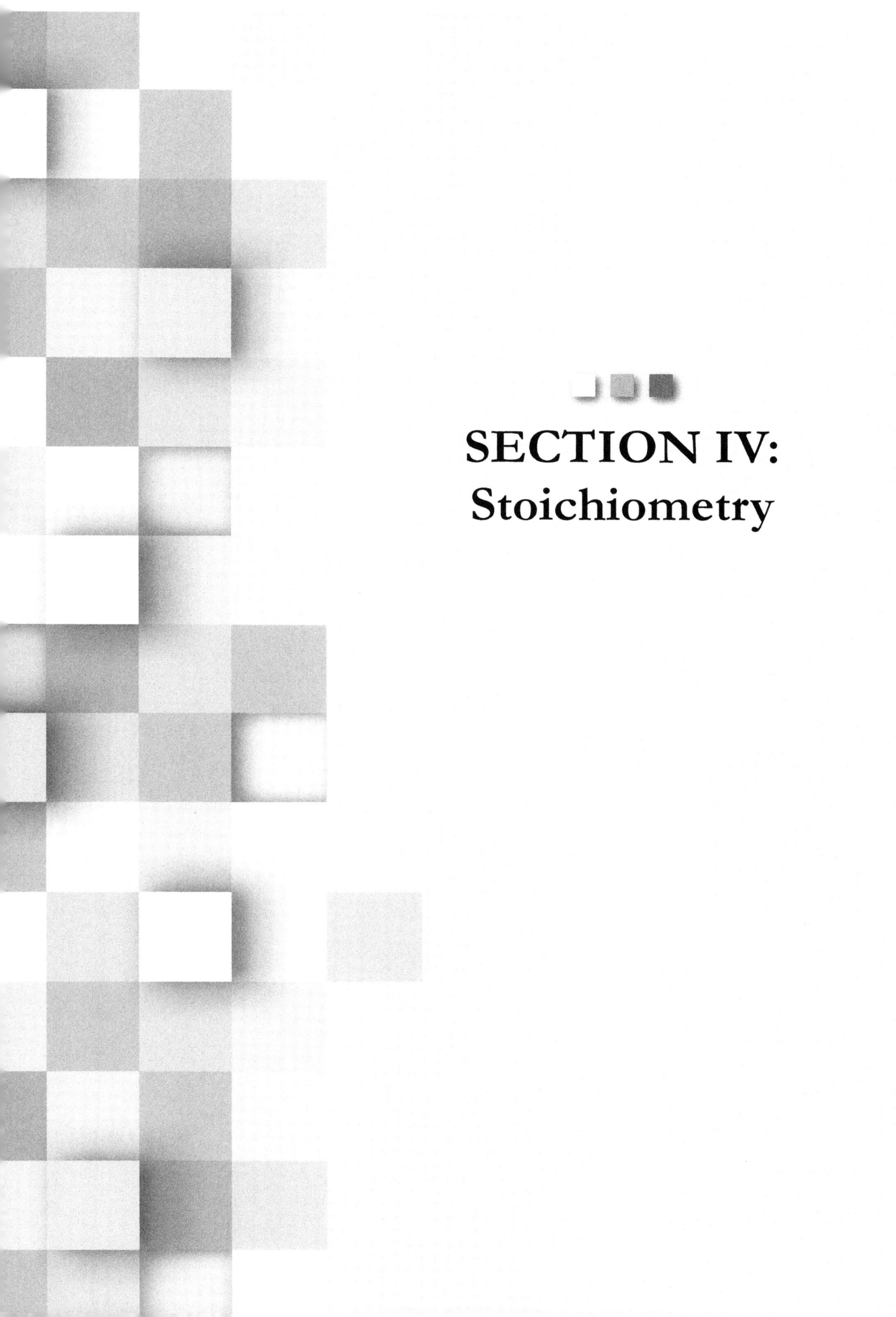

SECTION IV: Stoichiometry

Chapter 10: Mole Concept

10.1 Molar mass

Application of units of mass, volume, and moles to determine concentrations and dilutions of solutions.

As discussed briefly in Chapter 6.1, the amount of solute dissolved in a given amount of solvent is called **concentration.** Solutions that cannot dissolve any additional solute are referred to as **saturated solutions**.

Concentration may be expressed in a number of quantitative ways. Many times they are given as percent by mass or volume, molarity, molality, as well as part per million.

Percent by mass or volume expresses the amount of solute present as a percentage of the total solution present, either by mass or by volumes. This was reviewed in Chapter 6.1.

Example: The label on a 500 mL bottle of hydrogen peroxide, H_2O_2, says 3% by volume. How much hydrogen peroxide does it contain?

Solution: $$\text{\% by volume} = \frac{\text{volume of solute}}{\text{volume of solute + volume of solvent}} \times 100\%$$

(or total volume of solution)

volume of H_2O_2 = 0.03 (volume of solution) = 0.03 x 500 mL = 15 mL of H_2O_2 present.

Molarity, M, is an expression of concentration that compares the number of moles of solute present in the solution to the total volume in liters of the solution.

$$\text{Molarity} = \frac{\text{moles solute}}{\text{volume of solution in liters}}$$

Molarity is the most frequently used concentration unit in chemical reactions because it reflects the number of solute moles available. By using Avogadro's number, the number of molecules in a flask--a difficult image to conceptualize in the lab--is expressed in terms of the volume of liquid in the flask—a straightforward image to visualize and actually manipulate.

Example: What is the molarity of a 5.00 liter solution that was made with 10.0 moles of $CuCl_2$?

Solution: We can use the original formula. Note that in this particular example, where the number of moles of solute is given, the identity of the solute ($CaCl_2$) has nothing to do with solving the problem.

$$\text{Molarity} = \frac{\text{\# of moles of solute}}{\text{Liters of solution}}$$

Given: # of moles of solute = 10.0 moles
Liters of solution = 5.00 liters

$$\text{Molarity} = \frac{10.0 \text{ moles of } CaCl_2}{5.00 \text{ Liters of solution}} = 2.00 \text{ M}$$

Answer = 2.00 M

Example: A 250 ml solution is made with 0.50 moles of NaCl. What is the molarity of the solution?

Solution: In this case we are given ml, while the formula calls for L. We must change the ml to Liters as shown below:

$$250 \cancel{\text{ml}} \times \frac{1 \text{ liter}}{1000 \cancel{\text{ml}}} = 0.25 \text{ liters}$$

Now, solve the problem using the equation:

$$\text{Molarity} = \frac{\text{\# of moles of solute}}{\text{Liters of solution}}$$

Given: Number of moles of solute = 0.50 moles of NaCl
Liters of solution = 0.25 L of solution

$$\text{Molarity} = \frac{0.50 \text{ moles of NaCl}}{0.25 \text{ L}} = 2.0 \text{ M solution}$$

Answer = 2.0 M solution of NaCl

Example: What is the molality of a solution composed of 2.55 g of acetone, $(CH_3)_2CO$ dissolved in 200g of water?

Solution: molar mass of acetone = 58.0 g/mol
grams of acetone = 2.55 g

2.55 g × 1 mol/ 58 g = 0.0440 mol acetone

molality= mol solute/ kg solvent

0.0440 mol acetone/ .200 kg water = 0.220 m

Molality is used when calculating **colligative properties** such as boiling point elevation and freezing point depression. (See section 6.4).

Parts per million, ppm, is used frequently when very small amounts of solute are present, such as when dealing with contaminants in water. When dealing with small amounts of solute present, it is more convenient to use the expression parts per million or even parts per billion. A 1% saline or NaCl solution means that there is 1 part NaCl per one hundred. Remember, percent is based on one hundred so 1% is equal to one part per hundred.

Well, $\frac{1}{100} = \frac{10}{1000} = \frac{100}{10000} = \frac{1000}{100000} = \frac{10000}{1000000} = \frac{10000000}{1000000000}$

or 1 pph = 10 ppt = 10,000 ppm = 10,000,000 ppb

Example: What is the concentration of a solution in pph, ppm, and ppb that contains 10 g of NaCl dissolved in 90 grams of H_2O?

Solution The total mass of the solution is 10 g + 90 g = 100 g. Therefore:

Percent by weight = 10 g NaCl / 100 g solution = 0.1 × 100% = 10%

ppm = 10 g NaCl / 100 g solution = 0.1 × 1,000,000 = 100,000 ppm

ppb = 10 g NaCl / 100 g solution = 0.1 × 1,000,000,000 = 100,000,000 ppb

The **normality** (abbreviated N) of a solution is defined as the number of **equivalents** of a solute per liter of solution.

$$\text{Normality} = \frac{\text{equivalents solute}}{\text{volume of solution in liters}}$$

An equivalent is defined according to the type of reaction being examined, but the number of equivalents of solute is always a whole number multiple of the number of moles of solute, and so the normality of a solute is always a whole-number multiple of its molarity. An equivalent is defined so that one equivalent of one reagent will react with one equivalent of another reagent.

For acid-base reactions, an equivalent of an acid is the quantity that supplies 1 mol of H^+ and an equivalent of a base is the quantity reacting with 1 mol of H^+. For example, one mole of H_2SO_4 in an acid-base reaction supplies two moles of H^+. The mass of one equivalent of H_2SO_4 is half of the mass of one mole of H_2SO_4, and its normality is twice its molarity. In a redox reaction, an equivalent is the quantity of substance that gains or loses 1 mol of electrons.

A **mole fraction** is used to represent a component in a solution as a portion of the entire number of moles present. If you were able to pick out a molecule at random from a solution, the mole fraction of a component represents the probability that the molecule you picked would be that particular component. The mole fractions for all components must add up to one and mole fractions have no units.

$$\text{Mole fraction of a component} = \frac{\text{moles of component}}{\text{total moles of all components}}$$

See 11.2, Solve solution stoichiometry problems.

See Determination of rate laws from concentration and rate data in Factors affecting reaction rates

See Interpret graphical and numerical titration data in 7.3

10.2 Avogadro's number

The mole and Avogadro's number

The *mole* (abbreviated mol) is the SI measure *oy number of particles of a "chemical entity"*, which can be an atom, molecule, formula unit, electron or photon. One mol of anything is just Avogadro's number of that something. Or, if you think like a lawyer, you might prefer the official SI definition:

Amadeo Avogadro (1766-1856) originated the *concept* of this number, whose actual value was first estimated by Josef Loschmidt, an Austrian chemistry teacher, in 1895.

It is obvious that if we had some way of counting atoms by the dozen, we could take dozens of them in a ratio that is exactly equal to the desired atom ratio. Unfortunately, a dozen atoms or molecules are still much too small to work with, so we must find a still larger unit. The "dozen" of a chemist is called the mol. It is composed of 6.022×10^{23} objects.

1 dozen = 12 objects

1 mole = 6.022×10^{23}

The same reasoning that we can use with the dozen applies to the mole. The mole is simply a much larger collection.

1 mole of C atoms + 2 moles of O $\longrightarrow$ 1 mole of CO_2
(6.022×10^{23} atoms of C) + ($2 \times 6.022 \times 10^{23}$ atoms of O) $\longrightarrow$ (6.022×10^{23} molecules of CO_2)

Avogadro's number $N_A = 6.02 \times 10^{23}$, like any pure number, is dimensionless. However, it also defines the mole, so we can also express N_A as 6.02×10^{23} mol^{-1}; in this form, it is properly known as **Avogadro's constant**. This construction emphasizes the role of Avogadro's number as a *conversion factor* between number of moles and number of "entities".

Example: How many atoms of each of S, O, and Fe are present in 15.2 g of Fe SO_4?

Solution:

The molecular weight of Fe SO_4 is (56 + 32 + 4(16)) = 152 atomic mass units

The no. of moles of Fe SO_4 is 15.2 gr./152 = 0.1 moles

The no. of Fe atoms present in 15.2 gr. Fe SO_4 = $0.1 \times (6.02 \times 10^{23}) = 6.02 \times 10^{22}$

The no. of S atoms present in 15.2 gr. Fe SO_4 = $0.1 \times (6.02 \times 10^{23}) = 6.02 \times 10^{22}$

The no. of O atoms present in 15.2 gr. Fe SO_4 = $4 \times 0.1 \times (6.02 \times 10^{23}) = 6.02 \times 10^{22} = 2.408 \times 10^{23}$

Avogadro's hypothesis

Avogadro's hypothesis states that equal volumes of different gases at the same temperature and pressure contain equal numbers of molecules. **Avogadro's law** states that the volume of a gas at constant temperature and pressure is directly proportional to the quantity of gas, or:

$V \propto n$ where *n* is the number of moles of gas.

Avogadro's law and the combined gas law yield $V \propto \frac{nT}{P}$

The proportionality constant *R*—the **ideal gas constant**—is used to express this proportionality as the **ideal gas law**:

$$PV = nRT$$

The ideal gas law ($PV = nRT$) is useful because it contains all the information of Charles's, Avogadro's, Boyle's, and the combined gas laws in a single equality.

If pressure is given in atmospheres and volume is given in liters, a value for *R* of **0.08206 L-atm/(mol-K)** is used. If pressure is given in Pascals (newtons/m^2) and volume in cubic meters, then the SI value for *R* of **8.314 J/(mol-K)** may be used. This is because a joule is defined as a Newton-meter. A value for *R* of **8.314 m^3-Pa/(mol-K)** is the ideal gas constant that is used with joules.

Many problems are given at "**standard temperature and pressure**" or "**STP**." Standard conditions are *exactly* **1 atm** (101.325 kPa) and **0°C (273.15 K)**. At STP, one mole of an ideal gas has a volume of:

$$V = \frac{nRT}{P}$$

$$= \frac{(1 \text{ mole})\left(0.08206 \ \frac{\text{L-atm}}{\text{mol-K}}\right)(273 \text{ K})}{1 \text{ atm}} = 22.4 \text{ L}.$$

The value of 22.4 L is known as the **standard molar volume of any gas at STP**.

10.3 Empirical and molecular formulas

Empirical formula

The **empirical formula** of a chemical compound is the simplest positive integer ratio of atoms present in a compound. An example of this concept is that the empirical formula of sulfur monoxide, or SO, would simply be SO. This, however, is also the empirical formula of disulfur dioxide, S_2O_2. This means that sulfur monoxide and disulfur dioxide, while very different molecules, both have the same empirical formula.

An empirical formula does not specify the arrangement or number of atoms. It is standard for many ionic compounds, like $CaCl_2$, and for macromolecules, such as SiO_2.

The **molecular formula**, on the other hand, shows the **number** of each type of atom in a molecule and the **structural formula** shows the **arrangement** of the atoms in the molecule. It is therefore possible for different types of compounds to have equal empirical formulas.

An important example he CH_2O, which is the empirical formula for glucose ($C_6H_{12}O_6$), ribose ($C_5H_{10}O_5$), acetic acid ($C_2H_4O_2$), and formaldehyde (CH_2O). All have different molecular formulas but the same empirical formula: CH_2O is the molecular formula for formaldehyde, but acetic acid has double the number of atoms, ribose has five times the number of atoms, and glucose has six times the number of atoms.

N-hexane, which is a hydrocarbon, has the structural formula $CH_3CH_2CH_2H\ _2CH_2CH_3$, which shows that it has 6 carbon atoms arranged in a chain, and 14 hydrogen atoms. Hexane's molecular formula is C_6H_{14}, and its empirical formula is $CH\ _7$, showing a C:H ratio of 3:7.

Calculation

Suppose you are asked to calculate the empirical formula of methyl acetate, a solvent commonly used in paints, inks, and adhesives. When methyl acetate was chemically analyzed, it was discovered to have 48.64% carbon (C), 8.16% hydrogen (H), and 43.20% oxygen (O). For the purposes of determining empirical formulas, we assume that we have 100 g of the compound. If this is the case, the percentages will be equal to the mass of each element in grams.

Step 1

Change each percentage to an expression of the mass of each element in grams. That is, 48.64% C becomes 48.64 g C, 8.16% H becomes 8.16 g H, and 43.20% O becomes 43.20 g O.

Step 2

Convert the amount of each element from gram to mole by dividing each element percentage by its atomic mass.

$$\left(\frac{48.64 \text{ g C}}{1}\right)\left(\frac{1 \text{ mol}}{12.01 \text{ g C}}\right) = 4.049 \text{ mol}$$
$$\left(\frac{8.16 \text{ g H}}{1}\right)\left(\frac{1 \text{ mol}}{1.008 \text{ g H}}\right) = 8.095 \text{ mol}$$
$$\left(\frac{43.20 \text{ g O}}{1}\right)\left(\frac{1 \text{ mol}}{16.00 \text{ g O}}\right) = 2.7 \text{ mol}$$

Step 3

Divide each of the calculated values by the smallest of these values (2.7)

$$\frac{4.049 \text{ mol}}{2.7 \text{ mol}} = 1.5$$
$$\frac{8.095 \text{ mol}}{2.7 \text{ mol}} = 3$$
$$\frac{2.7 \text{ mol}}{2.7 \text{ mol}} = 1$$

Step 4

Since the obtained numbers are not all cardinal numbers we have to convert them all to cardinal numbers. So we multiply these numbers by integers (here two) in order to get whole numbers; if an operation is done to one of the numbers, it must be done to all of them.

$1.5 \times 2 = 3$ carbons
$3 \times 2 = 6$ hydrogens
$1 \times 2 = 2$ oxygens

Therefore, the empirical formula of methyl acetate is $C_3H_6O_2$. This formula also happens to be methyl acetate's molecular formula.

Empirical versus Molecular Formulas

The formulas we have calculated in the previous section express the simplest atomic ratio between the elements in the compound. Such formulas are called empirical formulas. An empirical formula does not necessarily represent the actual numbers of atoms present in a molecule of a compound; it represents only the ratio between those numbers. Actually, the number of atoms of each element found in the smallest freely existing unit or molecule of the compound is expressed by the molecular formula of the compound. The molecular formula of a compound may be the empirical formula, or it may be a multiple of the empirical formula. An example is the molecular formula of butene, C_4H_8, showing that each freely existing molecule of butene contains four atoms of carbon and eight atoms of hydrogen. Its empirical formula is CH_2. One molecule of ethylene (molecular formula C_2H_4) contains two atoms of carbon and four atoms of hydrogen. Its empirical formula is CH_2. Both have the same empirical formula, yet they are different compounds with different molecular formulas. Butene is C_4H_8, or four times the empirical formula; ethylene is C_2H_4, or twice the empirical formula.

If the formula weight, or molecular weight, is known, the molecular formula of a compound can be determined from the empirical formula.

In order to determine the molecular formula of a compound two kinds of data are needed:

1. Its composition, from which we can calculate its empirical formula, and
2. Its molecular weight, which will be a multiple of the empirical formula weight. This same multiple of the empirical formula is the multiple of each of the atoms in the molecular formula.

Determine empirical formulas from experimental data.

If we know the chemical composition of a compound, we can calculate an **empirical formula** for it. An empirical formula is the **simplest formula** using the smallest set of integers to express the **ratio of atoms** present in a molecule.

The **percent composition** of a substance is the **percentage by mass of each element**. Chemical composition is used to verify the purity of a compound in the lab. An impurity will make the actual composition vary from the expected one.

To determine percent composition from a formula, follow these steps:

Write down the **number of atoms each element contributes** to the formula.

1. Multiply these values by the molecular weight of the corresponding element to determine the **grams of each element in one mole** of the formula
2. Add the values from step 1 to obtain the **formula mass**.
3. Divide each value from step 2 by the formula mass from step 3 and multiply by 100% to obtain the **percent composition of each element.**

Example: What is the chemical composition of ammonium carbonate $(NH_4)_2CO_3$?

Solution:

1. One $(NH_4)_2CO_3$ contains 2 N, 8 H, 1 C, and 3 O.

2.

$$\frac{2 \text{ mol N}}{\text{mol }(NH_4)CO_3} \times \frac{14.0 \text{ g N}}{\text{mol N}} = 28.0 \text{ g N/mol }(NH_4)CO_3$$

$$8(1.0) = 8.0 \text{ g H/mol }(NH_4)CO_3$$

$$1(12.0) = 12.0 \text{ g C/mol }(NH_4)CO_3$$

$$3(16.0) = 48.0 \text{ g O/mol }(NH_4)CO_3$$

$$\text{Sum is} \quad 96.0 \text{ g }(NH_4)CO_3\text{/mol }(NH_4)CO_3$$

3) $\%N = \frac{28.0 \text{ g N/mol }(NH_4)_2CO_3}{96.0 \text{ g }(NH_4)_2CO_3\text{/mol }(NH_4)_2CO_3} = 0.292 \text{ g N/g }(NH_4)_2CO_3 \text{ X}100\% = 29.2\%$

4) $\%H = \frac{8.0}{96.0}\text{X}100\% = 8.3\%$ $\quad \%C = \frac{12.0}{96.0}\text{X}100\% = 12.5\%$ $\quad \%O = \frac{48.0}{96.0}\text{X}100\% = 50.0\%$

If we know the chemical composition of a compound, we can calculate an **empirical formula** for it. An empirical formula is the **simplest formula** using the smallest set of integers to express the **ratio of atoms** present in a molecule.

To determine an empirical formula from a percent composition, follow these steps:

1. Change the "%" sign to grams for a base sample of 100 g of the compound.
2. Determine the moles of each element in 100 g of the compound.
3. Divide the values from step 1 by the smallest value to obtain ratios.
4. Multiply by an integer if necessary to get a whole-number ratio.

Example: What is the empirical formula of a compound with a composition of 63.9% Cl, 32.5% C, and 3.6% H?

Solution:

1. We will use a basis of 100 g of the compound containing 63.9 g Cl, 32.5 g C, and 3.6 g H.

$$\text{In 100 g, there are: } 63.9 \text{ g Cl} \times \frac{\text{mol Cl}}{35.45 \text{ g Cl}} = 1.802 \text{ mol Cl}$$

$$32.5/12.01 = 2.706 \text{ mol C}$$

$$3.6/1.01 = 3.56 \text{ mol H}$$

2. Dividing these values by the smallest yields:

$$\frac{2.706 \text{ mol C}}{1.802 \text{ mol Cl}} = 1.502 \text{ mol C/mol Cl}$$

$$\frac{3.56 \text{ mol H}}{1.802 \text{ mol Cl}} = 1.97 \text{ mol H/mol Cl}$$

Therefore, the elements are present in a ratio of C:H:Cl=1.50:2.0:1

3. Multiply the entire ratio by 2 because you cannot have a fraction of an atom. This corresponds to a ratio of 3:4:2 for an empirical formula of $C_3H_4Cl_2$.

The **molecular formula** describing the **actual number of atoms in the molecule** might actually be $C_3H_4Cl_2$ or it might be $C_6H_8Cl_4$ or some other multiple that maintains a 3:4:2 ratio.

Chapter 11: Chemical Equations

11.1 Balancing of equations

Balancing chemical equations.

A properly written chemical equation must contain properly written formulas and must be **balanced**. Chemical equations are written to describe a certain number of moles of reactants becoming a certain number of moles of products. Chemical equations obey the law of conservation of mass in that no atoms are created or destroyed, so each element has the same number of total atoms on the left of the arrow as it has on the right of the arrow when the equation is balanced. The number of moles of each compound is indicated by its **stoichiometric coefficient**. The number of atoms of an element is determined by multiplying the coefficient by the subscript, with subscripts outside of parentheses being multiplied by subscripts inside the parenthesis. This is done for each element in each compound. Then all the atoms of that element on that side of the arrow are added together.

Example: In the reaction

$$2H_2(g) + O_2(g) \rightarrow 2H_2O(l)$$

hydrogen has a stoichiometric coefficient of two, oxygen has a coefficient of one, and water has a coefficient of two because 2 moles of hydrogen react with 1 mole of oxygen to form two moles of water.

The number of atoms of hydrogen on the left is 4 which is found by multiplying the coefficient of 2 by the subscript of 2. The number of atoms of hydrogen on the right is found in the same way to be 4. The number of atoms of oxygen on the left is 2 since the coefficient of 1 is multiplied by the subscript of 2. Using the same method on the right, the coefficient of 2 is multiplied by the subscript of 1 (no subscript means "1") to obtain a total of 2 atoms of oxygen. Therefore, this reaction is balanced.

In a balanced equation, the stoichiometric coefficients are chosen so that the equation contains an **equal number of each type of atom on each side**. In our example, there are four H atoms and two O atoms on both sides. Therefore, the equation is balanced with respect to atoms.

Balance properly written chemical equations.

Balancing equations is a multi-step process.

1. Determine the **correct formulas** for all compounds.
2. Write an **unbalanced equation**. This requires knowledge of the proper products which can be predicted from the reactants. Reactants are written on the left and products are written on the right.
3. Determine the **number of each type of atom on each side** of the equation to determine whether or not the equation is already balanced. Under the reactants, list all the elements in the reactants starting with metals, then nonmetals, listing oxygen last and hydrogen next to last. Under the products, list all the elements in the same order as those under the reactants – preferably, straight across from them.
4. Count the atoms of each element on the left side and list the numbers next to the elements. Repeat for products. Don't forget that subscripts outside a parenthesis multiply subscripts inside the parenthesis. If each element has the same number of atoms on the right that it has on the left, the equation is balanced. If not, proceed to Step 5.
5. For the first element in the list that has unequal numbers of atoms, use a coefficient (whole number to the left of the compound or element) on either the left of the arrow or the right of the arrow to give an equal number of atoms. NEVER change the subscripts to balance an equation.
6. Go to the next unbalanced element and balance it, moving down the list until all are balanced.
7. Start back at the beginning of the list and actually count the atoms of each element on each side of the arrow to make sure the number listed is the actual number. Re-balance and re-check as needed.

Example: Balance the chemical equation describing the combustion of methanol (CH_4O) in oxygen to produce only carbon dioxide and water.

Solution:

1. The structural formula of methanol is CH_3OH, so its molecular formula is CH_4O. The formula for carbon dioxide is CO_2. Therefore the unbalanced equation is:

   $$CH_4O + O_2 \longrightarrow CO_2 + H_2O$$

2. On the left there are 1 C, 4 H, and 3 O atoms. On the right, there are 1 C, 2 H, and 3 O atoms. The equation is close to being balanced, but there is still work to do.
3. Assuming that CH_4O has a stoichiometric coefficient of one means that the left side has 1C and 4H that also must be present on the right. Therefore the stoichiometric coefficient of CO_2 will be 1 to balance C and the stoichiometric coefficient of H_2O will be 2 to balance H. Now we have:

   $$CH_4O + O_2 \longrightarrow CO_2 + 2H_2O$$

4. and only oxygen remains unbalanced. There are 4 O on the right and one of these is accounted for by methanol, leaving 3 O to be accounted for by O_2. This gives a stoichiometric coefficient of 3/2 and a balanced equation:

$$CH_4O + 3/2O_2 \longrightarrow CO_2 + 2H_2O$$

5. Whole-number coefficients are achieved by multiplying the entire equation by two:

$$2CH_4O + 3O_2 \longrightarrow 2CO_2 + 4H_2O$$

Reactions among ions in aqueous solution may be represented in three ways. When solutions of hydrochloric acid and sodium hydroxide are mixed, a reaction occurs and heat is produced. The **molecular equation** for this reaction is:

$$HCl + NaOH \longrightarrow NaCl + H_2O.$$

This is a molecular equation because the **complete chemical formulas** of reactants and products are shown. But in reality, both HCl and NaOH are strong electrolytes and exist in solution as ions. This is represented by a **complete ionic equation** that shows all the dissolved ions:

$$H^+ + Cl^- + Na^+ + OH^- \longrightarrow Na^+ + Cl^- + H_2.$$

Because $Na^+(aq)$ and $Cl^-(aq)$ appear as both reactants and products, they play no role in the reaction. Ions that appear in identical chemical forms on both sides of an ionic equation are called **spectator ions** because they aren't part of the action. When spectator ions are removed from a complete ionic equation, the result is a **net ionic equation** that shows the actual changes that occur to the chemicals when these two solutions are mixed together**ct.**

$$H^+ + OH^- \longrightarrow H_2O$$

An additional requirement for **redox** reactions (see Chapter 8) is that the equation contains an **equal charge on each side**. In the example above, the positive charge on the H^+ ion cancels out the negative charge on the OH^- ion, giving an overall neutral charge for both sides of the equation. Redox reactions may be divided into half-reactions which either gain or lose electrons.

11.2 Stoichiometric calculations

The mole is the chemist's counting unit. Working with the mole should become second nature since moles are very important in stoichiometry. Balanced equations provide mole ratios, not mass ratios.

Solve mass-mass stoichiometry problems.

Mass to moles

First determine the molar mass of the substance by adding the molar masses for each element in the substance and multiplying by the number of atoms present.

Example: What is the molar mass of $CuSO_4$?

Solution:

(1 mole of Cu = 63.5 g) + (1 mol of S = 32 g) + (4 mol O = 4 × 16 = 48 g) $\longrightarrow$ 143.5 g/mol $CuSO_4$

Then determine the number of moles present using the molar mass conversion 1 mol = molar mass of substance, for example: 1 mol $CuSO_4$ = 143.5 g $CuSO_4$

Example: 315 g of $CuSO_4$ is how many moles of $CuSO_4$?
315 g × 1 mol/143.5 g = 2.20 mol $CuSO_4$

Mole to grams conversions are the reverse

Solving these problems is a three-step process:

1. Grams of the given compound (known mass) are converted to moles (known moles) using molar mass.
2. The moles of the given compound (known moles) are related to moles of the second compound (unknown moles) by relating their stoichiometric coefficients.
3. The moles of the second compound (unknown moles) are converted to grams (unknown mass) using the molar mass of the compound.

These steps are often combined in one series of multiplications, which may be described as **"grams to moles to moles to grams.**" Another way to look at them is **"grams of known to moles of known to moles of unknown to grams of unknown.**"

Example: What mass of oxygen is required to consume 95.0 g of ethane in the following reaction?

$$2C_2H_6 + 7O_2 \rightarrow 4CO_2 + 6H_2O$$

Solution:

$$\underbrace{95.0\text{ g } C_2H_6 \times \frac{1\text{ mol } C_2H_6}{30.1\text{ g } C_2H_6}}_{\text{step 1}} \times \underbrace{\frac{7\text{ mol } O_2}{2\text{ mol } C_2H_6}}_{\text{step 2}} \times \underbrace{\frac{32.0\text{ g } O_2}{1\text{ mol } O_2}}_{\text{step 3}} = 359\text{ g } O_2$$

By expressing the molar mass of the first or given compound as moles per gram, the grams of the known compound cancel. By putting the moles of the known compound on the bottom with the molecular coefficient of the unknown compound on top in the second conversion the moles of the known are canceled and converted to moles of unknown. And finally, by expressing the molecular mass of the second compound as grams per mole, its moles are converted to grams for the final answer and the units (by canceling) end up as grams of the unknown.

Solve mass-gas volume stoichiometry problems.

The progress of reactions that produce or consume a gas may be described by measuring gas volume instead of mass. If the gas is at STP (standard temperature and pressure), then the equivalency 22.4 liters = 1 mole can be used.

Example: How many liters of nitrogen are needed to produce 34.0 grams of NH_3?

Solution: Write and balance the equation: $N_2 + 3\ H_2 \longrightarrow 2\ NH_3$

List what is given and what is unknown:

Mass of NH_3: 34.0 g	Mass of N_2: unknown
Molar mass of NH_3: 17.0 g/mol	Molar mass of N_2: 28.0 g/mol
Moles of NH_3: 2 (coefficient)	Moles N_2: 1 (no coefficient)

Use the 3-step label canceling method:

$$34.0\ g\ NH_2 \times \frac{1\ mole\ NH_2}{17.0\ g\ NH_3} \times \frac{1\ mole\ N_2}{2\ moles\ NH_3} \times \frac{22.4\ liters}{1\ mole\ N_2} = 22.4\ liters\ N_2$$

For 34 L grams of NH_3 to be formed, 22.4 L of N_2 would be needed.

If you need to get the number of molecules or formula units, use Avogadro's number of 6.02×10^{23} particles per mol in the equation.

If the gas is not at STP, the best way to solve these problems is to use the ideal gas equation to convert between volume and moles.

$$n = \frac{PV}{RT} \text{ and } V = \frac{RT}{nP}$$

If a volume is given, the steps are "**volume of known to moles of known to moles of unknown to grams of unknown**." If a mass is given, the steps will be "**grams of known to moles of known to moles unknown to volume of unknown**."

Example: What volume of oxygen, in liters, is generated at 40° C and 1 atm by the decomposition of 280 g of potassium chlorate in the following reaction?

$$2KClO_3 \rightarrow 2KCl + 3O_2(g)$$

Solution: We are given a mass and are asked to find volume, so the steps in the solution will be "grams to moles to moles to liters."

Thus we convert grams to moles:

$$280\ g\ KClO_3 \times \frac{1\ mol\ KClO_3}{122.548\ g\ KClO_3} \times \frac{3\ mol\ O_2}{2\ mol\ KClO_3} = 3.427\ mol\ O_2$$

Then we use the ideal gas law and convert to volume, in liters.

$$V = \frac{RT}{nP} = \frac{\left(0.08206\ \frac{\text{L-atm}}{\text{mol-K}}\right)(273.15+40)\text{K}}{(3.427\ \text{mol}\ O_2)(1\ \text{atm})} = 7.50\ \text{L}\ O_2$$

Solve solution stoichiometry problems.

Solving stoichiometry problems involving a solution requires that the moles of solute for each solution first be determined. This is done using the molarity definition. Once the moles of solute are determined, the problem becomes a mole-mole stoichiometry problem or a limiting reactant problem.

Avogadro's law states that the volume of a gas, at constant temperature and pressure, will be proportional to the number of moles. All components of a reaction are usually in the same vessel at the same temperature and pressure. Therefore, **stoichiometric coefficients may be used directly for volumes just as they are used for moles.**

Example: How many m^3 of water vapor are produced when 8.2 m^3 of hydrogen are consumed in this reaction: $2H_2(g) + O_2(g) \rightarrow 2H_2O(g)$?

Solution:

$$8.2\ m^3\ H_2 \times \frac{2\ m^3\ H_2O}{2\ m^3\ H_2} = 8.2\ m^3\ H_2O$$

Example: How much of a 0.50 M $Pb(NO_3)_2$ solution is required to completely react with 25 g of KI?

$$Pb(NO_3)_2\ (aq) + 2\ KI\ (s) \longrightarrow PbI_2\ (s) + 2\ KNO_3\ (aq)$$

Solution:

1. Find the moles of KI reacting:

   25 g KI × 1 mol KI/g/mol = 0.15 mol KI

2. Use the mole ratio from the balanced equation to find the required number of moles:

   0.15 mole KI × 1 mol $Pb(NO_3)_2$ / 2 mol KI = 0.075 mol $Pb(NO_3)_2$

3. Use molarity to determine the volume of $Pb(NO_3)_2$ required.

   M = mol solute/L solution

   L solution = mol solute/M = 0.075 mol/0.50 M =0.15 L or 150 mL

11.4 Percent Yield

Calculate percent yield

The **yield of a reaction is the amount of product** obtained. This value is nearly always less than what would be predicted from a stoichiometric calculation because side reactions may produce different products, the reverse reaction may occur, and some material may be lost during the procedure. The yield from a stoichiometric calculation on the limiting reagent is called the theoretical yield.

Percent yield is the actual yield divided by the theoretical yield times 100%:

$$\text{Percent yield} = \frac{\text{Actual yield}}{\text{Theoretical yield}} \text{x}100\%$$

Example: 387 g $AlBr_3$ are produced by the reaction described in the previous example. What is the percent yield?

Solution: $\frac{387 \text{ g } AlBr_3}{445 \text{ g } AlBr_3} \text{x}100\% = 87.0\%$ yield

11.5 Limiting reactants

Solve stoichiometry problems with limiting reactants.

The **limiting reagent** of a reaction is the **reactant that runs out first**. This reactant **determines the amount of products formed**, and any **other reactants remain unconverted** to product and are called **excess reagents**.

Example: Consider the reaction $3H_2 + N_2 \rightarrow 2NH_3$ and suppose that 3 mol H_2 and 3 mol N_2 are available for this reaction. What is the limiting reagent?

Solution: The equation tells us that 3 mol H_2 will react with one mol N_2 to produce 2 mol NH_3. This means that H_2 is the limiting reagent because when it is completely used up, 2 mol of N_2 will still remain.

The limiting reagent may be determined by **dividing the number of moles of each reactant by its stoichiometric coefficient.** This determines the moles of reactant if each reactant were limiting. The **lowest result** will indicate the actual limiting reagent. Remember to use moles and not grams for these calculations.

Example: 50.0 g Al and 400 g Br_2 react according the following equation:

$$2Al + 3Br_2 \rightarrow 2AlBr_3$$

until the limiting reagent is completely consumed. Find the limiting reagent, the mass of $AlBr_3$ formed, and the excess reagent remaining after the limiting reagent is consumed.

Solution: First convert both reactants to moles:

$$50.0\text{ g Al} \times \frac{1\text{ mol Al}}{26.982\text{ g Al}} = 1.85\mathit{3}\text{ mol Al and } 400\text{ g Br}_2 \times \frac{1\text{ mol Br}_2}{159.808\text{ g Br}_2} = 2.05\mathit{3}\text{ mol Br}_2$$

The final digits in the intermediate results above are italicized because they are insignificant. Dividing by stoichiometric coefficients gives:

$$1.85\mathit{3}\text{ mol Al} \times \frac{\text{mol reaction}}{2\text{ mol Al}} = 0.92\mathit{65}\text{ moles of reactant if Al is limiting.}$$

$$2.50\mathit{3}\text{ mol Br}_2 \times \frac{\text{mol reaction}}{3\text{ mol Br}_2} = 0.834\mathit{3}\text{ moles of reactant if Br}_2\text{ is limiting.}$$

Br_2 is the lower value and is the limiting reagent.

The reaction is expected to produce:

$$2.50\mathit{3}\text{ mol Br}_2 \times \frac{2\text{ mol AlBr}_3}{3\text{ mol Br}_2} \times \frac{266.694\text{ g AlBr}_3}{\text{mol AlBr}_3} = 445\text{ g AlBr}_3.$$

The reaction is expected to consume:

$$2.50\mathit{3}\text{ mol Br}_2 \times \frac{2\text{ mol Al}}{3\text{ mol Br}_2} \times \frac{26.982\text{ g Al}}{\text{mol Al}} = 45.0\text{ g Al.}$$

50.0 g Al – 45.0 g Al = 5.0 g Al.

These 5.0 g of aluminum will remain after all of the limiting reagent has been used up.

SECTION IV PRACTICE QUESTIONS

CHAPTER 10

1. **15.00 grams of aluminum sulfide and 10.00 g of water react until the limiting reagent is used up. Here is the balanced equation for the reaction:**

$$Al_2S_3 + 6\ H_2O \longrightarrow 2\ Al(OH)_3 + 3\ H_2S$$

(A) Water
(B) Al_2S_3
(C) H_2S
(D) 2 $Al(OH)_3$
(E) The reaction is not limited

The correct answer is A.
Determine the moles of Al_2S_3 and H_2O
aluminum sulfide: 15.00 g × 1 mol ÷ 158 g/mol = 0.099895 mol
water: 10.00 g × 1 mol ÷ 11 g/mol = 0.555093 mol
Divide each mole amount by equation coefficient
aluminum sulfide: 0.099895 mol ÷ 1 mol = 0.099895
water: 0.555093 mol ÷ 6 mol = 0.0925155
The water is the lesser amount; it is the limiting reagent.

Questions 2-4 refer to the following

(A) 22.5 L
(B) 6.2×10^{23}
(C) CH_2O
(D) $C_6H_{12}O_6$
(E) none of the above

2. **This is the volume of water vapor at STP. The correct answer is A**

3. **This is the empirical formula for sucrose. The correct answer is C**

4. **This is the number of atoms in 3 mole of NH_3. The correct answer is E.**

5. **Cause/Effect**

Statement I. Electrons are gained in a reduction reaction

Statement II. Energy must be conserved.

Answer: Both statements are True and there is a cause/effect relationship.

CHAPTER 11

1. **What is the pH of a 0.16 M solution of hydrochloric acid?**

   (A) pH = 1.6
   (B) pH = .80
   (C) pH = .016
   (D) pH = 8.0
   (E) Cannot be determined

   The correct answer is B.
   HCl is a strong acid, so $[H_3O^+]$ = concentration of HCl = 0.16 M
   pH = -log (0.16) = 0.7959 = 0.80

Questions 2-4, refer to the following:

(A) **Strong acid**
(B) **pH**
(C) **Equivalence point**
(D) **Oxidation reaction**

2. **A reaction in which electrons are lost: The correct answer is D.**

3. **Dissociates completely in solution: The correct answer is A.**

4. **A measure of the [H+] in solution: The correct answer is B.**

5. **Cause and Effect**

   Statement I
   Electrons are gained in a reduction reaction.
   This statement is True.

   Statement II
   Energy must be conserved.
   This statement is True.

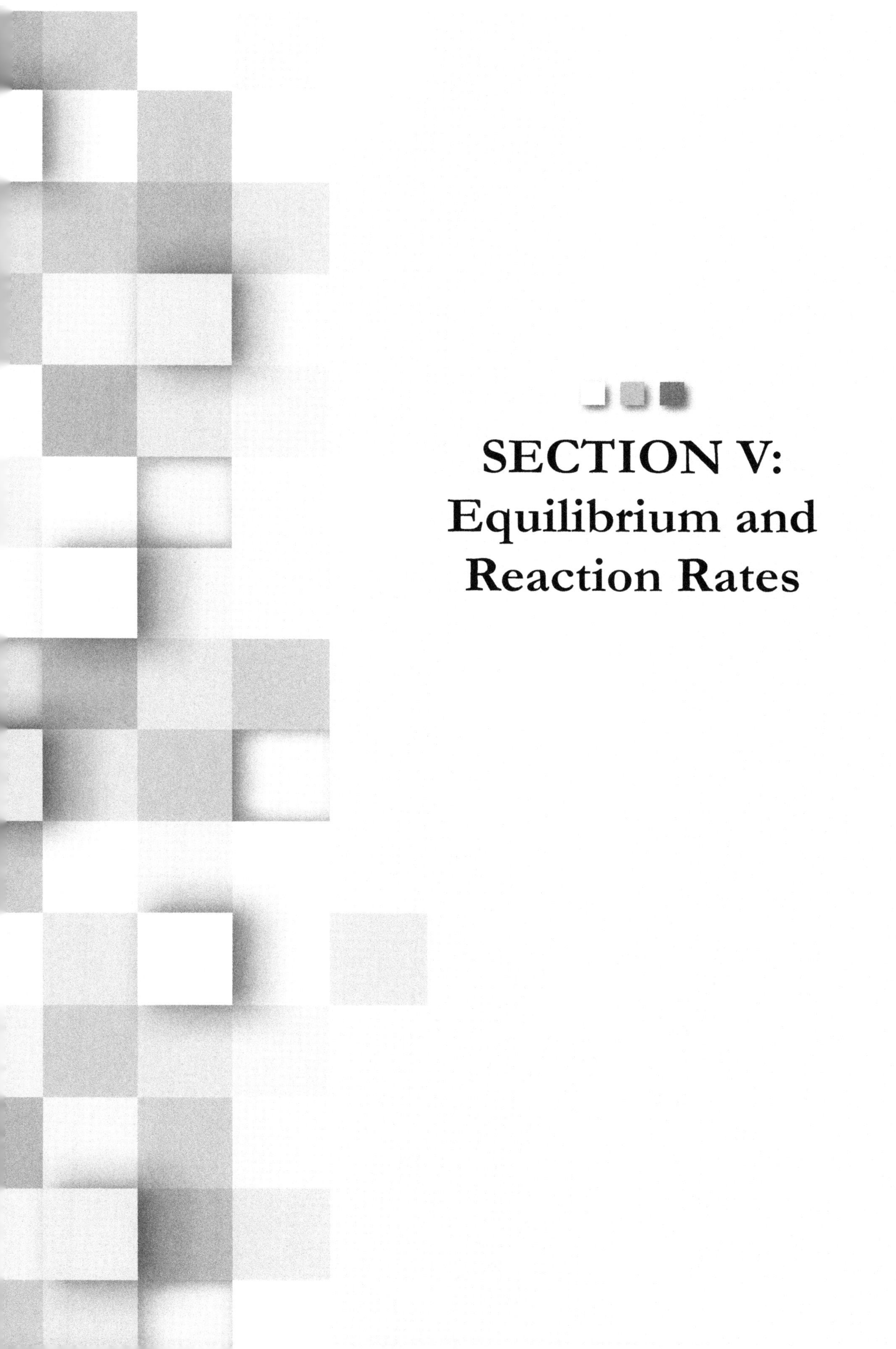

SECTION V: Equilibrium and Reaction Rates

Chapter 12: Equilibrium Systems

12.1 Le Châtelier's principle (factors affecting position of equilibrium (Le Châtelier's principle) in gaseous and aqueous systems)

Assess the effects of changes in concentration, temperature, or pressure on the state of a system initially at equilibrium (Le Châtelier's principle).

The state at which the concentration of reactants and products do not change with time is called a state of chemical equilibrium. The amount of unused reactants depends on the experimental conditions such as concentration, temperature, pressure and the nature of the reaction. At equilibrium, the rate of the forward reaction is equal to the rate of the reverse reaction. If equilibrium is disturbed by a change in concentration, pressure, or temperature, the state of balance is upset and the equilibrium will shift to achieve a new state of balance. **Le Châtelier's principle states that equilibrium will shift to partially offset the impact of an altered condition.**

Change in reactant and product concentrations

If a chemical reaction is at equilibrium, Le Châtelier's principle predicts that **adding a substance**—either a reactant or a product—will shift the reaction so that **a new equilibrium is established by consuming some of the added substance.** Removing a substance will cause the reaction to move in the direction that forms more of that substance.

Example: The reaction $CO + 2H_2 \longrightarrow CH_3O$ is used to synthesize methanol. Equilibrium is established, and then additional CO is added to the reaction vessel. Predict the impact on each reaction component after CO is added.

Solution: Le Châtelier's principle states that the reaction will shift to partially offset the impact of the added CO. Adding the CO will increase the concentration of CO. In response, CO will react in an attempt to restore the equilibrium concentration, and the reaction will "shift to the right." In the process, the concentration of H_2 will decrease, and the concentration of CH_3OH will increase.

Change in pressure for gases

If a chemical reaction is at equilibrium in the gas phase, Le Châtelier's principle predicts that **an increase in pressure** will shift the reaction so **a new equilibrium is established by decreasing the number of moles of gas present**. A decrease in the number of moles partially offsets this rise in pressure. Decreasing pressure will cause the reaction to move in the direction that forms more moles of gas. These changes in pressure might result from altering the volume of the reaction vessel at constant temperature.

Example: The reaction, $N_2 + 3H_2 \longrightarrow 2NH_3$ is used to synthesize ammonia. Equilibrium is established. Next, the reaction vessel is expanded at constant temperature. Predict the impact on each reaction component after this expansion occurs.

Solution: The expansion will result in a decrease in pressure (Boyle's Law). Le Châtelier's principle states that the reaction will shift to partially offset this decrease by increasing the number of moles present. There are 4 moles of gas on the left side of the equation and 2 moles of gas on the right, so the reaction will shift to the left. N_2 and H_2 concentration will increase. NH_3 concentration will decrease.

Change in temperature

Le Châtelier's principle predicts that **when heat is added** at constant pressure to a system at equilibrium, **the reaction will shift in the direction that absorbs heat** until a new equilibrium is established. For an endothermic process, the reaction will shift to the right towards product formation. For an exothermic process, the reaction will shift to the left towards reactant formation. If you understand the application of Le Châtelier's principle to concentration changes, then writing "heat" on the appropriate side of the equation will help you understand its application to changes in temperature.

Example: $N_2 + 3H_2 \longrightarrow 2NH_3$ is an exothermic reaction. First, equilibrium is established and then the temperature is decreased. Predict the impact of the lower temperature on each reaction component.

Solution: Since the reaction is exothermic, we may write it as:

$$N_2\ 3H_2 \longrightarrow 2NH_3 + \text{heat}$$

To find the impact of temperature on equilibrium processes, we may consider heat as if it were a reaction component. Le Châtelier's principle states that after a tempeecreaseincrease in an exothermic reaction, the reaction will shift to the left partially offset the impact of a gain of heat. Therefore, the reaction will shift to the left. N_2 and H_2 concentration will increase. NH_3 concentration will decrease.

12.2 Gaseous and aqueous systems

It is often desirable to be able to be able to predict how some disturbances imposed on an equilibrium system will influence the position of equilibrium, whether it is a gaseous phase equilibrium or aqueous phase equilibrium or even an equilibrium between a liquid and a gas. For instance we may wish to consider the following four equilibria:

$$A_{(g)} + B_{(g)} \longleftrightarrow C_{(g)} + D_{(g)}$$

$$A_{(l)} + B_{(g)} \longleftrightarrow C_{(l)} + D_{(g)}$$

$$\text{Liquids} \longleftrightarrow \text{vapor}$$

$$\text{Heat} + \text{liquid} \longleftrightarrow \text{vapor}$$

We may wish to predict, in a qualitative way, the conditions that favor the greatest production of products. Should we run the reaction at high or low temperature? Should the pressure on the system be high or low? These are the questions we would like to answer quickly without having to perform tedious computations. We can do this by applying Le Châtelier's principle which was discussed for gaseous systems in the previous section.

Evaluate the properties of buffer systems.

A **buffer solution** is a solution that **resists a change in pH** after addition of small amounts of an acid or a base. Buffer solutions require the presence of an acid to neutralize an added base and also the presence of a base to neutralize an added acid. These two components present in thfer also must also not neutralize each other.

A **conjugate acid-base pair is present in buffers** to fulfill these requirements. Buffers are prepared by mixing together **a weak acid or base and a salt of the acid or base** that provides the conjugate.

Consider the buffer solution prepared by mixing together acetic acid ($HC_2H_3O_2$) and sodium acetate ($C_2H_3O_2^-$) and containing Na^+ as a spectator ion. The equilibrium reaction for this acid/conjugate base pair is:

$$HC_2H_3O_2 \longleftrightarrow C_2H_3O_2^- + H^+$$

If H^+ ions from a strong acid are added to this buffer solution, Le Châtelier's principle predicts that the reaction will shift to the left and much of this H^+ will be consumed to create more $HC_2H_3O_2$ from $C_2H_3O_2^-$. If a strong base that consumes H^+ is added to this buffer solution, Le Châtelier's principle predicts that the reaction will shift to the right and much of the consumed H^+ will be replaced through the dissociation of $HC_2H_3O_2$. The net effect is that **buffer solutions prevent large changes in pH that occur when an acid or base is added to pure water** or to an unbuffered solution.

The amount of acid or base that a buffer solution can neutralize before a lare pH changes begins to occur is called its **buffering capacity**. Blood and seawater both contain several conjugate acid-base pairs to buffer the solution's pH and decrease the impact of acids and bases on living things.

12.3 Equilibrium constants

Special equilibrium constants

A few equilibrium constants are used often enough to have their own unique nomenclature.

The **solubility-product constant**, K_{sp}, is the equilibrium constant for an ionic solid in contact with a saturated aqueous solution. The two processes with equal rates in this case are dissolution and cystallization.

$$\text{ionic compound}(s) \longleftrightarrow p\ \text{Cation}^{+}(aq) + q\ \text{Anion}^{-}(aq)$$

$$K_{sp} = \left[\text{cation}^{+}\right]^{p}\left[\text{anion}^{-}\right]^{q}$$

Because this is an example of a heterogeneous equilibrium, the concentration of pure solid is not included as a variable. K_{sp} is a different quantity from solubility. K_{sp} is an equilibrium constant, and solubility is the mass of solid that is able to dissolve in a given quantity of water. See **Intermolecular forces such as hydrogen bonding, dipole-dipole forces, dispersion (London) forces in Section I** for details about solubility processes.

Example: Solid lead chloride $PbCl_2$ is allowed to dissolve in pure water until equilibrium has been reached and the solution is saturated. The concentration of Pb^{2+} is 0.016 M. What is the K_{sp} for $PbCl_2$?

Solution: For $PbCl_2\ (s) \longleftrightarrow Pb^{2+}(aq) + 2Cl^{-}(aq)$, $K_{sp} = [Pb^{2+}][Cl^{-}]^2$.
The only source of both ions in solution is $PbCl_2$, so the concentration of Cl^- must be twice that for Pb^{2+}, or 0.032 M. Therefore,

$K_{sp} = (0.016)(0.032)^2 = 1.6 \times 10^{-5}$

The **acid-dissociation constant**, K_a, is the equilibrium constant for the ionization of a weak acid to a hydrogen ion and its conjugate base.

$$HX(aq) \longleftrightarrow H^{+}(aq) + X^{-}(aq) \qquad K_a = \frac{[H^{+}][X^{-}]}{[HX]}$$

Polyprotic acids have unique values for each dissociation: K_{a1}, K_{a2}, etc.

Example: Hydrofluoric acid is dissolved in pure water until $[H^+]$ reaches 0.006 M. What is the concentration of undissociated HF? K_a for HF is 6.8×10^{-4}.

Solution:

$$HF(aq) \longleftrightarrow H^{+}(aq) + F^{-}(aq), \quad K_a = \frac{[H^{+}][F^{-}]}{[HF]}$$

The principle source of both ions is dissociation of HF (autoionization of water is negligible). Therefore,

$$\left[F^-\right]=\left[H^+\right]=0.006\ M \text{ and } \left[HF\right]=\frac{\left[H^+\right]\left[F^-\right]}{K_a}=\frac{(0.006)^2}{6.8\times10^{-4}}=0.05\ M$$

The **base-dissociation constant**, K_b, is the equilibrium constant for the addition of a proton to a weak base by water to form its conjugate acid and an OH^- ion. In these reactions, it is water that is dissociating as a result of reaction with the base:

$$\text{weak base}(aq)+H_2O(l)\longleftrightarrow\text{conjugate acid}(aq)+OH^-(aq)$$

$$K_b=\frac{\left[\text{conjugate acid}\right]\left[OH^-\right]}{\left[\text{weak base}\right]}$$

The concentration of water is nearly constant and is incorporated into the dissociation constant.

For ammonia (the most common weak base), the equilibrium reaction and base-dissociation constant are:

$$NH_3(aq)+H_2O(l)\longleftrightarrow NH_4^+(aq)+OH^-(aq)$$

$$K_b=\frac{\left[NH_4^+\right]\left[OH^-\right]}{\left[NH_3\right]}$$

Example: K_b for ammonia at 25°C is 1.8×10^{-5}. What is the concentration of OH^- in an ammonia solution at equilibrium containing 0.2 M NH_3 at 25°C?

Solution: Let $x=\left[OH^-\right]$. The principle source of both ions is NH_3 (autoionization of water is negligible). Therefore x = $[NH_3^+]$ also.

$$K_b=\frac{\left[NH_4^+\right]\left[OH^-\right]}{\left[NH_3\right]}=\frac{x^2}{0.2}=1.8\times10^{-5}$$

Solving for x yields: $x=\sqrt{(0.2)(1.8\times10^{-5})}=0.002\ M\ OH^-$

The **ion-product constant for water** (K_w) is the equilibrium constant for the dissociation of H_2O. Water molecules may donate protons to other water molecules in a process known as autoionization:

$$2\ H_2O(l)\longleftrightarrow H_3O^+(aq)+OH^-(aq)$$

A hydronium ion is often referred to as H^+ (*aq*), so the above equation may be rewritten as the following reaction that defines K_w. As with K_b, the concentration of wate is nearly constant.

$$2\ H_2O(l) \longleftrightarrow H^+\ (aq) + OH^-\ (aq)$$

$$Kw = [H^+][OH^-]$$

Example:

a) What is the concentration of OH^- in an aqueous solution with an H^+ oncentration of 2.5×10^{-6} M?

b) What is the concentration of H^+ when pure water reaches equilibrium?

Solution:

a) $K_w = [H^+][OH^-] = (2.5 \times 10^{-6})[OH^-] = 1.0 \times 10^{-14}$

Solving for $\left[OH^-\right]$ yields $\left[OH^-\right] = \dfrac{1.0\times10^{-14}}{2.5\times10^{-6}} = 4.0\times10^{-9}$.

b) The autoionization of pure water creates an equal concentration of the two ions:

$\left[H^+\right] = \left[OH^-\right]$. Therefore, $K_w = \left[H^+\right]^2 = 1.0\times10^{-14}$.

Solving for $[H^+]$ yields $[H^+] = 1.0 \times 10^{-7}$

Example: K_b for ammonia at 25°C is 1.8×10^{-5}. What is the concentration of OH^- in an ammonia solution at equilibrium containing 0.2 M NH_3 at 25°C?

Solution: Let $x = \left[OH^-\right]$. The principle source of both ions is NH_3 (auto-ionization of water is negligible).

Therefore $x = [NH_3^+]$ also.

$$K_b = \frac{\left[NH_4^+\right]\left[OH^-\right]}{[NH_3]} = \frac{x^2}{0.2} = 1.8\times10^{-5}$$

Solving for *x* yields: $x = \sqrt{(0.2)(1.8\times10^{-5})} = 0.002\text{ M OH}^-$

The **ion-product constant for water** (K_w) is the equilibrium constant for the dissociation of H_2O. Water molecules may donate protons to other water molecules in a process known as autoionization.

12.4 Equilibrium expressions

Calculate either the equilibrium constant or concentration of a reaction species at equilibrium (e.g., K_a, K_b, K_{sp}, K_w, K_{eq}).

The concentrations of reactants and products at equilibrium remain constant because the forward and reverse reactions take place at the same rate. This equilibrium responds to a perturbation in one concentration by altering every other concentration in a well-defined way until equilibrium is reestablished. The mathematical relationship between the concentrations of the reactants and products of a system is called the law of mass action. It governs equilibrium expressions.

Consider the general balanced reaction:

$$mA + nB \longrightarrow pR + qS$$

where m, n, p, and q are stoichiometric coefficients and A, B, R, and S are chemical species. An **equilibrium expression** relating the concentrations of chemical species at equilibrium is determined by the equation:

$$K_{eq} = \frac{[R]^p [S]^q}{[A]^m [B]^n}$$

where K_{eq} is a constant value called the **equilibrium constant**. Product concentrations raised to the power of their stoichiometric coefficients are placed in the numerator and reactant concentrations raised to the power of their coefficients are placed in the denominator.

Every reaction has a unique value of K_{eq} that varies only with temperature. Alternate subscripts are often given to the equilibrium constant. K_c or K with no subscript is often used instead of K_{eq} to represent the equilibrium constant. Other subscripts are used for specific reactions as described.

Example: Write the equilibrium expression for the following reaction.

$$2HI\ (g) \longrightarrow H_2\ (g) + I_2\ (g)$$

$$K_{eq} = \frac{[H_2]\,[I_2]}{[HI]^2}$$

The units associated with equilibrium constants in the expression above are molarity raised to the power of the stoichiometric coefficients of the reaction, but it is common practice to write these constants as dimensionless values. Multiplying or dividing the equilibrium expression by 1 M as needed achieves these dimensionless values. **The equilibrium expression for a reaction written in one direction is the reciprocal of the expression for the reaction in the reverse direction**.

For a heterogeneous equilibrium (a chemical equilibrium with components in different phases), reactants or products may be pure liquids or solids. The concentration of a pure liquid or solid in moles/liter cannot change. It is a constant property of the material, and these constants are incorporated into the equilibrium constant. Therefore, the concentrations of pure liquids and solids are absent from equilibrium expressions for heterogeneous equilibria.

Example: Write the equilibrium expression for the redox reaction between copper an silver.

$$Cu\ (s) + 2Ag^+\ (aq) \longleftrightarrow Cu^{2+}\ (aq) + 2Ag\ (s)$$

Solution: The solids do not appear in the equilibrium expression.

$$K_{eq} = \frac{[Cu^{2+}]}{[Ag^+]^2}$$

Calculation of unknown concentrations and concentration units

Many types of K_{eq} problems require the calculation of an unknown concentration at equilibrium from known quantities. These problems only require algebra to solve. Remember that equilibrium constants are dimensionless values. If all concentrations are represented in the same units, it is acceptable to be a little less cautious with units here than for other problems.

Example: The reaction:

$$N_2\ (g) + 3H_2\ (g) \longleftrightarrow 2NH_3\ (g)$$

achieves equilibrium in the presence of 0.27 M H_2 and 0.094 M N_2. What is the ammonia concentration, under these conditions, if the reaction at the given temperature has an equilibrium constant of $K_{eq} = 0.11$

Solution: First we will solve this problem ignoring units since all concentrations are given in M.

$$K_{eq} = \frac{[NH_3]^2}{[N_2]\,[H_2]^3} = \frac{[NH_3]^2}{(0.094)(0.27)^3} = 0.11$$

Solving for $[NH_3]$ yields:

$$[NH_3] = \sqrt{(0.11)(0.094)(0.27)^3} = 0.014\ M$$

This is the preferred method for solving these problems.

If units are to be treated rigorously, then a more explicit definition of K_{eq} should be written. This will achieve a dimensionless value for K_{eq} by repeatedly multiplying by 1 M.

$$K_{eq} = \frac{[NH_3]^2 (1\ M)^2}{[N_2]\,[H_2]^3} = \frac{[NH_3]^2 (1\ M)^2}{(0.094\ M)(0.27\ M)^3} = 0.11$$

Solving for $[NH_3]$ yields:

$$[NH_3] = \sqrt{(0.11)\frac{(0.094)(0.27)^3}{(1M)^2}} = 0.014\ M$$

Determining the directionality of a reaction

Some K_{eq} problems give every concentration value for a reaction that is not at equilibrium and ask which direction the reaction will proceed for a given equilibrium constant. Solving these problems is a two-step process:

1. Insert the non-equilibrium reaction concentrations into the equilibrium expression to obtain a **reaction quotient**, Q.
2. If $Q < K_{eq}$, there are too many reactant molecules for the products, and the reaction proceeds to the right. If $Q > K_{eq}$, there are too many product molecules for the reactants, and the reaction proceeds to the left. If $Q = K_{eq}$, the reaction is at equilibrium.

Example: Predict the direction in which the reaction

$$H_2\,(g) + I_2\,(g) \longrightarrow 2HI\,(g)$$

will proceed if initial concentrations are 0.004 mM H_2, 0.006 mM I_2, and 0.011 mM HI given K_{eq}=48.

Solution:
1) The reactant quotient is

$$Q = \frac{[HI]^2}{[H_2]\,[I_2]} = \frac{(0.011)^2}{(0.004)(0.006)} = 5$$

2) The reactant quotient, which is 5, is less than K_{eq}, which is 48. Therefore, the numerator (products) of the reaction quotient is too small for the denominator (reactants). Over time, the trend towards equilibrium will increase the numerator in relation to the denominator until a ratio of 48 is achieved. More reactants will turn into products and the reaction will proceed to the right.

Chapter 13: Rates of Reactions

13.1 Factors affecting reaction rates

Determine rate laws from concentration and rate data.

The rate of any process is measured by its change per unit time. The speed of a car is measured by its change in position over time using units of miles per hour. The speed of a chemical reaction is usually measured by a change in the concentration of a reactant or product over time using units of **molarity per second** (M/s). The molarity of a chemical is represented in mathematical equations using brackets.

The **average reaction rate** is the change in concentration of either reactant or product per unit time during a time interval:

$$\text{Average reaction rate} = \frac{\text{Change in concentration}}{\text{Change in time}}$$

Reaction rates are positive quantities. Product concentrations increase and reactant concentrations decrease with time, so a different formula is required depending on the identity of the component of interest:

$$\text{Average reaction rate} = \frac{[\text{product}]_{\text{final}} - [\text{product}]_{\text{initial}}}{\text{time}_{\text{final}} - \text{time}_{\text{iniial}}}$$

$$= \frac{[\text{reactant}]_{\text{initial}} - [\text{reactant}]_{\text{final}}}{\text{time}_{\text{final}} - \text{time}_{\text{iniial}}}$$

The **reaction rate** at a given time refers to the **instantaneous reaction rate**. This is found from the absolute value of the **slope of a curve of concentration vs. time**. An estimate of the reaction rate at time *t* may be found from the average reaction rate over a small time interval surrounding *t*. For those familiar with calculus notation, the following equations define reaction rate, but calculus is not needed for this concept:

$$\text{Reaction rate at time } t = \frac{d[\text{product}]}{dt} = -\frac{d[\text{reactant}]}{dt}$$

Example: The following concentration data describes the decomposition of N_2O_5 according to the reaction below.

$$N_2O_5 \longrightarrow 4NO_2 + O_2$$

Time (sec)	$[N_2O_5]$ (M)
0	0.0200
1000	0.0120
2000	0.0074
3000	0.0046
4000	0.0029
5000	0.0018
7500	0.0006
10000	0.0002

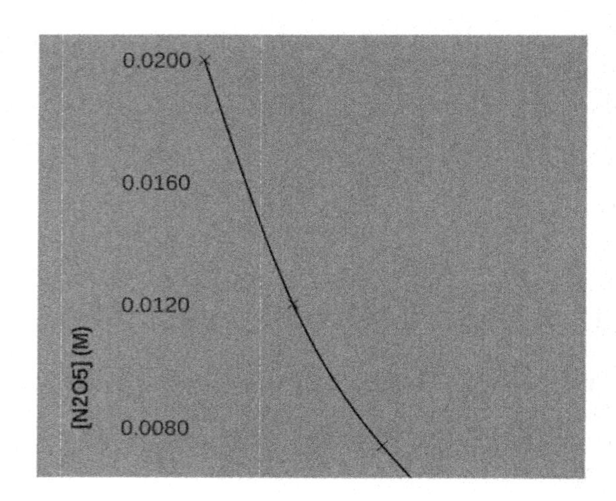

Determine the average reaction rate from 1000 to 5000 seconds and the instantaneous reaction rate at 0 and at 4000 seconds.

$$\frac{[\text{reactant}]_{\text{initial}} - [\text{reactant}]_{\text{final}}}{\text{time}_{\text{final}} - \text{time}_{\text{iniial}}} = \frac{0.0120 \text{ M} - 0.0018 \text{ M}}{5000 \text{ sec} - 1000 \text{ sec}} = 2.55 \times 10^{-6} \frac{\text{M}}{\text{s}}$$

Instantaneous reaction rates are found by drawing lines tangent to the curve, finding the slopes of these lines, and forcing these slopes to be positive values.

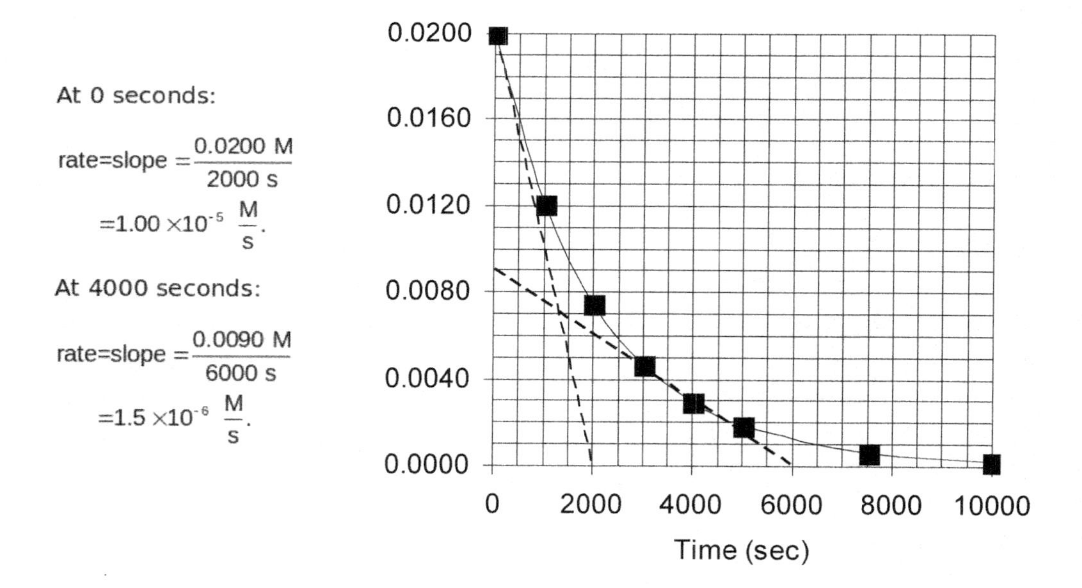

Deriving rate laws from reaction rates

A **rate law** is an **equation relating a reaction rate to concentration**. The rate laws for most reactions discussed in high-school level chemistry are of the form:

$$\text{Rate} = k\,[\text{reactant } 1]^a\,[\text{reactant } 2]^b \ldots$$

In the above general equation, K is called the **rate constant.** a and b are called **reaction orders**. Most reactions considered in introductory chemistry have a reaction order of zero, one, or two. The sum of all reaction orders for a reaction is called the **overall reaction order**. Rate laws cannot be predicted from the stoichiometry of a reaction. They must be determined by experiment or derived from knowledge of reaction mechanism.

If a reaction is zero order for a reactant, the concentration of that reactant has no impact on the rate as long as some reactant is present. If a reaction is first order for a reactant, the reaction rate is proportional to the reactant's concentration. For a reaction that is second order with respect to a reactant, doubling that reactant's concentration increases reaction rate by a factor of four. Rate laws are determined by finding the appropriate reaction order describing **the impact of reactant concentration on reaction rate**.

Reaction rates typically have units of M/s (moles/liter-sec) and concentrations have units of M (moles/liter). For units to cancel properly in the expression above, the units of the rate constant K must vary with overall reaction order as shown in the following table. The value of K may be determined by finding the slope of a plot charting a function of concentration against time. These functions may be memorized or computed using calculus.

Overall reaction order	Units of rate constant K	Method to determine K for rate laws with one reactant
0	M/sec	–(slope) of a chart of [reactant] vs. t
1	sec^{-1}	–(slope) of a chart of ln[reactant] vs. t
2	M^{-1}/sec^{-1}	slope of a chart of 1/[reactant] vs. t

As an alternative to using the rate constant K, the course of **first order reactions** may be expressed in terms of a **half-life**, $t_{halflife}$. The half-life of a reaction is the time required for reactant concentration to reach half of its initial value. First order rate constants and half-lives are inversely proportional:

$$t_{halflife} = \frac{\ln 2}{k_{first\ order}} = \frac{0.693}{k_{first\ order}}$$

Example: Derive a rate law for the reaction $2N_2O_5 \longrightarrow 4NO_2 + O_2$ using data from the previous example.

Solution: Three methods will be used to solve this problem.

1. In the previous example, we found the following two **instantaneous reaction rates**:

Time (sec)	$[N_2O_5]$ (M)	Reaction rate (M/sec)
0	0.0200	1.00×10^{-5}
4000	0.0029	1.5×10^{-6}

A decrease in reactant concentration to 0.0029/0.0200 = 14.5% of its initial value led to a nearly proportional decrease in reaction rate to 15% of its initial value. In other words, reaction rate remains proportional to reactant concentration. The reaction is first order:

$$\text{Rate} = k\,[N_2O_5]$$

We may estimate a value for the rate constant by dividing reaction rates by the concentration:

$$k_{\text{first order}} = \frac{\text{Rate}}{[N_2O_5]}$$

Time (sec)	$[N_2O_5]$ (M)	Reaction rate (M/sec)	K (sec^{-1})
0	0.0200	1.00×10^{-5}	5.00×10^{-4}
4000	0.0029	1.5×10^{-6}	5.2×10^{-4}

2. We could estimate this rate constant by finding **average reaction rates** in each small time interval and assuming this rate occurs halfway between the two concentrations:

Average rate (M/sec)	Halfway $[N_2O_5]$ (M)	k (sec^{-1})
8.00×10^{-6}	0.0160	5.00×10^{-4}
4.6×10^{-6}	0.0097	4.7×10^{-4}
2.8×10^{-6}	0.0060	4.7×10^{-4}
1.7×10^{-6}	0.0038	4.5×10^{-4}
1.1×10^{-6}	0.0024	4.7×10^{-4}
4.8×10^{-7}	0.0012	4.0×10^{-4}
2×10^{-7}	0.0004	4×10^{-4}

3. If **concentration data** are given then no rate data needs to be found in order to determine a rate constant. For a first order reaction, chart the natural logarithm of concentration against time and find the slope.

Time (sec)	$[N_2O_5]$ (M)	$\ln[N_2O_5]$
0	0.0200	-3.91
1000	0.0120	-4.41
2000	0.0074	-4.90
3000	0.0046	-5.37
4000	0.0029	-5.83
5000	0.0018	-6.30
7500	0.0006	-7.39
10000	0.0002	-8.46

The slope may be determined from a best-fit method or it may be estimated from

$$\frac{-8.46-(-3.91)}{10000} = -5\times10^{-4}.$$

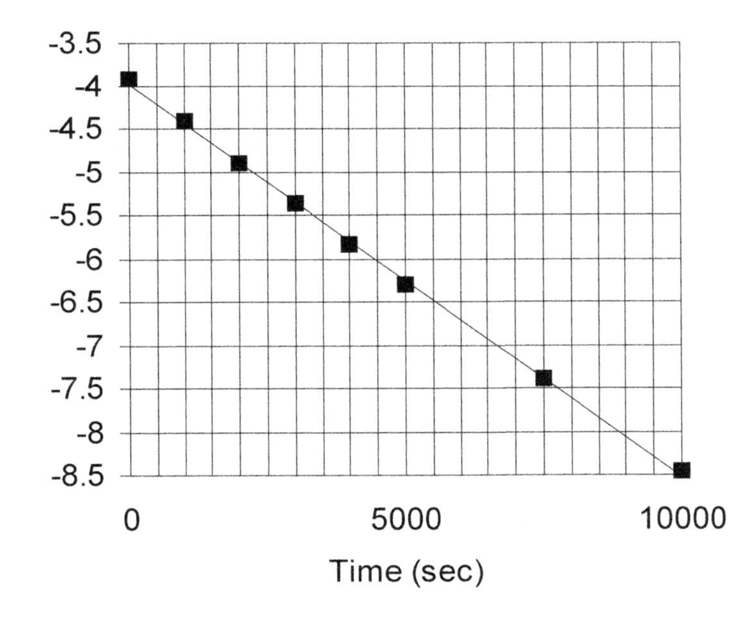

The rate law describing this reaction is:

$$\text{Rate} = \left(5 \times 10^{-4} \frac{\text{M}}{\text{sec}}\right)[N_2O_5]$$

Derive rate laws from simple reaction mechanisms

A **reaction mechanism** is a series of **elementary reactions** that explain how a reaction occurs. These elementary reactions are also called elementary processes or elementary steps. **A reaction mechanism cannot be determined from reaction stoichiometry.** Stoichiometry indicates the number of molecules of reactants and products in an **overall reaction**. Elementary steps represent a **single event**. This might be a collision between two molecules or a single rearrangement of electrons within a molecule.

The simplest reaction mechanisms consist of a single elementary reaction. The number of molecules required determines the rate laws for these processes. For a **unimolecular process:**

$$A \longrightarrow \text{products}$$

the number of molecules of A that decompose in a given time will be proportional to the number of molecules of A present. Therefore unimolecular processes are first order:

$$\text{Rate} = k[A]$$

For **bimolecular processes**, the rate law will be second order.

For $A + A \longrightarrow$ products, Rate $= k[A]^2$

For $A + B \longrightarrow$ products, Rate $= k[A][B]$

Most reaction mechanisms are multi-step processes involving **reaction intermediates**. Intermediates are chemicals that are formed during one elementary step and consumed during another, but they are not overall reactants or products. In many cases one elementary reaction in particular is the slowest and determines the overall reaction rate. This slowest reaction in the series is called the **rate-limiting step** or rate determining step.

Example: The overall reaction:

$$NO_2\,(g) + CO\,(g) \longrightarrow NO\,(g) + CO_2\,(g)$$

is composed of the following elementary reactions in the gas phase:

$$NO_2 + NO_2 \longrightarrow NO + NO_3$$

$$NO_3 + CO \longrightarrow NO_2 + CO_2$$

The first elementary reaction is very slow compared with the second. Determine the rate law for the overall reaction if NO_2 and CO are both present in sufficient quantity for the reaction to occur. Also name all reaction intermediates.

Solution: The first step will be rate limiting because it is slower. In other words, almost as soon as NO_3 is available, it reacts with CO, so the rate-limiting step is the formation of NO_3. The first step is bimolecular. Therefore, the rate law for the entire reaction is:

$$\text{Rate} = k\,[NO_2]^2$$

NO_3 is formed during the first step and consumed during the second. NO_3 is the only reaction intermediate because it is neither a reactant nor a product of the overall reaction.

Identify the characteristics of a chemical system in dynamic equilibrium.

A dynamic equilibrium consists of two **opposing reversible processes** that both occur at the **same rate**. *Balance* is a synonym for equilibrium. A system at equilibrium is stable; it does not change with time. Equilibria are drawn with a double arrow.

When a process at equilibrium is observed, it often doesn't seem like anything is happening, but **at the chemical level, two events are taking place that balance each other**. An example is presented on the right. Arrows in this diagram represent the movement of molecules. When water is placed in a closed container, the water evaporates until the air in the container is saturated. After this occurs, the water level no longer changes, so an observer at the macroscopic scale would say that evaporation has stopped, but the reality on a microscopic scale is that both evaporation and condensation are taking place at the same rate. All equilibria between different phases of matter have this dynamic character.

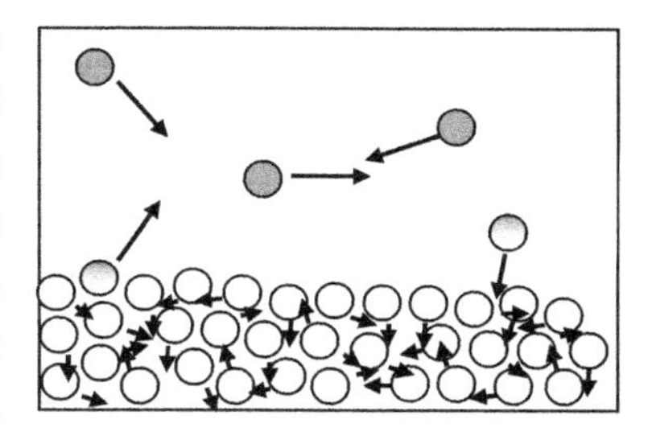

$$\textbf{Liquid} \overset{\text{Evaporation}}{\underset{\text{Condensation}}{\longleftrightarrow}} \textbf{Vapor}$$

Chemical reactions often do not "go to completion." Instead, products are generated from reactants up to a certain point when the reaction no longer seems to occur, leaving some reactant unaltered. At this point, the system is in a state of **chemical equilibrium** because **the rate of the forward reaction is equal to the rate of the reverse reaction**. An example is shown to the right. Arrows in this diagram represent the chemical reactions of individual molecules. An observer at the macroscopic scale might say that no reaction is taking place at equilibrium, but at the chemical level, both the forward and reverse reactions are occurring at the same rate.

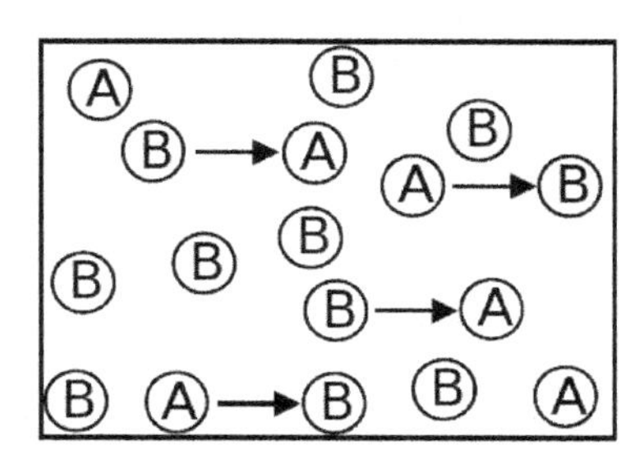

Homogeneous equilibrium refers to a chemical equilibrium among reactants and products that are all in the same phase of matter. **Heterogeneous equilibrium** takes place between two or more chemicals in different phases.

If a reaction at equilibrium is disturbed, changes occur to reestablish equilibrium. **Le Châtelier's principle** states that the equilibrium will be reestablished in a manner that counteracts the effect of the initial disturbance (see **12.1**).

A reaction at equilibrium contains a constant ratio of chemical species. The nature of this ratio is determined by an **equilibrium constant.** The mathematical relationship between the concentrations of the reactants and products of a system is called the law of mass action. It governs equilibrium expressions.

13.2 Potential energy diagrams

Distinguish between different forms of energy (e.g., thermal, electrical, nuclear).

Abstract concept it might be, but energy is one of the most fundamental concepts in our world. We use energy to move people and things from place to place, to heat and light our homes, to entertain ourselves, to produce food and goods and to communicate with each other. It is not some sort of magical invisible fluid, poured, weighed or bottled.

Technically, **energy is the ability to do work or supply heat.** Work is the transfer of energy to move an object a certain distance. It is the motion against an opposing force. Lifting a chair into the air is work; the opposing force is gravity. Pushing a chair across the floor is work; the opposing force is friction. Heat, on the other hand, is a form of energy transfer.

Energy, according to the **First Law of Thermodynamics**, is conserved. That means **energy is neither created nor destroyed.** Energy is merely changed from one form to another. Energy in all of its forms must be conserved. In any system, $\Delta E = q + w$ (E = energy, q = heat and w = work).

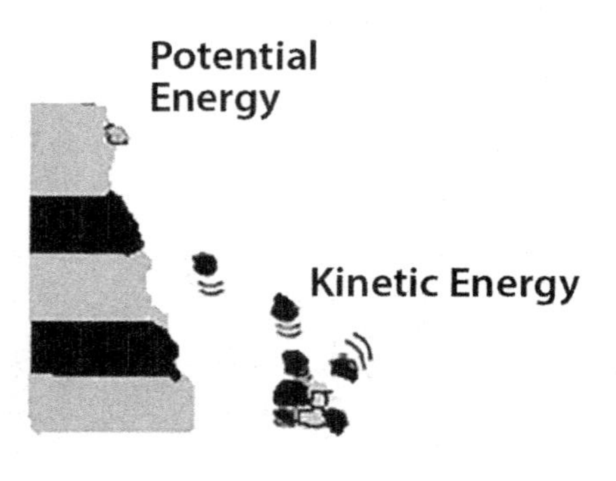

Energy exists in two basic forms, potential and kinetic. Kinetic energy is the energy of a moving object. Potential energy is the energy stored in matter due to position relative to other objects.

In any object, solid, liquid or gas, the atoms and molecules that make up the object are constantly moving (vibrational, translation and rotational motion) and colliding with each other. They are not stationary.

Due to this motion, the object's particles have varying amounts of kinetic energy. A fast moving atom can push a slower moving atom during a collision because it has energy. All moving objects have energy and that energy depends on the object's mass and velocity. Kinetic energy is calculated: $K.E. = \frac{1}{2} mv^2$, where m = mass of the object and v = velocity of the object.

The temperature of an object is proportional to the average kinetic energy of the particles in the substance. Increase the temperature of a substance and its particles move faster; their average kinetic energies increase. But temperature is NOT a type of energy, and it is not conserved.

The energy an object has due to its position or arrangement of its parts is called potential energy. Potential energy due to position is equal to the mass of the object times the gravitational pull on the object times the height of the object, or:

PE = mgh

Where PE = potential energy; m = mass of object; g = gravity; and h = height.

Heat is energy that is transferred between objects because of differences in their temperatures. Heat passes spontaneously from an object of higher temperature to one of lower temperature. This transfer continues until both objects reach the same temperature. Both kinetic energy and potential energy can be transformed into heat energy. When you step on the brakes in your car, the kinetic energy of the car is changed to heat energy by friction between the brake and the wheels. Other transformations can occur from kinetic to potential as well. Since most of the energy in our world is in a form that is not easily used, humans have developed some clever ways of changing one form of energy into another form that may be more useful.

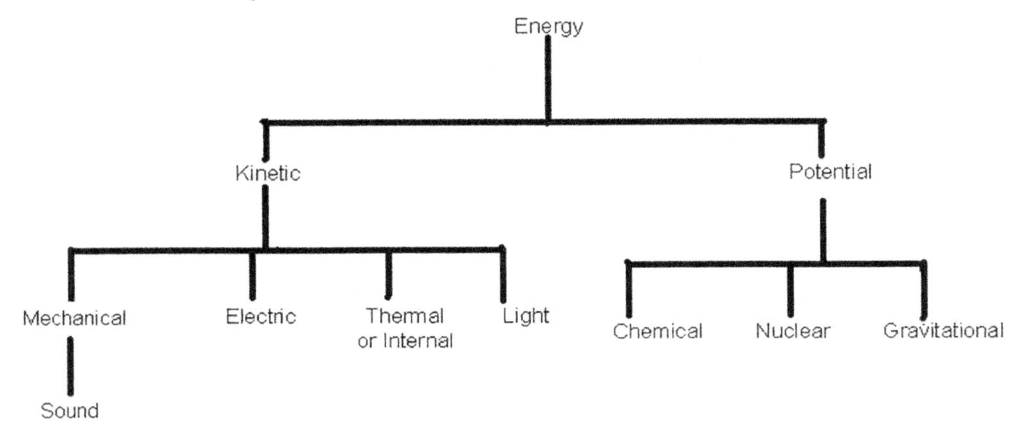

Gravitational Potential Energy

When something is lifted or suspended in air, work is done on the object against the pull of gravity. This work is converted to a form of potential energy called gravitational potential energy.

Nuclear Potential Energy

The energy trapped inside the atom is referred to as nuclear energy. When the atom is split, tremendous energy is released in the form of heat and light.

Chemical Potential Energy

The energy generated from chemical reactions in which the chemical bonds of a substance are broken and rearranged to form new substances is called chemical potential energy.

Electrical Kinetic Energy

The flow of electrons along a circuit is called electrical energy. The movement of electrons creates an electric current which gives us electricity.

Mechanical Kinetic Energy

Mechanical energy is the energy of motion doing work, like a pendulum moving back and forth in a grandfather clock.

Thermal Kinetic Energy

Thermal energy is defined as the energy that a substance has due to the chaotic motion of its molecules. Molecules are in constant motion, and always possess some amount of kinetic energy. It is also called internal energy and is not the same as heat.

Light or Radiant Kinetic Energy

Radiant energy comes from a light source, such as the sun. Radiant energy released from the sun is in the form of photons, tiny particles that can be seen by the human eye as light. Light is also a wave and different types of light are characterized by their different wavelengths, with shorter wavelength light being higher in energy than longer wavelength light.

Energy transformations make it possible for us to use energy to do work. Here are some examples of how energy is transformed to do work:

1. Different types of stoves are used to transform the chemical energy of the fuel (gas, coal, wood, etc.) into heat.
2. Solar collectors can be used to transform radiant energy from the sun into electrical energy.
3. Wind mills make use of the kinetic energy of the air molecules, transforming it into mechanical or electrical energy.
4. Hydroelectric plants transform the kinetic energy of falling water into electrical energy.
5. A flashlight converts chemical energy stored in batteries to light energy and heat. Most of the energy is converted to heat; only a small amount is actually changed into light energy.

13.3 Activation energies

Analyze the effects of concentration, temperature, pressure, surface area, and the presence or absence of catalysts on the rates of reaction.

In order for one species to be converted to another during a chemical reaction, the reactants must collide. The collisions between the reactants determine how fast the reaction takes place. However, during a chemical reaction, only a fraction of the collisions between the appropriate reactant molecules convert them into product molecules. This occurs for two reasons:

1. Not all collisions occur with a **sufficiently high energy** for the reaction to occur.
2. Not all collisions **orient the molecules properly** for the reaction to occur.

The **activation energy**, E_a, of a reaction is the **minimum energy needed to overcome the barrier to the formation of products** and allow the reaction to occur.

At the scale of individual molecules, a reaction typically involves a very small period of time when old bonds are broken and new bonds are formed. During this time, the molecules involved are in a **transition state** between reactants and products. A threshold of maximum energy is crossed when the arrangement of molecules is in an unfavorable intermediate state between reactants and products known as the **activated complex**. Formulas and diagrams of activated complexes are often written within brackets to indicate they are transition states that are present for extremely small periods of time.

The activation energy, E_a, is the difference between the energy of reactants and the energy of the activated complex. The energy change during the reaction, ΔE, is the difference between the energy of the products and the energy of the reactants. The activation energy of the reverse reaction is $E_a + \Delta E$. These energy levels are represented in an **energy diagram** such as the one shown below.

$$NO_2 + CO \rightarrow NO + CO_2$$

This is an exothermic reaction because products are lower in energy than reactants.

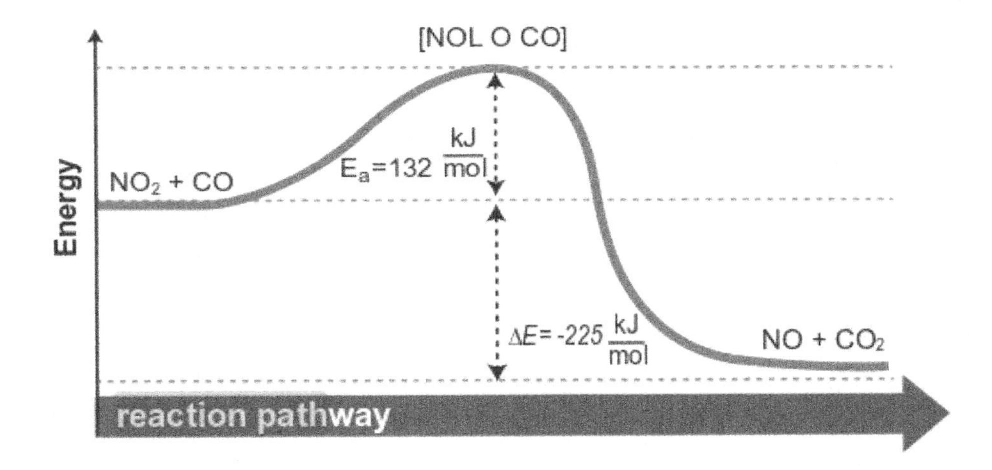

An energy diagram is a conceptual tool, so there is some variability in how its axes are labeled. The y-axis of the diagram is usually labeled energy (E), but it is sometimes labeled "enthalpy (H)" or (rarely) "free energy (G)." There is an even greater variability in how the x-axis is labeled. The terms "reaction pathway," "reaction coordinate," "course of reaction," or "reaction progress" may be used on the x-axis, or the x-axis may remain without a label.

The energy diagrams of an endothermic and exothermic reaction (See 13.2 Potential energy diagrams) are compared below.

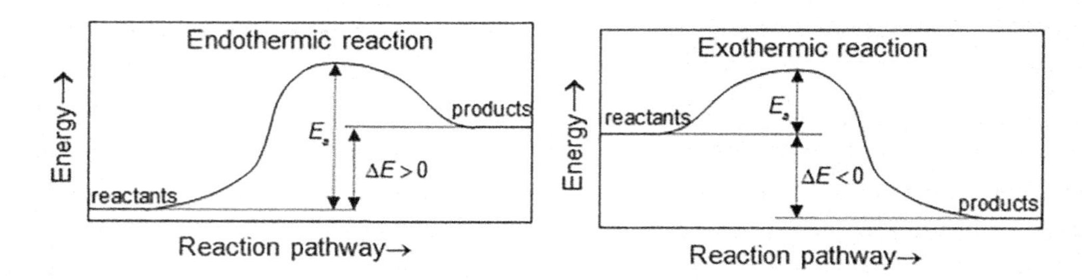

The rate of most simple reactions **increases with temperature** because a **greater fraction of molecules have the kinetic energy** required to overcome the reaction's activation energy. The chart below shows the effect of temperature on the distribution of kinetic energies in a sample of molecules. These curves are called **Maxwell-Boltzmann distributions**. The shaded areas represent the fraction of molecules containing sufficient kinetic energy for a reaction to occur. This area is larger at a higher temperature; so more molecules are above the activation energy and more molecules react per second.

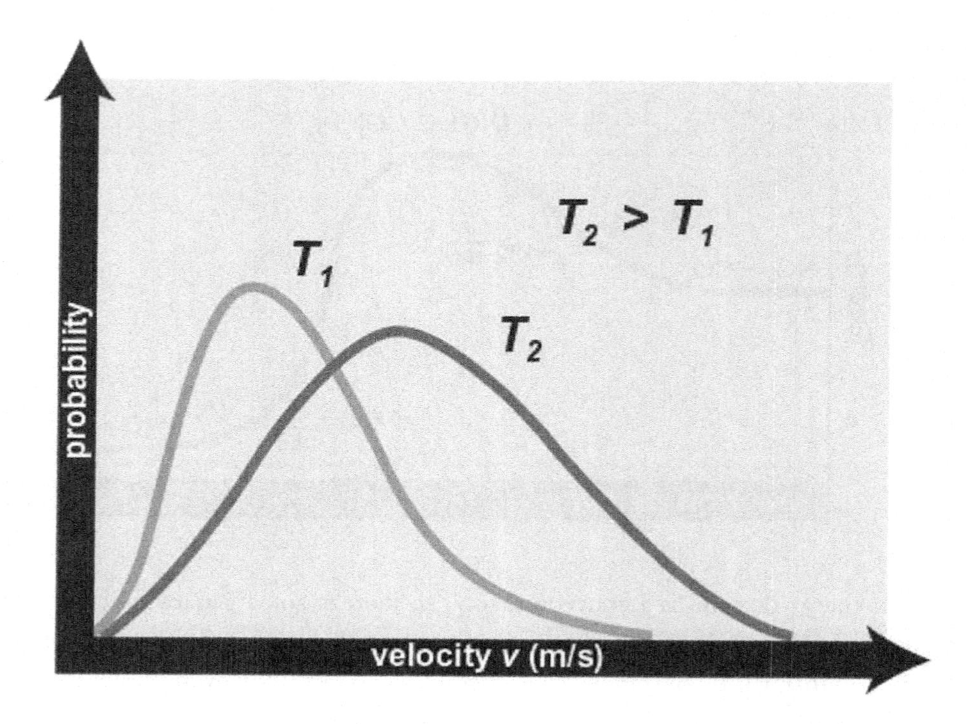

http://www.mhhe.com/physsci/chemistry/essentialchemistry/flash/activa2.swf provides an animated audio tutorial on energy diagrams.

Kinetic molecular theory may be applied to reaction rates in addition to physical constants like pressure. **Reaction rates increase with reactant concentration** because more reactant molecules are present and more are likely to collide with one another in a certain volume at higher concentrations. The nature of these relationships determines the rate law for the reaction. For ideal gases, the concentration of a reactant is its molar density, and this varies with pressure and temperat in.

Kinetic molecular theory also predicts that **reaction rate constants (values for *k*) increase with temperature** because of two reasons:

1. More reactant molecules will collide with each other per second.
2. These collisions will each occur at a higher energy that is more likely to overcome the activation energy of the reaction.

A **catalyst** is a material that increases the rate of a chemical reaction without changing itself permanently in the process. Catalysts provide an alternate reaction mechanism for the reaction to proceed in the forward and in the reverse direction. Therefore, **catalysts have no impact on the chemical equilibrium** of a reaction. They will not make a less favorable reaction more favorable.

Catalysts reduce the activation energy of a reaction. This is the amount of energy needed for the reaction to begin. Molecules with such low energies that they would have taken a long time to react will react more rapidly if a catalyst is present.

The impact of a catalyst may also be represented on an energy diagram (as shown below). **A catalyst increases the rate of both the forward and reverse reactions by lowering the activation energy for the reaction.** Catalysts provide a different activated complex for the reaction at a lower energy state.

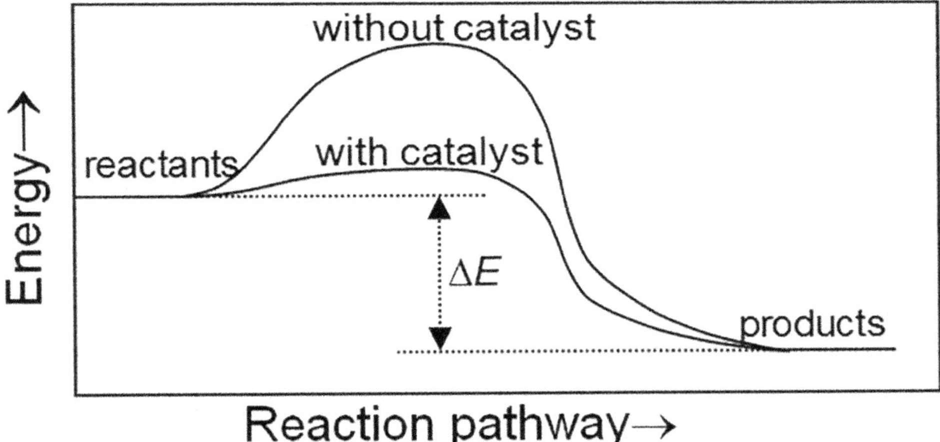

There are two types of catalysts: **Homogeneous catalysts** are in the same physical phase as the reactants. Biological catalysts are called **enzymes**, and most are homogeneous catalysts. A typical homogenous catalytic reaction mechanism involves an initial reaction with one reactant followed by a reaction with a second reactant and release of the catalyst:

$$A + C \longrightarrow AC$$

$$B + AC \longrightarrow AB + C$$

Net reaction: $A + B \xrightarrow{\text{catalyst C}} AB$

Heterogeneous catalysts are present in a different physical state from the reactants. A typical heterogeneous catalytic reaction involves a solid surface onto which molecules in a fluid phase temporarily attach themselves in a way that favors a rapid reaction. Catalytic converters in cars utilize heterogeneous catalysis to break down harmful chemicals in exhaust.

SECTION V PRACTICE QUESTIONS

CHAPTER 12

1. **A buffer solution is one that**

   (A) prevents the pH from going too high to maintain a neutral or acidic solution
   (B) resists a change in pH
   (C) contains enough water to keep the solution neutral
   (D) is always at pH = 7
   (E) None of the above.

   The correct answer is B.
   A buffer solution is a solution that resists change in pH after the addition of small amounts of an acid or base.

Questions 2-4 refer to the following terms

$$2SO(g) + O_2(g) \longrightarrow 2SO_2(g) + \text{heat}$$

(A) shift to the right (products)
(B) shift to the left (reactants)
(C) increase eEquilibrium
(D) increase Q > K

2. **If we add more SO_2 gas to the reaction chamber: The corect answer is B.**

3. **If we decrease the volume of the reaction chamber: The corect answer is A.**

4. **K = 10, and the reaction presently has a Q = 5. What must happen in order for the reaction to reach equilibrium? : The corect answer is A.**

5. **Cause and Effect**

   Statement I
   Every reaction has a unique value of K_{eq} that varies only with temperature.
   This statement is True.

   Statement II
   When calculating K_{sp} the concentration of pure solid is not included as a variable.
   This statement is True.
   Both statements are TRUE but this is not a cause and effect relationship.

CHAPTER 13

1. **The first Law of Thermodynamics states that**

   (A) Heat moves from hot objects to cold objects
   (B) The entropy of the universe is increasing
   (C) Energy is neither created nor destroyed
   (D) There is no temperature change during a phase change.
   (E) The temperature of an object is proportional to the average kinetic energy of the particles

   The correct answer is C.
   While all the statements are true, only C is the definition of the First Law of Thermodynamics.

Questions 2-4 refer to the following terms

(A) 1st order
(B) second order
(C) zero order
(D) unable to determine
(E) variable

2. **The rate order if the concentration of that reactant has no impact on the rate: The correct answer is C.**

3. **The rate order if doubling that reactant's concentration increases reaction rate by a factor of four: The correct answer is B.**

4. **The rate order if a plot of ln[reactant] vs.** *t is linear*: **The correct answer is A.**

5. **Cause and Effect**

   Statement I
   According to the Kinetic Molecular Theory, Reaction rates increase with reactant concentration
   This statement is True.

   Statement II
   If more reactant molecules are present they are likely to collide with one another in a certain volume at higher concentrations.
   This statement is True.
   Both statements are TRUE and there IS a cause and effect relationship.

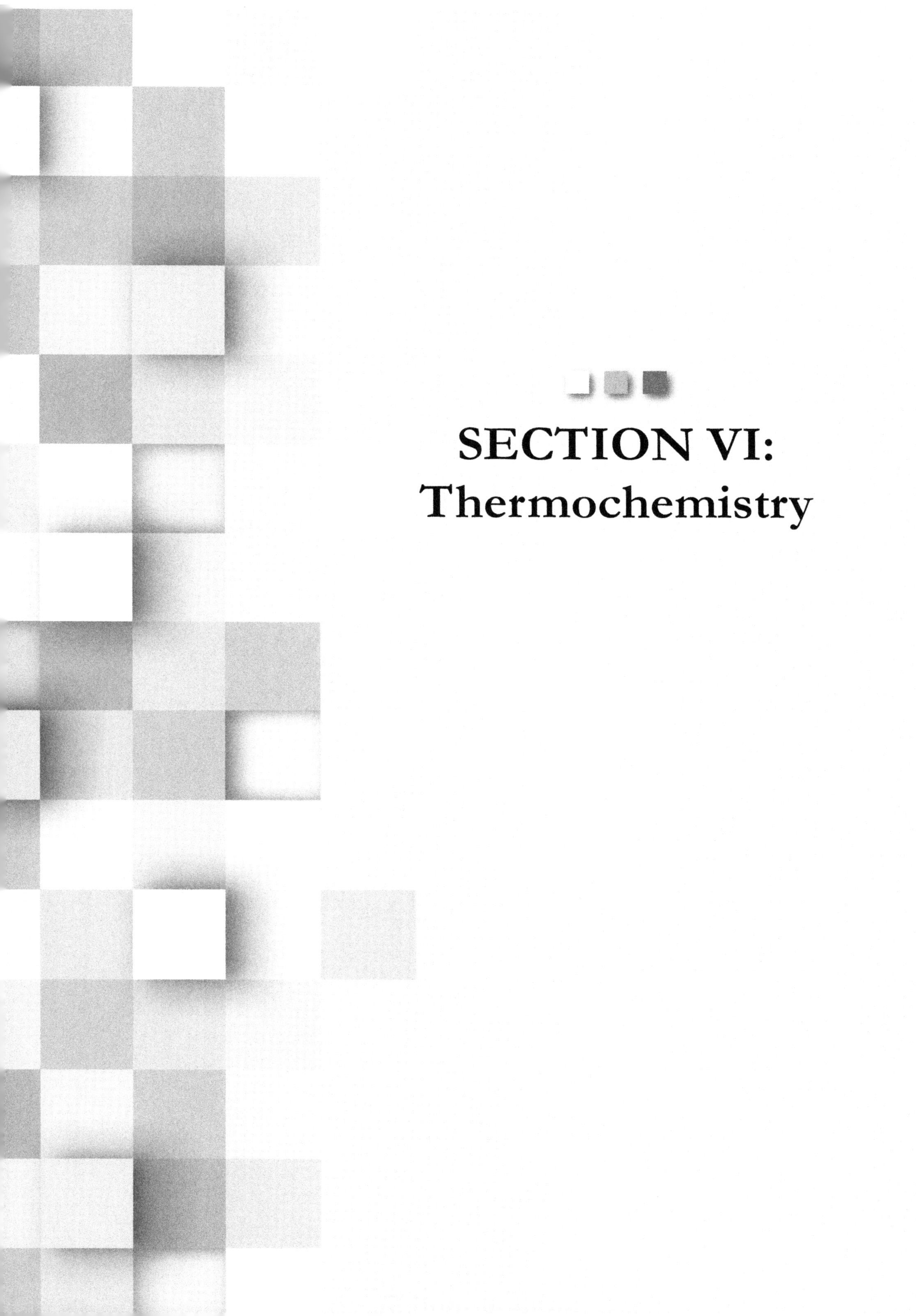

SECTION VI: Thermochemistry

Chapter 14: Thermochemistry

14.1 Conservation of energy

Conservation of Energy can be discussed from two perspectives:

1. The Law of Conservation of Energy, or Hess's Law, from the thermochemical point of view of chemical reactions. This theory explained the conservation law by proposing that all matter is made up of atoms which are never created or destroyed in chemical reactions.
2. The Law of Conservation of Energy from the point of view of an isolated system in which there is no net change in energy and where energy is neither created nor destroyed. However, even though energy can be neither created nor destroyed, energy can change forms; for example, potential to kinetic energy.

Hess's Law: The Principle of Conservation of Energy

Henri Hess's Law is very important in thermochemical studies. It states that *the heat evolved or absorbed in a chemical process is the same whether this process takes place in one or in several steps*. This law is also known as **the law of constant heat summation**. Hess's law is illustrated in the following thermal equations:

Thermal equations
$A + B = AB$, ΔH_1
$AB + B = AB_2$, ΔH_2
then,
$A + 2B = AB_2$, $\Delta H_{12} = \Delta H_1 + \Delta H_2$

Chemical energy and Hess's law

Standard enthalpy of reaction and *standard enthalpy of formation* are very useful chemical properties. Using these properties, we can calculate enthalpies of reactions that are difficult to measure. We have already mentioned some basic rules regarding the quantities ΔH, $\Delta H°$, and ΔH_f in the preceding equations.

If the number of moles is altered on both sides of the equations by multiplying by the same factor, then ΔH, $\Delta H°$, or ΔH_f for the equation should be multiplied by the same factor, since they are natural quantities per equation as written. Thus, the equation

$C\ (graphite) + 0.5O_2 \longrightarrow CO,\ \Delta H° = -110 kJ/mol.$

can be written in either of the following forms:

$2C\ (graphite) + O_2 \longrightarrow 2CO,\ \Delta H° = -220 kJ/mol$ (multiplied by 2)

$6C\ (graphite) + 3O_2 \longrightarrow 6CO,\ \Delta H° = -660 kJ/mol$ (multiplied by 6)

To write the reverse reaction, we have to change the signs of these quantities (multiply by -1) which gives the following equations:

$CO \longrightarrow C\ (graphite) + 0.5O_2,\ \Delta H° = 110\ kJ/mol$

$2CO \longrightarrow 2\ C\ (graphite) + O_2,\ \Delta H° = 220 kJ/mol$

According to Hess's law, the amount of energy depends only on the states of the reactants and of the products, not on the intermediate steps. In chemical reactions, the energy changes (enthalpy changes) will be the same regardless of whether the reaction proceeds in one step or several steps. In a chemical reaction, the total energy change will be the sum of the energy changes of all the steps making up the overall reaction.

For example, we look at the oxidation of carbon into CO and CO_2 in the following diagram. The enthalpy of oxidation of carbon (graphite) into CO_2 is -393 kJ/mol while the enthalpies of oxidation of carbon into CO and then CO_2 are, respectively, -110 and -283 kJ/mol. Therefore, the sum of enthalpy in the overall steps is exactly -393 kJ/mol; the same as the one-step reaction.

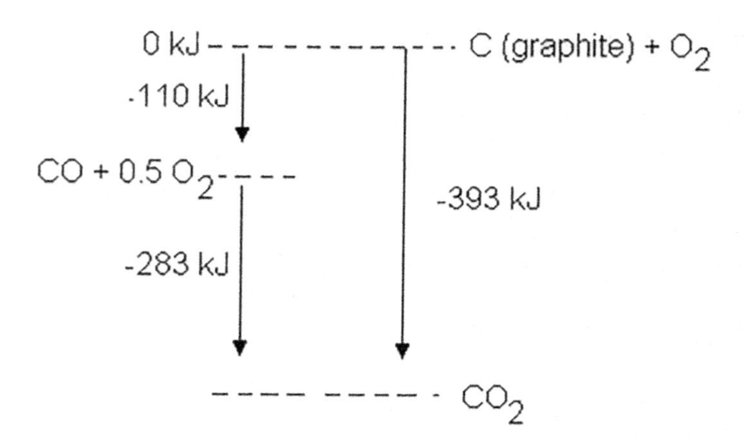

The two-step reactions are:

$C + 12O_2 \longrightarrow CO,\ \Delta H° = -110\ kJ/mol$

$CO + 12O_2 \longrightarrow CO_2,\ \Delta H° = -283\ kJ/mol$

Adding the two equations together and canceling out the intermediate, CO, on both sides leads to

$C + O_2 \longrightarrow CO_2,\ \Delta H° = (-110) + (-283) = -393\ kJ/mol$

Application of Hess's Law

Based on Hess's law, we can calculate enthalpies of reactions that are difficult to measure. For example in the above example it would be very difficult to obtain pure CO from oxidation of graphite. But the enthalpy of oxidation of graphite to CO_2 is easily measured, as is the oxidation of CO to CO_2. Hess's law allows us to calculate the enthalpy of formation of CO because:

$$C + O_2 \longrightarrow CO_2,\ \Delta H° = -393\ kJ/mol$$

$$CO + ½O_2 \longrightarrow CO_2,\ \Delta H° = -283\ kJ/mol$$

By subtracting the second equation from the first we find:

$$C + ½O_2 \longrightarrow CO,\ \Delta H° = -393 - (-283) = -110 kJ/mol$$

The equation shows the standard enthalpy of formation of CO to be -110 kJ /mol.

Also we are able to calculate ΔH, $\Delta H°$, and ΔH_f for chemical reactions that are impossible to measure, provided that we have all the data for related reactions.

Example:

The standard enthalpies of formation of SO_2 and SO_3 are respectively -297 and -396 kJ/mol. Calculate the standard enthalpy of reaction for the reaction:

$$SO_2 + ½O_2 \longrightarrow SO_3$$

Solution:

We write the equations according to the given data:

$$SO_2\ (g) \longrightarrow S(s) + O_2\ (g)\ \Delta H = 297 kJ$$

$$S(s) + 3/2O_2 \longrightarrow SO_3\ \Delta H = -396 kJ$$

By adding the two equations, the enthalpy of the desired reaction is:

$$SO_2\ (g) + ½O_2 \longrightarrow SO_3\ \Delta H = -99 kJ$$

Conservation of Energy of an Isolated system

This aspect of the law of Conservation of Energy discusses an isolated system in which there is no net change in energy and where energy is neither created nor destroyed, but can change forms, like potential to kinetic energy. In other words, the sum of the total energy (E) for a specific isolated system would be the sum of the potential energy (V) and the kinetic energy (T).

$$E = T + V$$

Another way in which energy can change forms is heat (q) and work (w). When heat is applied to a closed system, the system does work by increasing its volume.

$$w = P_{ext}\ \Delta V$$

Here P_{ext} is the external pressure, and delta V is the change of volume. When we add heat to a cylinder, the pressure inside the cylinder will increase and hence the piston rises to relieve the pressure difference between inside and outside of the cylinder. Therefore, by increasing the volume of the cylinder, the piston has done work.

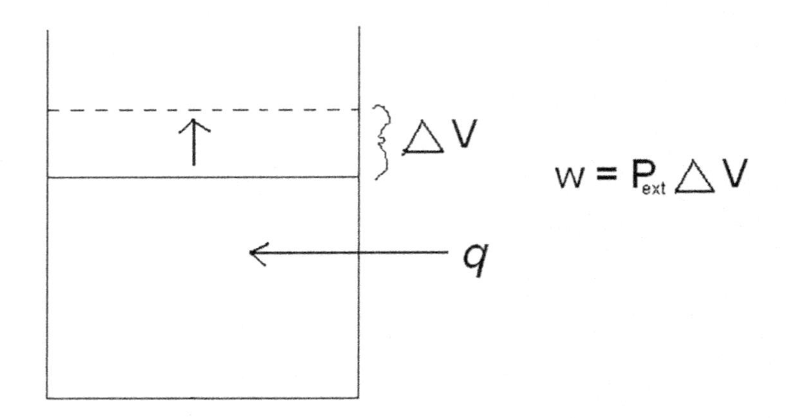

The sum of heat and work is the change in internal energy, ΔU.

In an isolated system, q = –w. Therefore, ΔU=0.

14.2 Calorimetry and Specific Heat

Specific Heat

A substance's molar **heat capacity** is the heat required to **change the temperature of one mole of the substance by one degree**. Heat capacity has units of joules per mol-kelvin or joules per mol-°C. The two units are interchangeable because we are only concerned with differences between one temperature and another. A Kelvin degree and a Celsius degree are the same size.

The **specific heat** of a substance (also called specific heat capacity) is the heat required to **change the temperature of one gram or kilogram by one degree Celsius.** Specific heat has units of joules per gram or joules per kilogram.

These terms are used to solve thermochemistry problems involving a change in temperature by applying the formula:

A substance's **enthalpy of fusion** (ΔH_{fusion}) is the heat required to **change a substance from a solid to a liquid** by melting. This is also the heat released from the substance when it changes from a liquid to a solid (freezes).

$q = n \times C \times \Delta T$ where $q \Rightarrow$ heat added (positive) or evolved (negative)

$n \Rightarrow$ amount of material

$C \Rightarrow$ molar heat capacity if n is in moles, specific heat if n is a mass

$\Delta T \Rightarrow$ change in temperature $T_{final} - T_{initial}$

Calorimetry: Measuring energy changes in chemical reactions

Calorimetry is an experimental technique that uses temperature to determine the amount of heat exchanged/transferred in a chemical system during processes such as phase transitions, physical changes, and chemical reactions. The apparatus that is used to measure the heat of reaction is called a calorimeter (from the words "calorie meter", although we now express the heat unit in joules rather than calories). A calorimeter is a well-insulated device in which a reaction takes place. No heat can leave or enter the system. The system is put in thermal contact with a bath composed of a substance with a known heat capacity. Examples include water, which is an excellent absorber of heat and has one of the highest specific heats (4.184 J/g°C). Since the heat capacity of the bath substance is known, changes in temperature are utilized to estimate the amount of heat exchanged between the system and the heat bath. If the **heat bath increases** in temperature then the **energy of the system has decreased**. If the **heat bath decreases** in temperature, it has decreased in energy—it has transferred energy to the chemical system being measured, and **the energy of the system has increased** by the same amount.

There are different designs for this apparatus, and one called a bomb calorimeter is shown in Fig 1.1. This kind of calorimeter is normally used to study exothermic reactions that do not begin until they are initiated by applying heat – reactions such as the combustion of CH_4 or the reaction of H_2 with O_2. The apparatus consists of a strong steel container (the bomb) into which the reactants are placed. The bomb is then immersed in an insulated bath that is fitted with a stirrer and a thermometer. The initial temperature of the bath is measured and then the reaction is set off by a small heater wire within the bomb. The heat given in the reaction is absorbed by the bomb and the bath, causing the temperature of the entire apparatus to rise.

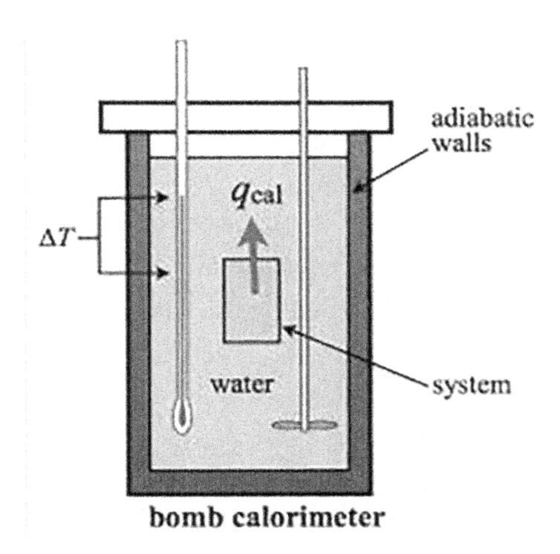

Fig 1.1 – A bomb calorimeter

Example: Measuring the heat of reaction with a bomb calorimeter

In an experiment, 0.200 g of H_2 and 1.600 g of O_2 were compressed into a 1.00 L bomb, which then was placed into a calorimeter with the capacity of 1.816×10^5 J/°C. The initial calorimeter temperature was measured to be 26°C and after the reaction took place, the calorimeter's final temperature was 26.155 °C. Calculate the amount of heat given off in the reaction of H_2 and O_2 to form H_2O, expressed (a) in kilojoules and (b) in kilojoules per mole of H_2O formed.

Solution:

To calculate the heat evolved we have to multiply the heat capacity (1.816×10^5 J/°C) by the temperature change in °C. The temperature change is 0.155. Therefore,

$(1.816 \times 10^5 \text{ J/1 °C}) \times 0.155 \text{ °C} = 2.82 \times 10^4 \text{ J} = 28.2$ KJ. So this is the amount of heat given off by the reactants.

(b) The amount of heat released in a reaction depends on the amount of the reactants used. Large amounts of reactants produce large energy changes. The heat of reaction for different reactions is usually expressed on a "per mole "basis.

The reaction between H_2 and O_2 to form water is as follows

$$2\,H_2 + O_2 \longrightarrow 2\,H_2O$$

From 2 moles of H_2, 2 moles of H_2O is produced. Therefore,
the number of moles of H_2O produced = + 0.2/2 g H_2 = 0.1 moles

To get units of kilojoules per mole of water, we take the ratio of the number of kilojoules to the number of moles.

28.2 KJ/0.1 moles of water = 282 KJ/mol H_2O

Heat Capacity of Calorimeter

In calorimetry we need to know the heat capacity of the calorimeter itself rather than the heat capacity of the whole calorimeter system (calorimeter and water). The heat (q) released by a process or reaction is absorbed by the calorimeter and any substances in the calorimeter. If water is the only other substance in the calorimeter, the following energy balance exists:

$$q = q_{cal} + q_w$$

where q_{cal} is the heat flow for the calorimeter and q_w is the heat flow for the water.

Both of these separate heat flows can be related to the heat capacity and temperature change for the substance.

$$q_{cal} = C_{cal}\Delta T$$

$$q_w = C_w \Delta T$$

Here, C_{cal} is the heat capacity of the calorimeter and C_w is the heat capacity of water. Because the water and calorimeter are in thermal equilibrium, they both have the same temperature and thus ΔT is the same for both. The consequence is that the heat capacity of the entire system (C) is the sum of the heat capacities for the individual components.

$$C = C_{cal} + C_w$$

Since the heat capacity is an extensive property, it depends upon the amount of substance present. The calorimeter exists as a fixed unit, so its heat capacity is a fixed value. The amount of water in the calorimeter, however, can vary, and therefore the heat capacity of the water can vary. When dealing with variable amounts of material, one often prefers to use an intensive measure of the heat capacity. Another common version of the heat capacity is the specific heat capacity (s), which is defined as the heat capacity of one gram of a substance.

$$s_w = \frac{C_w}{m_w}$$

Because the mass of water (m_w) and the specific heat capacity of water are both known, one can easily calculate the heat capacity of the water. The joule (J) is defined based upon the specific heat capacity of water:

$$s_w = 4.184 \text{ J } {}^{o}\text{C}^{-1} \text{ g}^{-1}$$

Overall, one can write

$$C = C_{cal} + s_w m_w$$

Quantity	Symbol	Unit	Meaning
heat	q	J	Energy transfer; producing difference in temperature
temperature	T	°C or K	Kinetic energy; measure of the molecular motion
temperature change	ΔT	°C or K	Difference between the final and initial temperatures of a process
mass	m	g	Amount of material
heat capacity	C	J ^{o}C^{-1} or J K^{-1}	Heat required to change the temperature of a substance one degree
specific heat capacity	s	J ^{o}C^{-1} g^{-1} or J K^{-1} g^{-1}	Heat required to change (raise or lower) the temperature of one gram of a substance one degree

14.3 Enthalpy (heat) changes associated with phase changes and chemical reactions

Calculate thermal changes in chemical reactions, such as heats of reaction, heats of formation, and/or heats of combustion, from data.

Energy is the **driving force for change**. Energy has units of joules (J). Temperature remains constant during phase changes, so the **speed** of molecules and their **translational kinetic energy do not change** during a change in phase.

The **internal energy** of a material is the **sum of the total kinetic energy** of its molecules and the **potential energy** of interactions between those molecules. Total kinetic energy includes the contributions from translational motion and other components of motion such as rotation. The potential energy includes **energy stored in the form of resisting intermolecular attractions** between molecules.

The **enthalpy** (*H*) of a material is the **sum of its internal energy and the mechanical work** it can do by driving a piston. We usually don't deal with mechanical work in high school chemistry, so the differences between internal energy and enthalpy are not important. The key concept is that a change in the **enthalpy** of a substance is the total **energy** change caused by **adding/removing heat** at constant pressure.

When a material is heated and experiences a phase change, **thermal energy is used to break the intermolecular bonds** holding the material together. Similarly, bonds are formed with the release of thermal energy when a material changes its phase during cooling. Therefore, **the energy of a material increases during a phase change that requires heat and decreases during a phase change that releases heat**. For example, the energy of H_2O increases when ice melts and decreases when liquid water freezes.

Hess's law states that energy changes are state functions. The amount of energy depends only on the states of the reactants and the state of the products, but not on the intermediate steps. Energy (enthalpy) changes in chemical reactions are the same, regardless of whether the reactions occur in one or several steps. The total energy change in a chemical reaction is the sum of the energy changes in its many steps leading to the product of the overall reaction.

A **standard** thermodynamic value occurs with all components at 25 °C and 100 kPa. This *thermodynamic standard state* is slightly different from the *standard temperature and pressure* (STP) often used for gas law problems (0 °C and 1 atm = 101.325 kPa). Standard properties of common chemicals are listed in tables.

The **heat of formation, ΔH_f,** of a chemical is the heat required (positive) or emitted (negative) when elements react to form the chemical. It is also called the enthalpy of formation. The **standard heat of formation** ΔH_f° is the heat of formation with all reactants and products at 25 °C and 100 kPa.

Elements in their **most stable form** are assigned a value of ΔHf° = 0 kJ/mol. Different forms of an element in the same phase of matter are known as **allotropes**.

Example: The heat of formation for carbon as a gas is: ΔH_f° for C(*g*) = 718.4 $\frac{\text{kJ}}{\text{mol}}$.

Carbon in the solid phase exists in three allotropes. A C_{60} *buckyball* (one face is shown to the left), contains C atoms linked with aromatic bonds and arranged in the shape of a soccer ball. C_{60} was discovered in 1985. *Diamonds* (below middle) contains single C–C bonds in a three dimensional network. The most stable form of carbon at 25 °C is *graphite* (below right). Graphite is composed of C atoms with aromatic bonds in sheets.

$$\Delta H_f^\circ \text{ for } C_{60}(\textit{buckminsterfullerene or buckyball}) = 38.0\ \frac{kJ}{mol}$$

$$\Delta H_f^\circ \text{ for } C_\infty(\textit{diamond}) = 1.88\ \frac{kJ}{mol}$$

$$\Delta H_f^\circ \text{ for } C_\infty(\textit{graphite}) = 0\ \frac{kJ}{mol}.$$

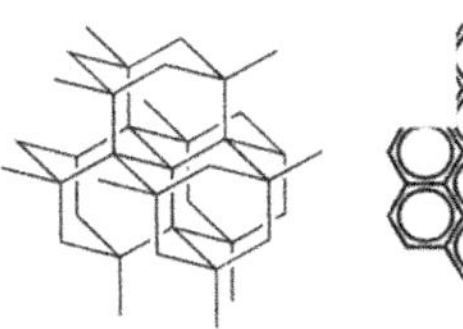

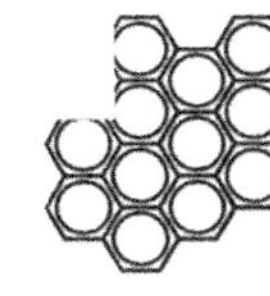

Heat of combustion ΔH_c (also called enthalpy of combustion) is the heat of reaction when a chemical **burns in O_2** to form completely oxidized products such as **CO_2 and H_2O**. It is also the heat of reaction for **nutritional molecules that are metabolized** in the body. The standard heat of combustion **ΔH_c°** takes place at 25 °C and 100 kPa. **Combustion is always exothermic**, so the negative sign for values of ΔHc is often omitted. If a combustion reaction is used in Hess's Law, the value must be negative.

14.4 Heating and cooling curves

Interpret a heating/cooling curve of a substance.

Kinetic models are used to describe gases, liquids, and solids. **Phase changes** occur when the relative importance of kinetic energy and intermolecular forces is altered sufficiently for a substance to change its state. See Chapters 5.3 and 5.4 for more on heating curves and phase diagrams.

The transition from gas to liquid is called **condensation** and from liquid to gas is called **vaporization**. The transition from liquid to solid is called **freezing** and from solid to liquid is called **melting**. The transition from gas to solid is called **deposition** and from solid to gas is called **sublimation**.

Phase transitions, such as ice to liquid water, require (absorb) a particular amount of standard enthalpy:

Standard Enthalpy of Vaporization (ΔH°_{vap}) is the energy that must be supplied as heat at constant pressure per mole of molecules vaporized (liquid to gas).

Standard Enthalpy of Fusion (ΔH°_{fus}) is the energy that must be supplied as heat at constant pressure per mole of molecules melted (solid to liquid).

Standard Enthalpy of Sublimation (ΔH°_{sub}) is the energy that must be supplied as heat at constant pressure per mole of molecules converted to vapor from a solid.

$$\Delta H^\circ_{sub} = \Delta H^\circ_{fus} + \Delta H^\circ_{vap}$$

The enthalpy of condensation is the reverse of the enthalpy of vaporization and the enthalpy of freezing is the reverse of the enthalpy of fusion. The enthalpy change of a reverse phase transition is the negative of the enthalpy change of the forward phase transition. Also the enthalpy change of a complete process is the sum of the enthalpy changes for each of the phase transitions incorporated in the process.

A substance's **enthalpy of vaporization** ($\Delta H_{vaporization}$) is the heat required to **change a substance from a liquid to a gas** or the heat released by condensation.

A substance's **enthalpy of sublimation** ($\Delta H_{sublimation}$) is the heat required to change the substance directly from a solid to a gas by sublimation or the heat released by deposition (transition from gas to solid).

These three values are also called "heats" or "latent heats" of fusion, vaporization, and sublimation. They have units of joules per mole, and are negative values when heat is released.

$$\text{Solid} \xrightarrow[\text{melting}]{\Delta H_{fusion}} \text{Liquid} \xrightarrow[\text{vaporization}]{\Delta H_{vaporization}} \text{Gas} \qquad \text{Solid} \xrightarrow[\text{sublimation}]{\Delta H_{sublimation}} \text{Gas}$$

$$\text{Gas} \xrightarrow[\text{condensation}]{-\Delta H_{vaporization}} \text{Liquid} \xrightarrow[\text{freezing}]{-\Delta H_{fusion}} \text{Solid} \qquad \text{Gas} \xrightarrow[\text{deposition}]{-\Delta H_{sublimation}} \text{Solid}$$

These terms are used to solve thermochemistry problems involving a change of phase by applying the formula:

where $q \Rightarrow$ heat added (positive) or evolved (negative)

$n \Rightarrow$ amount of material

$\Delta H_{change} \Rightarrow$ enthalpy of fusion, vaporization, or sublimation for heat added

$\Rightarrow$ –(enthalpy of fusion, vaporization, or sublimation) for heat evolved

where $q \Rightarrow$ heat added (positive) or evolved (negative)

$n \Rightarrow$ amount of material

$\Delta H_{change} \Rightarrow$ enthalpy of fusion, vaporization, or sublimation for heat added

$\Rightarrow$ –(enthalpy of fusion, vaporization, or sublimation) for heat evolved

The important thing to remember is that there is no temperature change during a phase change.

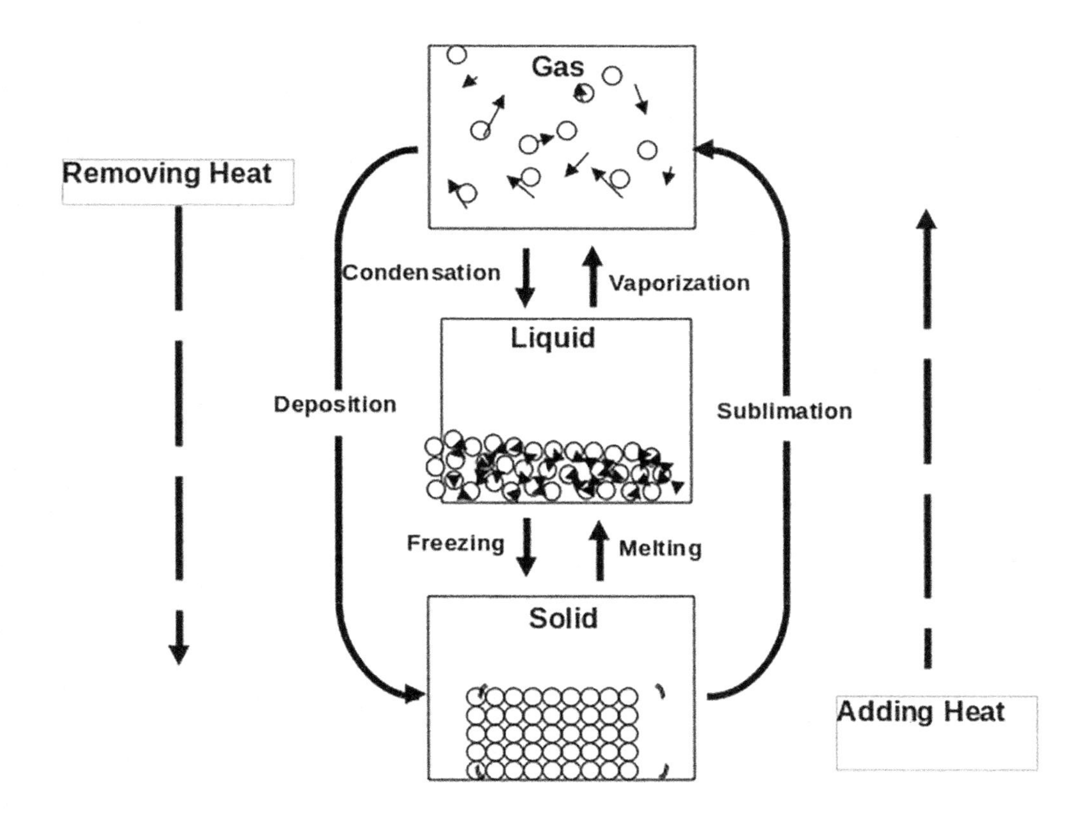

14.5 Entropy

Entropy (S) may be thought of as **the disorder in a system** or as a measure of the **number of states a system may occupy**. Changes due to entropy occur in one direction with no driving force. For example, a small volume of gas released into a large container will expand to fill it, but the gas in a large container never spontaneously collects itself into a small volume. This occurs because a large volume of gas has more disorder and has more places for gas molecules to be. Entropy has units of J/K.

If two different chemicals are at the same temperature, in the same state of matter, and they have the same number of molecules, their entropy difference will depend mostly on the number of ways the atoms within the two chemicals can rotate, vibrate, and flex. Most of the time**, the more complex molecule will have the greater entropy** because there are more energetic and spatial states in which it may exist.

In the solid phase, each molecule may vibrate, but it is otherwise locked into place in an ordered position and may only be in a relatively small number of locations. In the gas phase, however, each molecule could be almost anywhere and there is greater disorder. Liquids are intermediate in terms of movement of molecules. Therefore, **the entropy of a material increases during a phase change that raises the freedom of molecular motion and decreases during a phase change that prevents molecular motion.** Entropy also increases with temperature because molecules experience more disorder when they have a wider range of energy states to occupy.

Based on this principle, gases have greater entropy than liquids, liquids have greater entropy than solids, and matter in the same state increases in entropy with temperature.

Entropy is also an extensive property of matter. **A greater number of moles will have a larger entropy.**

Entropy also increases when there is an increase in volume (at constant temperature). The gas molecules are now able to move around a larger space. Review the gas laws in Chapter 4 for more details.

The standard molar entropy, S°, is the absolute entropy of a chemical at 1 atm and 25 °C. The molar entropy of a substance is expressed in units of J/mol-K. (See 5.3 Phase changes.)

Predict spontaneity of a chemical process given initial and final values of free energy, temperature, enthalpy, and/or entropy.

A reaction with a negative ΔH and a positive ΔS causes a decrease in energy and an increase in entropy. **These reactions will always occur spontaneously**. A reaction with a positive ΔH and a negative ΔS causes an increase in energy and a decrease in entropy. These reactions never occur to an appreciable extent because the reverse reaction takes place spontaneously.

Whether reactions with the remaining two possible combinations (ΔH and ΔS both positive or both negative) occur depends on the temperature. If $\Delta H - T\Delta S$ (known as **Gibbs Free Energy, ΔG**) is negative, the reaction will take place. If it is positive, the reaction will not occur to an appreciable extent. If $\Delta H - T\Delta S = 0$ exactly, then at equilibrium there will be 50% reactants and 50% products.

This chart will help identify four possible scenarios for the values of ΔH, ΔS and ΔG to predict spontaneity of the reaction.

ΔH	ΔS	ΔG
-	+	- (spontaneous)
-	-	- if $\lvert T\Delta S\rvert < \lvert\Delta H\rvert$ (depends on temperature)
+	+	- if $\lvert T\Delta S\rvert > \lvert\Delta H\rvert$ (depends on temperature)
+	-	+ not spontaneous

A spontaneous reaction is called *exergonic*. A non-spontaneous reaction is known as *endergonic*. These terms are used much less often than *exothermic* and *endothermic*.

Example (part 1): Determine the standard heat of formation ΔHf° for ethylene:

$$2C(graphite) + 2H_2(g) \rightarrow C_2H_4(g)$$

Use the heat of combustion for ethylene:

$$\Delta H_c^\circ = 1411.2 \frac{\text{kJ}}{\text{mol } C_2H_4} \quad \text{for } C_2H_4(g) + 3O_2(g) \rightarrow 2CO_2(g) + 2H_2O(l)$$

and the following two heats of formation for CO_2 and H_2O:

$$\Delta H_f^\circ = -393.5 \frac{\text{kJ}}{\text{mol C}} \quad \text{for} \quad C(graphite) + O_2(g) \rightarrow CO_2(g)$$

$$\Delta H_f^\circ = -285.9 \frac{\text{kJ}}{\text{mol } H_2} \quad \text{for} \quad H_2(g) + \frac{1}{2}O_2(g) \rightarrow H_2O(l).$$

Also find the standard change in entropy ΔS° for the formation of C_2H_4 given:

$$S^\circ(C(graphite)) = 5.7 \frac{\text{J}}{\text{mol K}} \quad S^\circ(H_2(g)) = 130.6 \frac{\text{J}}{\text{mol K}} \quad S^\circ(C_2H_4(g)) = 219.4 \frac{\text{J}}{\text{mol K}}.$$

Example (part 2): Will graphite and hydrogen gas react to form C_2H_4 at 25°C and 100 kPa?

Solution: Use Hess's Law after rearranging the given reactions so they cancel to yield the reaction of interest. Combustion is exothermic, so ΔH for this reaction is negative. We are interested in C_2H_4 as a product, so we consider the opposite (endothermic) reaction. The given ΔH values are multiplied by stoichiometric coefficients to give the reaction of interest as the sum of the three:

$$2CO_2(g) + 2H_2O(l) \rightarrow C_2H_4(g) + 3O_2(g) \quad \Delta H = 1411.2 \frac{\text{kJ}}{\text{mol reaction}}$$

$$2C(graphite) + 2O_2(g) \rightarrow 2CO_2(g) \quad \Delta H = -787.0 \frac{\text{kJ}}{\text{mol reaction}}$$

$$2H_2(g) + O_2(g) \rightarrow 2H_2O(l) \quad \Delta H = -571.8 \frac{\text{kJ}}{\text{mol reaction}}$$

$$2C(graphite) + 2H_2(g) \rightarrow C_2H_4(g) \quad \Delta H_f^\circ = 52.4 \frac{\text{kJ}}{\text{mol}}$$

$$2CO_2(g) + 2H_2O(l) \rightarrow C_2H_4(g) + 3O_2(g) \quad \Delta H = 1411.2 \frac{\text{kJ}}{\text{mol reaction}}$$

$$2C(graphite) + 2O_2(g) \rightarrow 2CO_2(g) \quad \Delta H = -787.0 \frac{\text{kJ}}{\text{mol reaction}}$$

$$2H_2(g) + O_2(g) \rightarrow 2H_2O(l) \quad \Delta H = -571.8 \frac{\text{kJ}}{\text{mol reaction}}$$

$$2C(graphite) + 2H_2(g) \rightarrow C_2H_4(g) \quad \Delta H_f^\circ = 52.4 \frac{\text{kJ}}{\text{mol}}$$

The value -787.0 kJ/mol is found by multiplying 2 × 393.5 kJ/mol because 2 moles react. The same is true for the -571.8 kJ/mol value; 2 moles react so it becomes 2 × -285.9 kJ/mol. The value for the first equation is not multiplied by 2 because the ΔH is for the equation as it is written.

The entropy change is found from:

$$\Delta S^{\circ} = S^{\circ}(C_2H_4) - 2S^{\circ}(C) - 2S^{\circ}(H_2) = 219.4\frac{J}{mol\ K} - 2 \times 5.7\frac{J}{mol\ K} - 2 \times 130.6\frac{J}{mol\ K}$$

$$= -53.2\frac{J}{mol\ K}.$$

This reaction is **endothermic** with a **decrease in entropy**, so it is **endergonic.** Graphite and hydrogen gas will not react to form C_2H_4.

In the diagram that follows, we look at the oxidation of carbon into CO and CO_2. The direct oxidation of carbon (graphite) into CO_2 yields an enthalpy of -393 kJ/mol. When carbon is oxidized into CO and then CO is oxidized to CO_2, the enthalpies are -110 and -283 kJ/mol respectively.

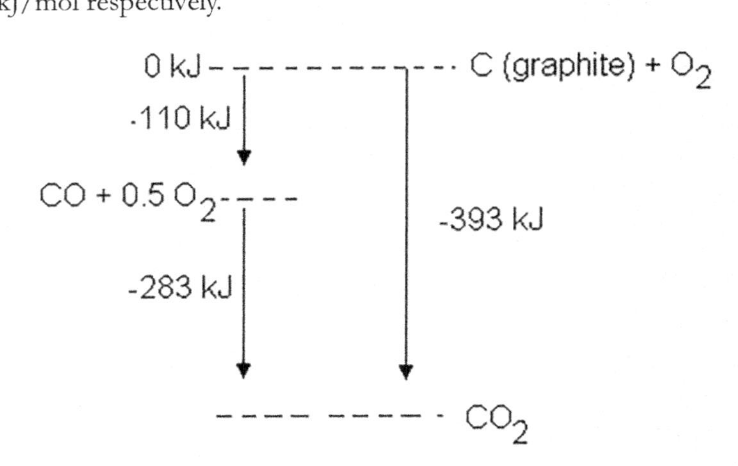

The sum of enthalpy for the two steps is -393 kJ/mol, exactly the same as that for the one-step reaction.

The two reaction steps are:

$C + \frac{1}{2}O_2 \longrightarrow CO$, $\Delta H^{\circ} = -110$ kJ/mol

$CO + \frac{1}{2}O_2 \longrightarrow CO_2$, $\Delta H^{\circ} = -283$ kJ/mol.

Adding the two equations together and canceling out the intermediate, CO, on both sides leads to

$C + O_2 \longrightarrow CO_2$, $\Delta H^{\circ} = (-110) + (-283) = -393$ kJ/mol.

Application of Hess's law enables us to calculate ΔH, ΔH°, and ΔH_f for chemical reactions that are impossible to measure, providing that we have all the data for related reactions.

The enthalpy of combustion for H_2, C (graphite) and CH_4 are -285.8, -393.5, and -890.4 kJ/mol respectively. Calculate the standard enthalpy of formation ΔH_f for CH_4.

Using the equations and their ΔH values:

1. $H_2(g) + 0.5\ O_2(g) \longrightarrow H_2O(l)$ $\Delta H = -285.8$ kJ/mol
2. $C(graphite) + O_2(g) \longrightarrow CO_2(g)$ $\Delta H = -393.5$ kJ/mol
3. $CH_4(g) + 2O_2(g) \longrightarrow CO_2(g) + 2H_2O(l)$ $\Delta H = -890.4$ kJ/mol

Find: $C + 2H_2 \longrightarrow CH_4$

$2\ (H_2(g) + 0.5\ O_2(g) \longrightarrow H_2O(l)$ $\quad \Delta H = -285.8\ kJ/mol)$

or $2H_2(g) + O_2\ (g) \longrightarrow 2H_2O(l)$ $\quad \Delta H = -571.6\ kJ$

$+\ \ C(graphite) + O_2(g) \longrightarrow CO_2(g)$ $\quad \Delta H = -393.5\ kJ$

$+\ \ CO_2(g) + 2H_2O(l) \longrightarrow CH_4(g) + 2O_2(g)$ $\quad \Delta H = +890.4\ kJ$

$C(graphite) + 2H_2(g) \longrightarrow CH_4(g)$ $\quad \Delta H = -74.7\ kJ$

From these data, we can construct an energy level diagram for these chemical combinations as follows:

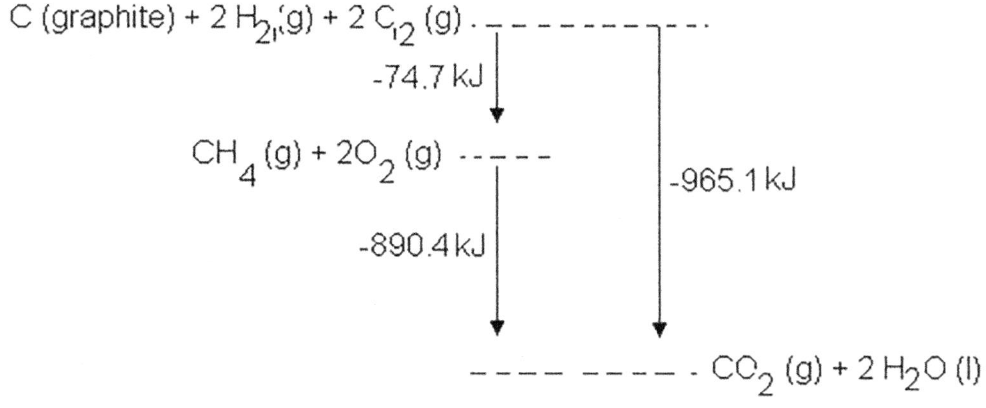

$C\ (graphite) + 2\ H_2\ (g) + 2\ O_2\ (g)$

-74.7 kJ

$CH_4\ (g) + 2O_2\ (g)$

-965.1 kJ

-890.4 kJ

$CO_2\ (g) + 2\ H_2O\ (l)$

Chemical energy is an important source of energy. It is the energy stored in substances by virtue of the arrangement of atoms within the substance. When atoms are rearranged during chemical reactions, energy is either released or consumed. It is the energy released from chemical reactions that fuels our economy and powers our bodies. Most of the electricity produced comes from chemical energy released by the burning of petroleum, coal, and natural gas. ATP is the molecule used by our body to carry chemical energy from cell to cell.

The energy within molecules is located in the bonds between the atoms in the molecule. To break these bonds requires energy. Once broken apart, the atoms, ions or molecules rearrange themselves to form new substances, making new bonds. Making new bonds releases energy.

If, during a chemical reaction, more energy is needed to break the reactant bonds than is released when product bonds form, the reaction is **endothermic** and heat is absorbed from the environment, which becomes colder.

On the other hand, if more energy is released due to product bonds forming than is needed to break reactant bonds, the reaction is **exothermic** and the excess energy is released to the environment as heat. The temperature of the environment goes up.

The total energy absorbed or released in the reaction can be determined by using bond energies. The total energy change of the reaction is equal to the total energy of all of the bonds of the products minus the total energy of all of the bonds of the reactants.

For example, propane , C_3H_8, is a common fuel used in heating homes and backyard grills. When burned, excess energy from the combustion reaction is released and used to cook our food.

$$C_3H_8\ (g) + 5\ O_2\ (g) \longrightarrow 3\ CO_2\ (g) + 4\ H_2O(l)$$

The total energy of the products came from the bonds found in the carbon dioxide molecules and the water molecules.

$$3\ O{=}C{=}O + 4\ H{-}O{-}H$$

or 6 C=O bonds and 8 H–O bonds.

A table of bond energies gives the following information:

C=O 743 kJ/mol
H–O 463 kJ/mol

So for these molecules there would be: (6 × 743 kJ) + (8 × 463 kJ) = 8162 kJ of energy released when these molecules form.

The reactants are these:

$$H{-}\underset{H}{\overset{H}{C}}{-}\underset{H}{\overset{H}{C}}{-}\underset{H}{\overset{H}{C}}{-}H + 5\ O{=}O$$

or

2 C–C bonds
8 C–H bonds and
5 O=O bonds

These bonds require the following energy to break:

C–C 348 kJ/ mol
C–H 412 kJ/mol
O=O 498 kJ/mol

The total energy required for the reactants would be

(2 × 348 kJ) + (8 × 412 kJ) +(5 × 498 kJ) = 6482 kJ of energy.

The total energy change that occurs during the combustion of propane is then:

8162 kJ – 6482 kJ = 680 kJ of energy is released for every mole of propane that burns in excess oxygen.

SECTION VI PRACTICE QUESTIONS

CHAPTER 14

1. **Which is correct regarding an endothermic reaction?**

    (A) The reaction does not occur.
    (B) The reaction liberates heat.
    (C) The enthalpy of the reactants is higher than the enthalpy of the products.
    (D) The sign of the ΔH is positive.
    (E) The temperature increases when the reaction is carried out in a calorimeter.

    The correct answer is D.
    For an endothermic process, Delta H is positive.

Questions 2-4 refer to the following terms

(A) Endothermic
(B) Exothermic
(C) Spontaneous
(D) Enthalpy
(E) Entropy

2. **Is TRUE when Delta G is negative: The correct answer is C.**

3. **Is TRUE when the energy of the products is greater than the energy of the reactants: The correct answer is A**

4. **Increases when a solid melts: The correct answer is E.**

5. **Cause Effect**

    Statement I
    The transition from gas to liquid is called **condensation**
    This statement is True.

    Statement II
    The enthalpy of condensation is the reverse of the enthalpy of vaporization
    This statement is True.
    Both statements are TRUE but there is no cause/effect relationship.

Practice Questions

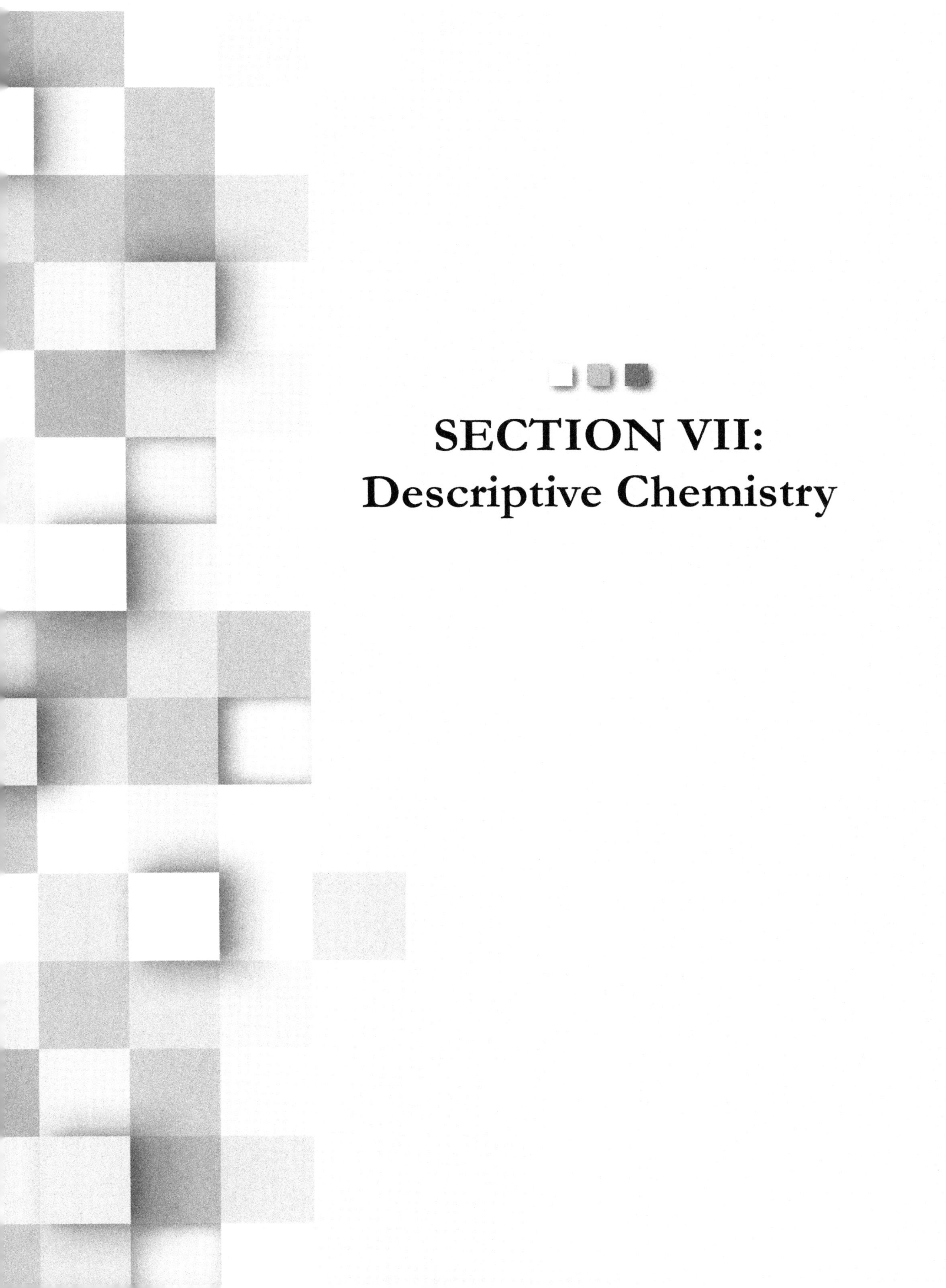

SECTION VII: Descriptive Chemistry

Chapter 15: Descriptive Chemistry

15.1 Nomenclature of ions and compounds

Identify an inorganic chemical formula (ionic or molecular), given the name.

The IUPAC is the **International Union of Pure and Applied Chemistry**, an organization that formulates naming rules for chemical elements and compounds. **Organic compounds contain carbon**, and they have a separate system of nomenclature but some of the simplest molecules containing carbon also fall within the scope of inorganic chemistry.

Naming rules depend on whether the chemical is an ionic compound or a molecular compound containing only covalent bonds. There are special rules for naming acids. The rules below describe a group of traditional "semi-systematic" names accepted by IUPAC.

Ionic compounds: Cation

Ionic compounds are named with the **cation (positive ion) first**. Nearly all cations in inorganic chemistry are **monatomic**, meaning they just consist of one atom (like Ca^{2+}, the calcium ion.) This atom will be a **metal ion**. For common ionic compounds, the **alkali metals always have a 1+ charge** and the **alkali earth metals always have a 2+ charge.**

Many metals may form cations of more than one charge. In this case, a Roman numeral in parenthesis after the name of the element is used to indicate the ion's charge in a particular compound. This Roman numeral method is known as the **Stock system**. An older nomenclature used the suffix *–ous* for the lower charge and *–ic* for the higher charge and is still used occasionally.

Example: Fe^{2+} is the iron(II) ion and Fe^{3+} is the iron(III) ion.

The only common inorganic **polyatomic cation** is **ammonium: NH_4^+**.

Ionic compounds: Anion

The **anion** (negative ion) is named and written last. Monatomic anions are formed from nonmetallic elements and are named by **replacing the end of the element's name with the suffix –*ide*.**

Examples: Cl^- is the chloride ion, S^{2-} is the sulfide ion, and N^{3-} is the nitride ion.

These anions also end with *–ide*:

C_2^{2-}	N_3^-	O_2^{2-}	O_3^-	S_2^{2-}	CN^-	OH^-
carbide or acetylide	azide	peroxide	ozonide	disulfide	cyanide	hydroxide

Oxoanions (also called oxyanions) **contain one element in combination with oxygen.** Many common polyatomic anions are oxoanions that **end with the suffix** *–ate*. If an element has two possible oxoanions, the one with the element at a lower oxidation state **ends with** *–ite*. This anion will also usually have **fewer oxygens per non-oxygen atom.** See **3.2.a Recognition of oxidation-reduction reactions** for a discussion of oxidation numbers. Additional oxoanions are named with the prefix *hypo-* if they have a lower oxidation number than the *–ite* form and the prefix *per–* if they have a higher oxidation number than the *–ate* form. Note the chlorate series of hypochlorite (ClO^-), chlorite (ClO_2^-), chlorate (ClO_3^-), and perchlorate (ClO_4^-).

Common examples:

				CO_3^{2-}	carbonate		
		SO_3^{2-}	sulfite	SO_4^{2-}	sulfate		
		PO_3^{3-}	phosphite	PO_4^{3-}	phosphate		
$N_2O_2^{2-}$	hyponitrite	NO_2^-	nitrite	NO_3^-	nitrate		
ClO^-	hypochlorite	ClO_2^-	chlorite	ClO_3^-	chlorate	ClO_4^-	perchlorate
BrO^-	hypobromite	BrO_2^-	bromite	BrO_3^-	bromate	BrO_4^-	perbromate
				MnO_4^{2-}	manganate	MnO_4^-	permanganate
				CrO_4^{2-}	chromate	CrO_8^{3-}	perchromate

Note that manganate/permanganate and chromate/perchromate are exceptions to the general rules because there are *–ate* ions but no *–ite* ions and because the charge changes.

Other polyatomic anions that end with *–ate* are:

$C_2O_4^{2-}$	$Cr_2O_7^{2-}$	SCN^-	HCO_2^-	$CH_3CO_2^-$
oxalate	dichromate	thiocyanate	formate	acetate

HCO_2^- and $CH_3CO_2^-$ are condensed **structural formulas** because they show how the atoms are linked together. Their molecular formulas would be CHO_2^- and $C_2H_3O_2^-$

If an H atom is added to a polyatomic anion with a negative charge greater than one, the word *hydrogen* or the prefix *bi-* are used for the resulting anion. If two H atoms are added, *dihydrogen* is used.

Examples: bicarbonate or hydrogen carbonate ion: HCO_3^-
dihydrogen phosphate ion: $H_2PO_4^-$

Ionic compounds: Hydrates

Water molecules often occupy positions within the lattice of an ionic crystal. These compounds are called **hydrates**, and the water molecules are known as **water of hydration**. The water of hydration is added after a centered dot in a formula. In a name, a number-prefix (listed below for molecular compounds) indicating the number of water molecules is followed by the root *–hydrate*.

Ionic compounds: Putting it all together

We now have the tools to name most common salts given a formula and to write a formula for them given a name. To determine a formula given a name, the number of anions and cations that are needed to achieve a neutral charge must be found.

Example: Determine the formula of cobalt(II) phosphite octahydrate.

Solution: For the cation, find the symbol for cobalt (Co) and recognize that it is present as Co^{2+} ions from the Roman numerals. For the anion, remember that the phosphite ion is PO_3^{3-}. A neutral charge is achieved with 3 Co^{2+} ions for every 2 PO_3^{3-} ions. Add eight H_2O for water of hydration for the answer:

$$Co_3(PO_3)_2 \cdot 8H_2O$$

Molecular compounds

Molecular compounds (compounds making up molecules with a neutral charge) are usually composed entirely of nonmetals and are named by placing the **less electronegative atom first**. See **3.4 Relationships of bonding to properties and structures: Intermolecular Forces** for the relationship between electronegativity and the periodic table. The suffix *–ide* is added to the second, more electronegative atom, and prefixes indicating numbers are added to one or both names if needed.

Prefix	*mono-*	*di-*	*tri-*	*tetra-*	*penta-*	*hexa-*	*hepta-*	*octa-*	*nona-*	*deca-*
Meaning	1	2	3	4	5	6	7	8	9	10

The final "o" or "a" may be left off these prefixes for oxides.

The electronegativity requirement is the reason the compound with two oxygen atoms and one nitrogen atom is called nitrogen dioxide, NO_2 and *not* dioxygen nitride O_2N. The hydride of sodium is NaH, sodium hydride, but the hydride of bromine is HBr, hydrogen bromide (or hydrobromic acid if it's in aqueous solution). Oxygen is only named first in compounds with fluorine such as oxygen difluoride, OF_2, and fluorine is never placed first because it is the most electronegative element.

Examples: N_2O_4, dinitrogen tetroxide (or tetraoxide)
Cl_2O_7, dichlorine heptoxide (or heptaoxide)
ClF_5 chlorine pentafluoride

Acids

There are special naming rules for acids that correspond with the **suffix of their corresponding anion** if hydrogen were removed from the acid. Anions ending with *–ide* correspond to acids with the prefix *hydro–* and the suffix *–ic*. Anions ending with *–ate* correspond to acids with no prefix that end with *–ic*. Oxoanions ending with *–ite* have associated acids with no prefix and the suffix *–ous*. The *hypo–* and *per–* prefixes are maintained. Some examples are shown in the following table:

anion	anion name	acid	acid name
Cl^-	chloride	$HCl(aq)$	hydrochloric acid
CN^-	cyanide	$HCN(aq)$	hydrocyanic acid
CO_3^{2-}	carbonate	$H_2CO_3(aq)$	carbonic acid
SO_3^{2-}	sulfite	$H_2SO_3(aq)$	sulfurous acid
SO_4^{2-}	sulfate	$H_2SO_4(aq)$	sulfuric acid
ClO^-	hypochlorite	$HClO(aq)$	hypochlorous acid
ClO_2^-	chlorite	$HClO_2(aq)$	chlorous acid
ClO_3^-	chlorate	$HClO_3(aq)$	chloric acid
ClO_4^-	perchlorate	$HClO_4(aq)$	perchloric acid

Example: What is the molecular formula of phosphorous acid?

Solution: If we remember that the *–ous* acid corresponds to the *–ite* anion, and that the *–ite* anion has one less oxygen than (or has an oxidation number 2 less than) the *–ate* form, we only need to remember that phosphate is PO_4^{3-}. Then we know that phosphite is PO_3^{3-} and phosphorous acid is H_3PO_3.

Select the name of an inorganic chemical compound (ionic or molecular), given its formula.

Properly written and named formulas

Proper formulas will follow the rules of the previous subject. Here are some ways to identify improper formulas that are emphasized below by underlining them.

In all common names for **ionic compounds**, **number prefixes are not used** to describe the number of anions and cations.

Examples:

$CaBr_2$ is calcium bromide, *not calcium dibromide.*
$Ba(OH)_2$ is barium hydroxide, *not barium dihydroxide.*
Cu_2SO_4 is copper(I) sulfate, *not dicopper sulfate or copper(II) sulfate or dicopper sulfur tetroxide.*

All ionic compounds must have a **neutral charge in their formula** representations.

Example: **MgBr is an improperly written formula** because Mg ion always exists as 2^+ and Br ion is always a 1^- ion. $MgBr_2$, magnesium bromide, is correct.

Proper oxoanions and acids use the correct prefixes and suffixes.

Example: HNO_3 is nitric acid because NO_3^- is the nitrate ion.

In both ionic and molecular compounds, the **less electronegative element comes first**.

Example: **CSi is an improperly written formula** because Si is below C on the periodic table and therefore less electronegative. SiC, silicon carbide, is correct.

Identify properly named formulas for simple organic compounds.

The IUPAC is the **International Union of Pure and Applied Chemistry**. **Organic compounds contain carbon,** and have their own branch of chemistry because of the huge number of carbon compounds in nature, including nearly all the molecules in living things. The 1979 IUPAC organic nomenclature is used and taught most often today, and it is the nomenclature described here.

The simplest organic compounds are called **hydrocarbons** because they **contain only carbon and hydrogen**. Hydrocarbon molecules may be divided into the classes of **cyclic** and **open-chain** depending on whether they contain a ring of carbon atoms. Open-chain molecules may be divided into **branched** or **straight-chain** categories.

Hydrocarbons are also divided into classes called **aliphatic** and **aromatic**. Aromatic hydrocarbons are related to benzene and are always cyclic. Aliphatic hydrocarbons may be open-chain or cyclic. Aliphatic cyclic hydrocarbons are called **alicyclic**. Aliphatic hydrocarbons are one of three types: alkanes, alkenes, and alkynes.

Alkanes

Alkanes contain only single bonds. Alkanes have the maximum number of hydrogen atoms possible for their carbon backbone, so they are called **saturated**. Alkenes, alkynes, and aromatics are **unsaturated** because they have fewer hydrogens.

Straight-chain alkanes are also called **normal alkanes**. These are the simplest hydrocarbons. They consist of a linear chain of carbon atoms. The names of these molecules contain the suffix *–ane* and a **root based on the number of carbons in the chain** according to the table on the following page. The first four roots, *meth–*, *eth–*, *prop–*, and *but–* have historical origins in chemistry, and the remaining alkanes contain common Greek number prefixes. Alkanes have the general formula C_nH_{2n+2}.

A single molecule may be represented in multiple ways. Methane and ethane in the table are shown as three-dimensional structures with dashed wedge shapes attaching atoms behind the page and thick wedge shapes attaching atoms in front of the page.

Number of carbons	Name	Formula	Structure
1	Methane	CH_4	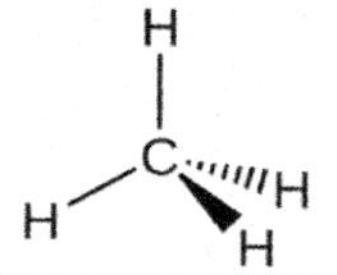 H C H H H
2	Ethane	C_2H_6	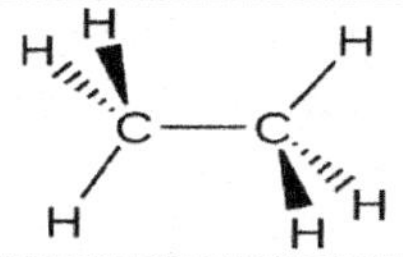 H H H C C H H H
3	Propane	C_3H_8	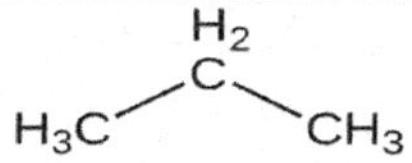 H_2 C H_3C CH_3
4	Butane	C_4H_{10}	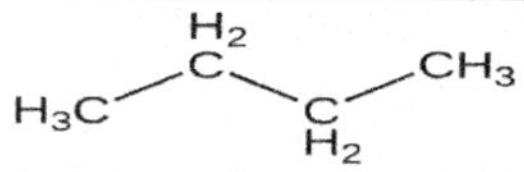 H_2 C CH_3 H_3C C H_2
5	Pentane	C_5H_{12}	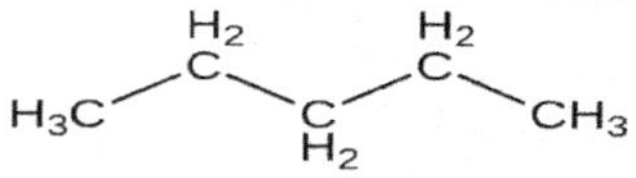 H_2 H_2 C C H_3C C CH_3 H_2
6	Hexane	C_6H_{14}	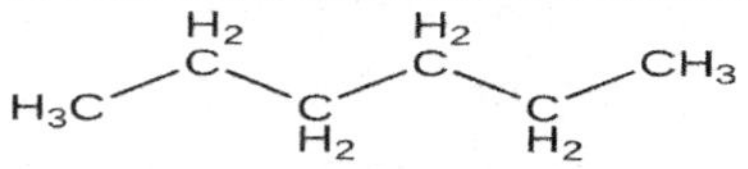 H_2 H_2 C C CH_3 H_3C C C H_2 H_2
7	Heptane	C_7H_{16}	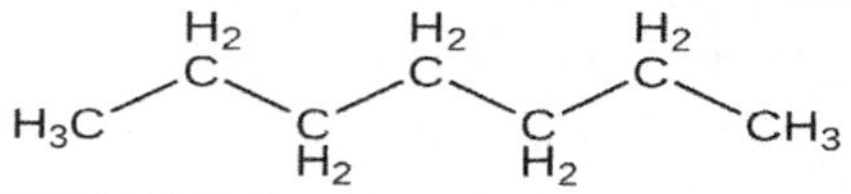 H_2 H_2 H_2 C C C H_3C C C CH_3 H_2 H_2
8	Octane	C_8H_{18}	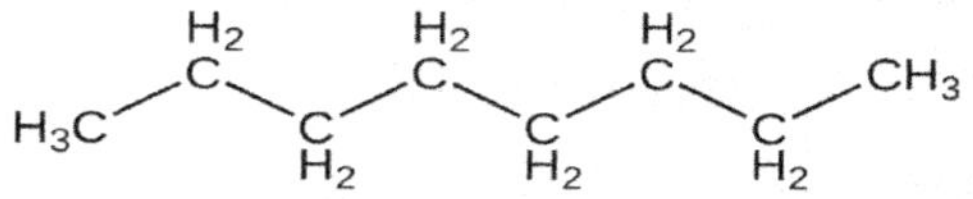 H_2 H_2 H_2 C C C CH_3 H_3C C C C H_2 H_2 H_2

Additional ways that pentane might be represented are:

n-pentane (the *n* represents a *normal* alkane)

$CH_3CH_2CH_2CH_2CH_3$

$CH_3(CH_2)_3CH_3$

If one hydrogen is removed from an alkane, the residue is called an **alkyl** group. The *–ane* suffix is replaced by an *–yl–* infix when this residue is used as **functional group**. Functional groups are used to systematically build up the names of organic molecules.

Branched alkanes are named using a four-step process:

1. Find the longest continuous carbon chain. This is the parent hydrocarbon.
2. Number the atoms on this chain beginning at the end near the first branch point, so the lowest locant numbers are used. Number functional groups from the attachment point.
3. Determine the numbered locations and names of the substituted alkyl groups. Use *di–*, *tri–*, and similar prefixes for alkyl groups represented more than once. Separate numbers by commas and groups by dashes.

4. List the locations and names of alkyl groups in alphabetical order by their name (ignoring the *di–*, *tri–* prefixes) and end the name with the parent hydrocarbon.

Example: Name the following hydrocarbon:

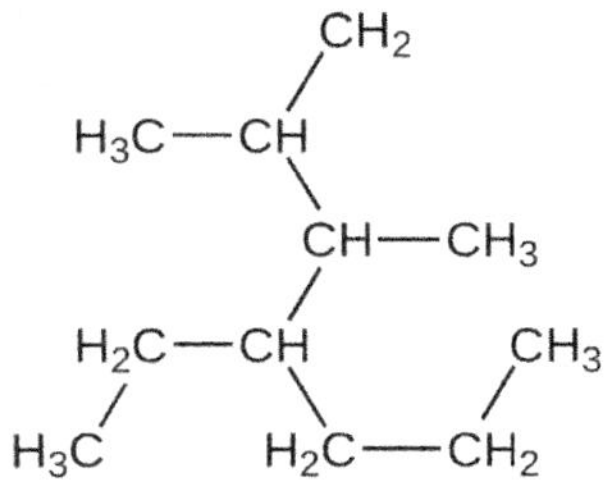

CH_2
$H_3C—CH$
$CH—CH_3$
$H_2C—CH$ CH_3
H_3C $H_2C—CH_2$

Solution:

(A) The longest chain is seven carbons in length, as shown by the bold lines below. This molecule is a heptane.

(B) The atoms are numbered from the end nearest the first branch as shown:

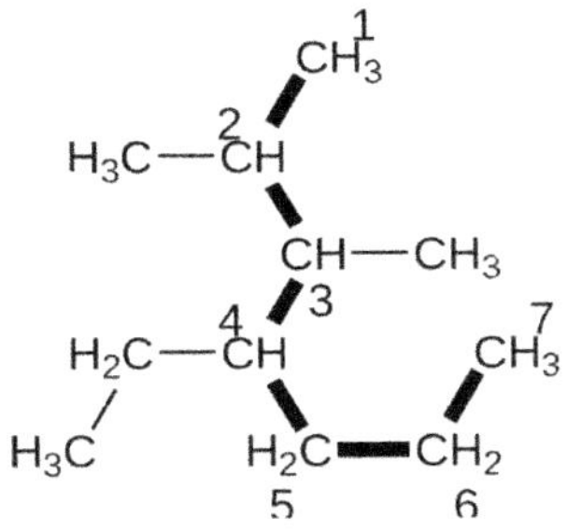

$\overset{1}{CH_3}$
$H_3C—\overset{2}{CH}$
$CH—CH_3$ 3
$H_2C—\overset{4}{CH}$ $\overset{7}{CH_3}$
H_3C $\underset{5}{H_2C}—\underset{6}{CH_2}$

(C) Methyl groups are located at carbons 2 and 3 (2,3-dimethyl), and an ethyl group is located at carbon 4.

(D) "Ethyl" precedes "methyl" alphabetically. The hydrocarbon name is: 4-ethyl-2,3-dimethylheptane.

The following branched alkanes have IUPAC-accepted common names:

Structure	Systematic name	Common name
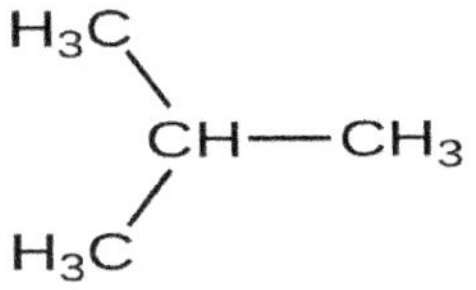 H_3C / $CH—CH_3$ / H_3C	2-methylpropane	isobutane
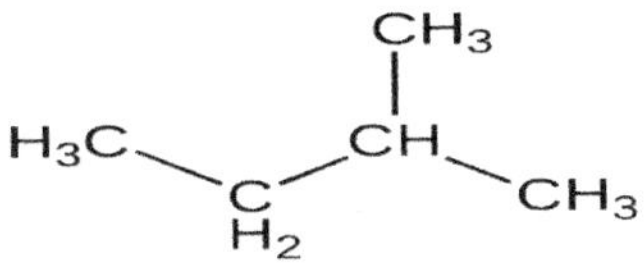 CH_3 / H_3C $\underset{H_2}{C}$ CH CH_3	2-methylbutane	isopentane
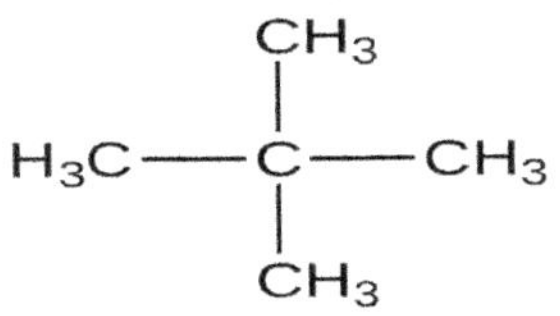 CH_3 / $H_3C—C—CH_3$ / CH_3	2,2-dimethylpropane	neopentane

The following alkyl groups have IUPAC-accepted common names. The systematic names assign a locant number of 1 to the attachment point:

Structure	Systematic name	Common name
H_3C–CH(–)–CH_3	1-methylethyl	isopropyl
$(H_3C)_2$CH–CH_2–	2-methylpropyl	isobutyl
H_3C–CH_2–CH(–)–CH_3	1-methylpropyl	*sec*-butyl
H_3C–C(–)(CH_3)–CH_3	1,1-dimethylethyl	*tert*-butyl

Alkenes

Alkenes contain one or more double bonds. Alkenes are also called olefins. The suffix used in the naming of alkenes is *–ene*, and the number roots are those used for alkanes of the same length.

A number preceding the name shows the location of the double bond for alkenes of length four and above. Alkenes with one double bond have the general formula C_nH_{2n}. Multiple double bonds are named using *–diene*, *–triene*, etc. The suffix *–enyl–* is used for functional groups after a hydrogen is removed from an alkene. Ethene and propene have the common names **ethylene** and **propylene**. The ethenyl group has the common name **vinyl** and the 2-propenyl group has the common name **allyl**.

Examples:

$H_2C{=}CH_2$ is ethylene or ethene. $H_2C{=}CH$– is a vinyl or ethenyl group.

$H_2C{=}CH{-}CH_3$ is propylene or propene. $H_2C{=}CH{-}CH_2$– is an allyl or 2-propenyl group.

$H_3C{-}CH{=}CH{-}CH_2{-}CH_2{-}CH_3$ is 2-hexene.

$H_2C{=}CH{-}C(CH_3){=}CH_2$ is 2-methyl-1,3-butadiene (common name: isoprene).

Cis-trans isomerism is often part of the complete name for an alkene. Note that isoprene contains two adjacent double bonds, so it is a **conjugated** molecule.

Alkynes and alkenynes

Alkynes contain one or more triple bonds. They are named in a similar way to alkenes. The suffix used for alkynes is *–yne*. Ethyne is often called **acetylene**. Alkynes with one triple bond have the general formula C_nH_{2n-2}. Multiple triple bonds are named using *–diyne*, *–triyne*, etc. The infix *–ynyl–* is used for functional groups composed of alkynes after the removal of a hydrogen atom.

Hydrocarbons with **both double and triple bonds are known as alkenynes**. The locant number for the double bond precedes the name, and the locant for the triple bond follows the suffix *–en–* and precedes the suffix *–yne*.

Examples: $HC{\equiv}CH$ is acetylene or ethyne.

$HC{\equiv}C{-}CH_2{-}CH_3$ is 1-butyne.

$HC{\equiv}C{-}C{\equiv}C{-}CH_3$ is 1,3-pentadiyne.

$H_3C{-}C{\equiv}C{-}CH_2{-}CH_2{-}CH_2{-}$ is a 4-hexynyl group.

$H_2C{=}CH{-}C{\equiv}CH$ is 1-buten-3-yne. This compound has the common name of vinylacetylene.

Cycloalkanes, –enes, and –ynes

Alicyclic hydrocarbons use the prefix *cyclo–* before the number root for the molecule. The structures for these molecules are often written as if the molecule lay entirely within the plane of the paper even though in reality, these rings dip above and below a single plane. When there is more than one substitution on the ring, numbering begins with the first substitution listed in alphabetical order.

Cis-trans isomerism is often part of the complete name for a cycloalkane

Examples:

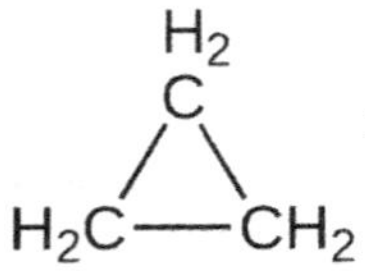

is cyclopropane

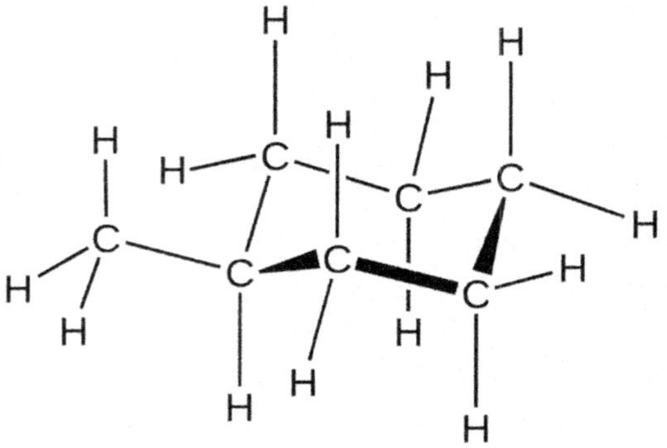

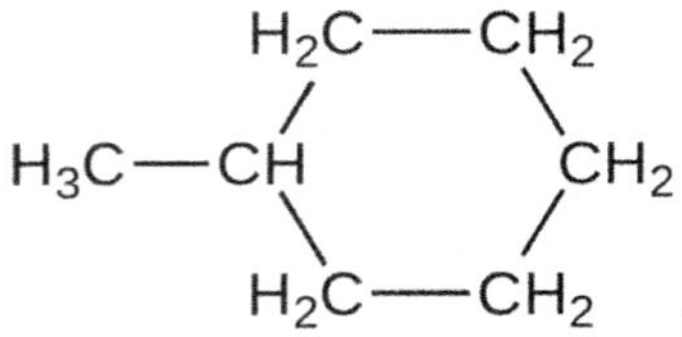

is methylcyclohexane.

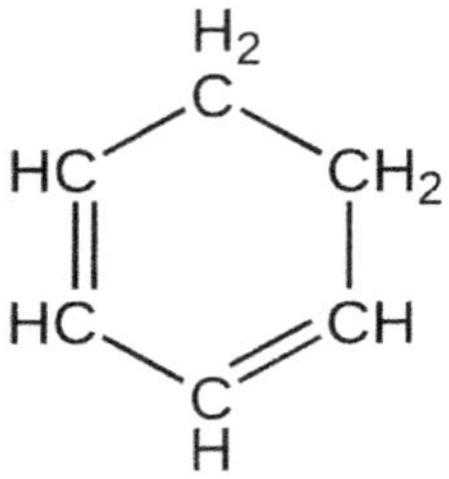

is 1,3-cyclohexadiene.

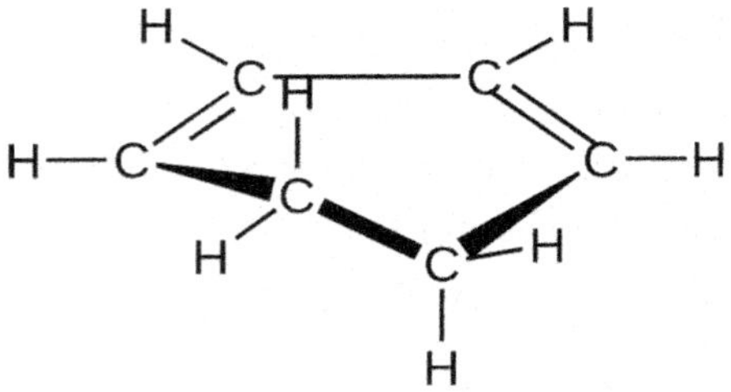

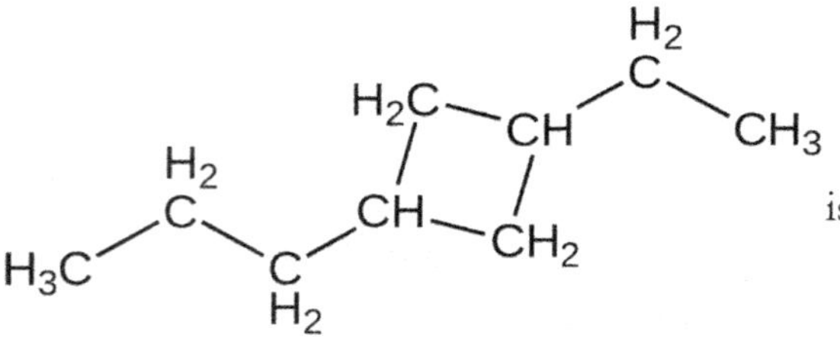

is 1-ethyl-3-propylcyclobutane.

Aromatic hydrocarbons

Aromatic hydrocarbons are structurally related to benzene or made up of benzene molecules fused together. These molecules are called **arenes** to distinguish them from alkanes, alkenes, and alkynes. All atoms in arenes lie in the same plane. In other words, aromatic hydrocarbons are flat. Aromatic molecules have electrons in delocalized π orbitals that are free to migrate throughout the molecule.

Substitutions onto the benzene ring are named in alphabetical order using the lowest possible locant numbers. The prefix *phenyl–* may be used for C_6H_5– (benzene less a hydrogen) attached as a functional group to a larger hydrocarbon residue. Arenes in general form aryl functional groups. A phenyl group may be represented in a structure by the symbol Ø. The prefix *benzyl–* may used for $C_6H_5CH_2$– (methylbenzene with a hydrogen removed from the methyl group) attached as a functional group.

Examples: or 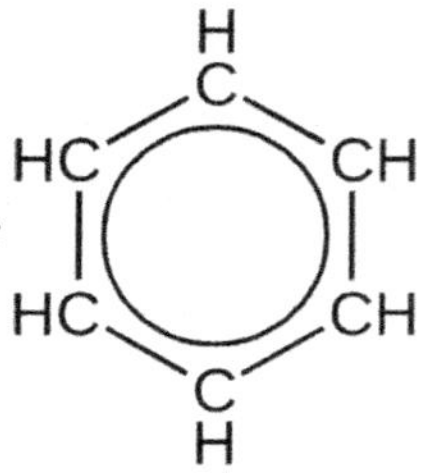 is benzene.

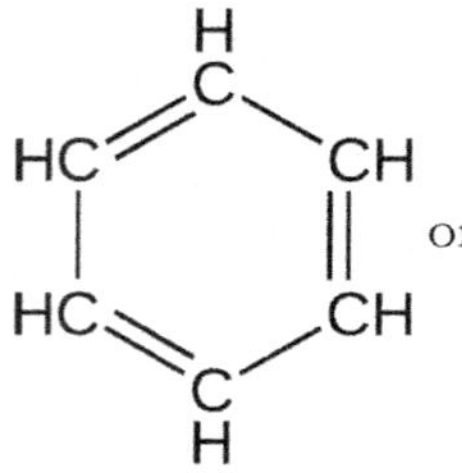

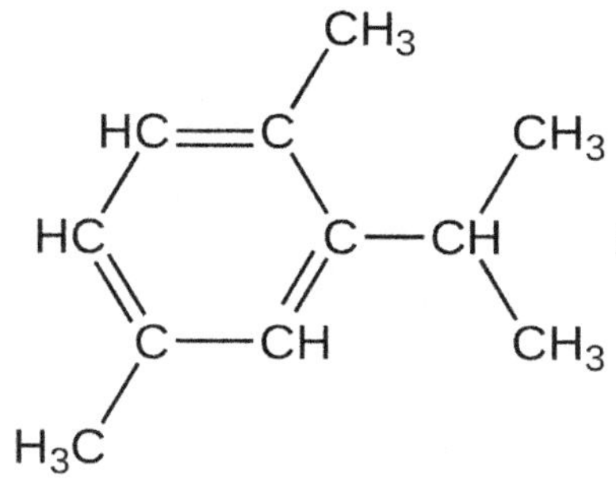

CH_3
HC=C CH_3
HC C—CH
C—CH CH_3
H_3C

is 2-isopropyl-1,4- dimethylbenzene.

is 3-phenyloctane or (1-ethylhexyl)benzen.

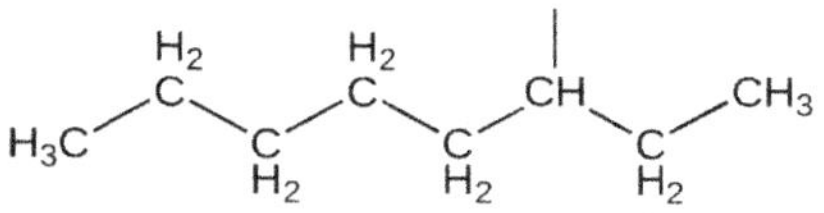

H_2 H_2
H_3C C C C CH C CH_3
H_2 H_2 H_2

The most often used common names for aromatic hydrocarbons are listed in the following table. Naphthalene is the simplest molecule formed by fused benzene rings.

Structure	Systematic name	Common name
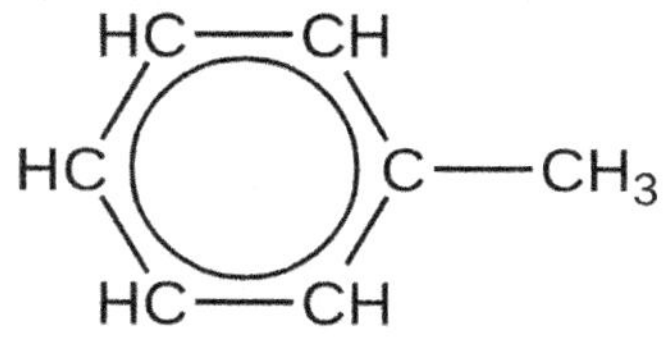 HC—CH HC C—CH_3 HC—CH	methylbenzene	toluene
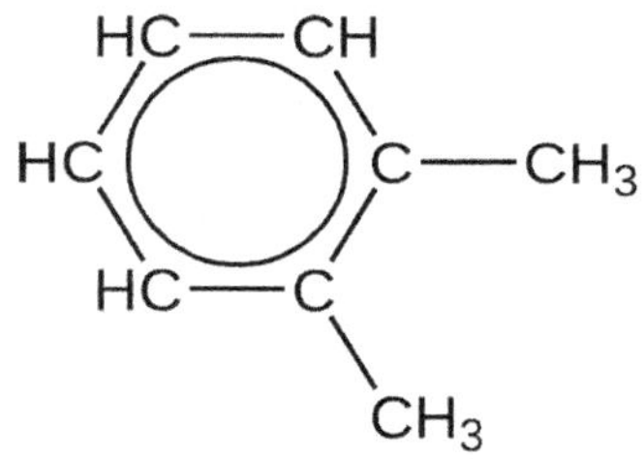 HC—CH HC C—CH_3 HC—C CH_3	1,2-dimethylbenzene	*ortho*-xylene or *o*-xylene
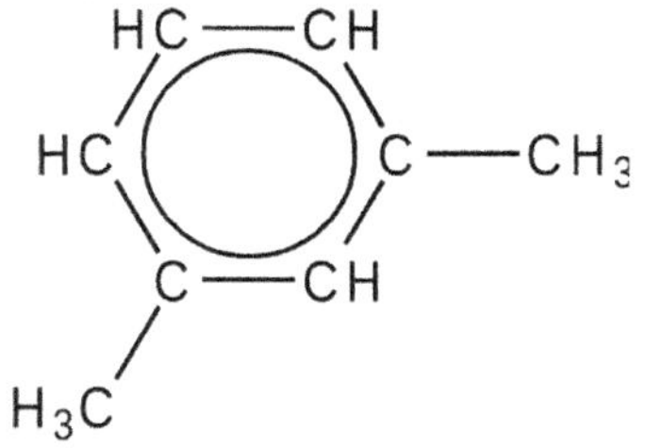 HC—CH HC C—CH_3 C—CH H_3C	1,3-dimethylbenzene	*meta*-xylene or *m*-xylene
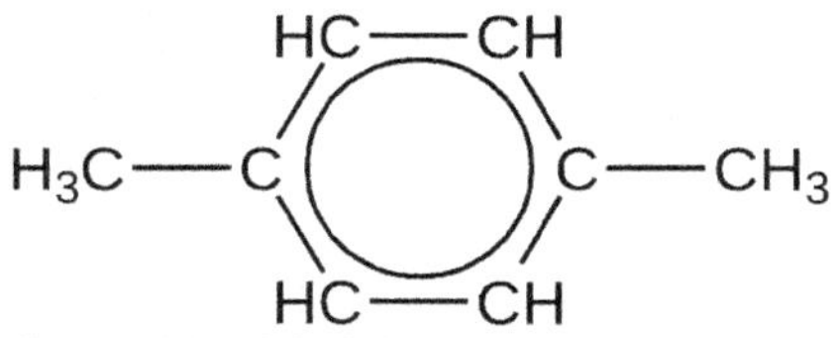 HC—CH H_3C—C C—CH_3 HC—CH	1,4-dimethylbenzene	*para*-xylene or *p*-xylene
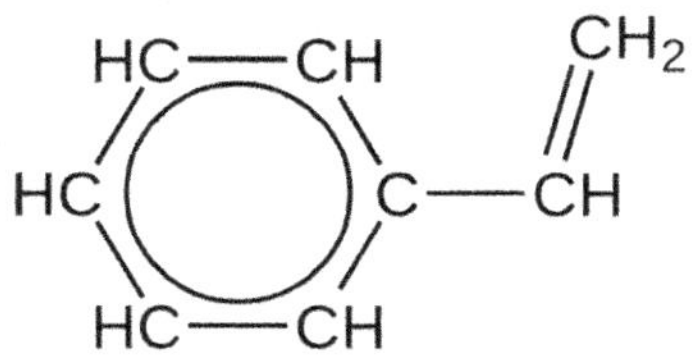 HC—CH CH_2 HC C—CH HC—CH	ethenylbenzene	styrene

Structure	Systematic name	Common name
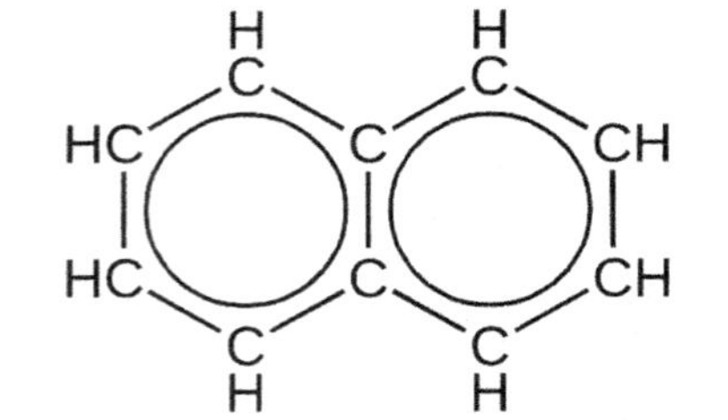 H H C C HC C CH HC C CH C C H H	Bicyclo[4.4.0]deca-1,3,5,7,9-pentene	naphthalene

Identify common organic functional groups.

Hydrocarbons consist entirely of nonpolar C-H bonds with no unpaired electrons. These compounds are relatively unreactive. The substitution of one or more atoms with unpaired electrons into the hydrocarbon backbone creates a **hydrocarbon derivative**. The unpaired electrons result in polar or charged portions of these molecules. These atoms fall into categories known as **functional groups**, and they create **local regions of reactivity**. Alkyl, alkenyl, alkynyl, and aryl groups may also be considered functional groups in some circumstances.

Two types of names are often used for the same hydrocarbon derivative. For **substitutive names**, the hydrocarbon name is written out and the correct prefix or suffix is added for the derivative group. For **functional class names**, the hydrocarbon is written as a functional group and the derivative is added as a second word. Chapter 15.5 also contains information on functional groups.

Examples: CH_3Br may be called bromomethane (substitutive name) or methyl bromide (functional class name). CH_3CH_2OH is ethanol (substitutive name) or ethyl alcohol (functional class name).

In the tables on the following pages, the symbols R and R' represent hydrocarbons in covalent linkage to the functional group. Many derivatives are named in a similar manner to alkenes and alkynes, but the location and suffix of the functional group is used in place of *–ene* and *–yne*.

The **acyl group** is

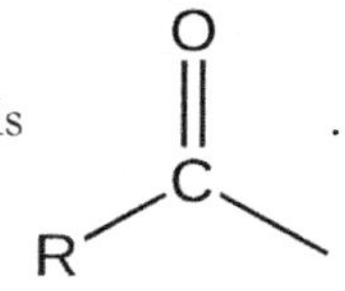

O
||
C
R

.

There are non-systematic number roots for hydrocarbon derivatives containing acyl groups or derived from acyl groups. These are the ketones, carboxylic acids, esters, nitriles, and amides:

Number of carbons (including the acyl carbon)	Systematic prefix	Accepted prefix for acyl and nitrile functional groups
1	meth–	form–
2	eth–	acet–
3	prop–	propion–
4	but–	butyr–
5	pent–	pent–
6 and above	same for larger numbers	

Several functional groups utilize oxygen.

Class of molecule	Functional group	Structure	Affix	Example
Alcohol	Hydroxyl —OH	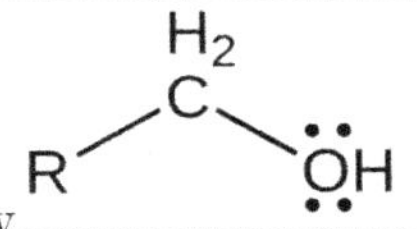 primary: 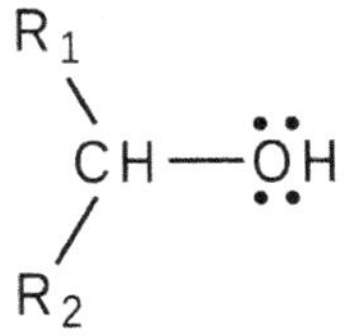 $R-CH_2-\ddot{O}H$ secondary: $R_1R_2CH-\ddot{O}H$ tertiary: $R_1R_2R_3C-OH$ 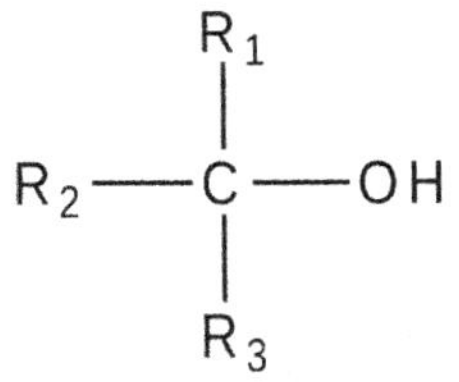	*–ol*	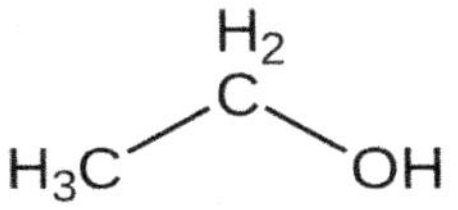 H_3C-CH_2-OH ethanol
Ether	Oxy 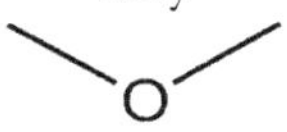 —O—	 $R_1-\ddot{O}-R_2$	*–oxy–*	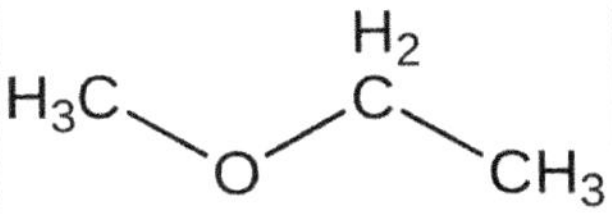 $H_3C-O-CH_2-CH_3$ methoxyethane or ethyl methyl ether

Class of molecule	Functional group	Structure	Affix	Example
Aldehyde	Carbonyl $-C(=O)-$	$R{-}C(=O){-}H$	*–al*	$O{=}CH{-}CH_2{-}CH_3$ propionaldehyde or propanal
Ketone		$R_1{-}C(=O){-}R_2$	*–one*	$H_3C{-}C(=O){-}CH_3$ acetone
Carboxylic acid	Carboxyl $-C(=O){-}OH$	$R{-}C(=O){-}OH$	*–oic acid*	$HC(=O){-}OH$ formic acid or methanoic acid
Ester	Oxycarbonyl $-C(=O){-}O-$	$R_2{-}C(=O){-}O{-}R_1$	*–yl –oate*	$CH_3{-}H_2C{-}CH_2{-}C(=O){-}O{-}CH_3$ methyl butyrate or methyl butanoate
Acid anhydride	Carbonyloxy-carbonyl $-C(=O){-}O{-}C(=O)-$	$R_1{-}C(=O){-}O{-}C(=O){-}R$	*–oic anhydride*	$H_3C{-}C(=O){-}O{-}C(=O){-}CH_3$ acetic anhydride or ethanoic anhydride

Derivatives utilizing other atoms are also common.

Class of molecule	Functional group	Structure	Affix	Example
Nitrile	Cyanide —C≡N	R—C≡N:	*–nitrile*	$H_3C—C≡N$ acetonitrile or ethanonitrile
Amine	Amino —N—	primary: R—N(H)—H; secondary: R_1—N(R_2)—H; tertiary: R_1—N(R_2)—R_3	*–amine*	H_3C—CH_2—NH_2 ethanamine
Amide	Aminocarbonyl —C(=O)—N—	primary: R—C(=O)—N(H)—H; secondary: R_2—C(=O)—N(R_1)—H; tertiary: R_3—C(=O)—N(R_2)—R_1	*–amide*	H_3C—CH_2—C(=O)—NH_2 propionamide or propanamide
Alkyl halide	Halide —X (where X is F, Cl, Br, or I)	R—X:	*fluoro– chloro– bromo– iodo–*	H_3C—CH(Br)—CH_2—CH_3 2-bromobutane

Hydrocarbon derivatives and functional groups may be identified using these tables. One way to memorize the derivatives utilizing oxygen is to divide them into pairs with a hydrogen atom in the first element of the pair replaced by a hydrocarbon in the second element. These pairs are:

1. alcohol/ether.
2. aldehyde/ketone.
3. carboxylic acid/ester.

A final word about IUPAC organic nomenclature

"IUPAC nomenclature" has become a synonym for "correct nomenclature," but the most recent IUPAC recommendations for organic nomenclature are ignored in most textbooks and curricula in the United States. This situation is similar to the nomenclature for inorganic chemistry in that the older nomenclature is "systematic enough" to be taught and to avoid most errors, but is a poor choice to name new compounds in an unambiguous way.

Two important differences between the 1979 and more recent nomenclature involve the use of locant numbers:

1. The 1979 nomenclature usually places locant numbers before the name of the entire compound, but the 1993 recommendations suggest placing locant numbers immediately before the group that they represent.
2. The 1979 nomenclature always assigns a locant number of 1 to the attachment point of groups, but the 1993 nomenclature recommends stating the attachment point explicitly before "*–yl–*." This decreases the complexity of hydrocarbon residues.

The 1993 nomenclature permits the 1979 nomenclature as "naming method a" and presents the newer method as a "method b." In provisional 2004 recommendations, the newer "method b" names are usually explicitly preferred. The newer method is more commonly used in Europe and elsewhere. These differences are shown below for some of the hydrocarbons described in 15.2 **Identify properly named formulas for simple organic compounds**.

Identify common organic functional groups.

Structure	Systematic name (1979)	Systematic name (2004)	Common name
$H_3C-CH=CH-CH_2-CH_2-CH_3$	2-hexene	hex-2-ene	
$H_2C=HC-C\equiv CH$	1-buten-3-yne	but-1-en-3-yne	vinylacetylene
$H_3C-CH(-)-CH_3$	1-methylethyl	propan-2-yl	isopropyl
$(H_3C)_2CH-CH_2-$	2-methylpropyl	2-methylpropan-1-yl	isobutyl
$H_3C-CH_2-CH(-)-CH_3$	1-methylpropyl	butan-2-yl	*sec*-butyl
$H_3C-C(CH_3)(-)-CH_3$	1,1-dimethylethyl	2-methylpropan-2-yl	*tert*-butyl
$H_3C-CH_2-CH_2-CH_2-CH_2-CH(ø)-CH_2-CH_3$	3-phenyloctane or (1-ethylhexyl) benzene	octan-3-ylbenzene	

Provisional IUPAC recommendations from 2004 discuss a single "preferred IUPAC name" for organic compounds, and this nomenclature may be taught in the classrooms of the future, but the rules in **15.5** and **Identify common organic functional groups** in 15.2 corresponding to 1979 conventions are what nearly everyone means by IUPAC organic nomenclature today, and they are what you should learh while being aware of the more recent changes.

Many IUPAC rules for organic chemistry may be found on-line here: http://www.acdlabs.com/iupac/nomenclature/.

15.2 Reactivity of elements and prediction of products of chemical reactions

Given common chemical species and reaction conditions, predict probable reaction products.

Chemical reactions occur all around us everyday-- so many chemical reactions that it would be very difficult to understand all of the different ones. However, the millions of chemical reactions that take place each and every day fall into only a few basic categories. Using these categories can help predict products of reactions that are unfamiliar or new.

Once we have an idea of the **reaction type**, we can make a good prediction about the products of chemical equations, and also balance the reactions. **General reaction types** are listed in the following table. Some reaction types have multiple names.

Reaction type	General equation	Example
Combination	$A + B \rightarrow C$	$2H_2 + O_2 \rightarrow 2H_2O$
Synthesis		
Decomposition	$A + B \rightarrow C$	$2KClO_3 \rightarrow 2KCl + 3O_2$
Single substitution	$A + BC \rightarrow AB + B$	$Mg + 2HCl \rightarrow MgCL_2 + H_2$
Single displacement		
Single replacement		
Double substitution	$AC + BD \rightarrow AD + BC$	$HCl + NaOH \rightarrow NaCl + H_2O$
Double displacement		
Double replacement		
Ion exchange		
Metathesis		
Isomerization	$A \rightarrow A$	C_3H_6 (cyclopropane: H_2C–CH_2 ring with H_2C) $\rightarrow$ C_3H_6 (propene: H_3C–$CH=CH_2$)

Example: Determine the products of a reaction between Cl_2 and a solution of NaBr.

Solution: The first step is to write

$$Cl_2 + NaBr(aq) \rightarrow ?$$

Now examine the possible choices from the table. Decomposition and isomerization reactions require only one reactant. It also can not be a double substitution reaction because one of the reactants is an element.

A synthesis reaction to form some NaBrCl compound would require very unusual valences! The most likely reaction is the remaining one: Cl replaces Br in aqueous solution with Na:

$$Cl_2 + NaBr(aq) \rightarrow NaCl(aq) + Br_2$$

After balancing, the equation is:

$$Cl_2 + 2NaBr(aq) \rightarrow 2NaCl(aq) + Br_2$$

Many **specific reaction types** also exist. Always determine the complete ionic equation for reactions in solution. This will help you determine the reaction type. The most common specific reaction types are summarized in the following table:

Reaction type	General equation	Example
Precipitation	Net ionic: $A^{+}(aq) + D^{-}(aq) \quad AD(s \text{ or } g)$	Net ionic: $Ni^{2+}(aq) + S^{2-}(aq) \quad NiS(s)$
Acid-base neutralization	Arrhenius: $H^{+} + OH^{-} \quad H_2O$	Arrhenius: $HNO_3 + NaOH \quad NANO_3 + H_2O$ $H^{+} + OH^{-} \quad H_2O$ (net ionic)
	Brønsted-Lowry: $HA + B \quad HB + A$	Brønsted-Lowry: $HNO_3 + KCN \quad HCN + KNO_3$ $H^{+} + CN^{-} \quad HCN$ (net ionic)
Redox	Half reactions: $A \quad C + e^{-}$ and $e^{-} + B \quad D$	$Ni \quad Ni^{2+} + 2e^{-}$ $2e^{-} + Cu^{2+} \quad Cu$
Combustion	organic module + O_2 $CO_2 + H_2O$ + heat	Combustion of methane: $CH_4(g) + 2\,O_2(g) \rightarrow CO_2(g) + 2\,H_2O(g)$

Whether precipitation occurs among a group of ions—and which compound will form the precipitate—may be determined by the solubility rules on the following page. (There is also a simplified list of solubility rules in 3.5)

If protons are available for combination or substitution, then they are likely being transferred from an acid to a base. An unshared electron pair on one of the reactants may form a bond in a Lewis acid-base reaction.

The possibility that oxidation numbers may change among the reactants indicates an electron transfer and a redox reaction. (See 3.2.c Oxidation numbers for redox reactions) Combustion reactants consist of an organic molecule and oxygen.

Ion	Solubility	Exception
All compounds containing alkali metals	All are soluble	
All compounds containing ammonium, NH_4^+	All are soluble	
All compounds containing nitrates, NO^{3-}	All are soluble	
All compounds containing chlorates, ClO^{3-} and perchlorates, ClO_4^-	All are soluble	
All compounds containing acetates, $C_2H_3O^{2-}$	All are soluble	
Compounds containing Cl^-, Br^-, and I^-	Most are soluble	Except halides of Ag^+, Hg_2^{2+}, and Pb^{2+}
Compounds containing F^-	Most are soluble	Except fluorides of Mg^{2+}, Ca^{2+}, Sr^{2+}, Ba^{2+}, and Pb^2
Compounds containing sulfates, SO_4^{2-}	Most are soluble	Except sulfates of Mg^{2+}, Ca^{2+}, Sr^{2+}, B^{a2}+, and Pb^2
All compounds containing carbonates, CO_3^{2-}	Most are insoluble	Except those containing alkali metals or NH_4^+
All compounds containing phosphates, PO_4^{3-}	Most are insoluble	Except those containing alkali metals or NH_4^+
All compounds containing oxalates, $C_2O_4^{2-}$	Most are insoluble	Except those containing alkali metals or NH_4^+
All compounds containing chromates, CrO_4^{2-}	Most are insoluble	Except those containing alkali metals or NH_4^+
All compounds containing oxides, O^{2-}	Most are insoluble	Except those containing alkali metals or NH_4^+
All compounds containing sulfides, S^{2-}	Most are insoluble	Except those containing alkali metals or NH_4^+
All compounds containing sulfites, SO_3^{2-}	Most are insoluble	Except those containing alkali metals or NH_4^+
All compounds containing silicates, SiO_4^{2-}	Most are insoluble	Except those containing alkali metals or NH_4^+
All compounds containing hydroxides	Most are insoluble	Except those containing alkali metals or NH_4^+, or Ba^{2+}

Example: Determine the products, and write a balanced equation for the reaction between sodium iodide and lead(II) nitrate in aqueous solution.

Solution: We know that sodium iodide is NaI and lead(II) nitrate is $Pb(NO_3)_2$. The reactants of the complete ionic equation are:

$$Na^+(aq) + Pb^{2+}(aq) + I^-(aq) + NO_3^-(aq) \rightarrow ?$$

The solubility rules indicate that lead iodide is insoluble and will form as a precipitate. The unbalanced net ionic equation is then:

$$Pb^{2+}(aq) + I^-(aq) \rightarrow PbI_2(s)$$

and the unbalanced molecular equation is:

$$NaI(aq) + Pb(NO_3)_2(aq) \rightarrow PbI_2(s) + NaNO_3(aq)$$

Balancing the equation yields:

$$2NaI(aq) + Pb(NO_3)_2(aq) \rightarrow PbI_2(s) + 2NaNO_3(aq)$$

15.3 Examples of simple organic compounds and compounds of environmental concern

Identify characteristics of simple organic compounds.

Many organic molecules contain functional groups which are groups of atoms in a particular arrangement that gives the molecule certain characteristics. Functional groups are named according to the composition of the group. The carboxyl group is the arrangement of -COOH atoms that tends to make a molecule exhibit acidic properties.

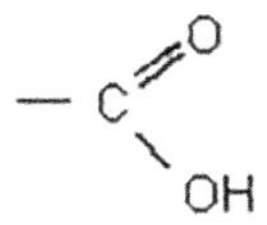

$$-C(=O)-OH$$

Some functional groups are polar and can ionize. For example, the hydrogen atom in the –COOH group can be removed (providing H^+ ions to a solution). When this occurs, the oxygen retains both the electrons it shared with the hydrogen and the molecule then has a negative charge.

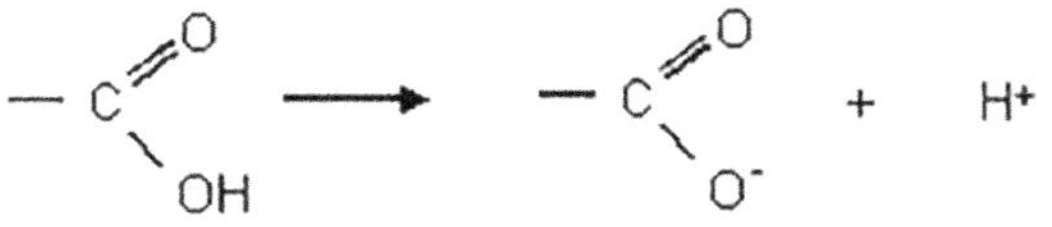

$$-C(=O)-OH \rightarrow -C(=O)-O^- + H^+$$

If polar or ionizing functional groups are attached to otherwise hydrophobic molecules, the molecule may become hydrophilic due to the functional group. Some ionizing functional groups are: -COOH, -OH, -CO, and $-NH_2$.

Some common functional groups include:

Hydroxyl group:

The hydroxyl group, -OH, is the functional group identifying alcohols. The hydroxyl group makes the molecule polar which tends to increase the solubility of the compound in polar solvents.

Carbonyl group:

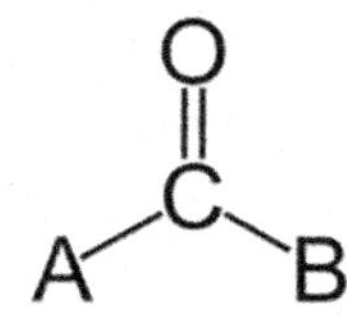

The carbonyl group is a -C=O. It is found in aldehydes (carboxyl group on end of carbon chain) and ketones (carboxyl group in middle of chain).

Aldehyde

ketone

If the carbon is attached to two carbon chains, the molecule is a ketone.

The double bonded oxygen atom is highly electronegative so it creates a polar molecule and will exhibit properties of polar molecules.

Carboxyl group:

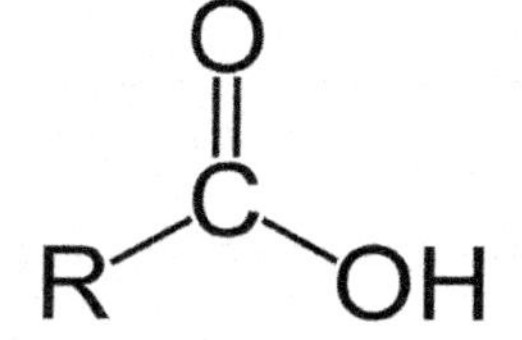

The –COOH group has the ability to donate a proton, thereby conferring acidic properties to the molecule.

Amino group:

An amino group contains an ammonia-like functional group composed of a nitrogen and two hydrogen atoms covalently bonded. The nitrogen atom has unshared electrons and so can bind free H^+ ions (proton). This gives the molecule basic properties. An organic compound that contains an amino group is called an amine. The amines are weak bases because the unshared electron pair of the nitrogen atom can form a coordinate bond with a proton. Another group of molecules that contains the amino group are amino acids. Every amino acids contains both a $-NH_2$ group and a –COOH group.

This is the amino acid glycine. H–CH–COOH with NH_2 on the central carbon

Sulfhydryl group:

A thiol is a compound that contains the functional group composed of a sulfur atom and a hydrogen atom (-SH). This functional group is referred to either as a *thiol group* or a *sulfhydryl group*. More traditionally, thiols have been referred to as *mercaptans*.

The small difference in electronegativity between the sulfur and the hydrogen atom produces a non-polar covalent bond. This in turn provides for no hydrogen bonding, giving thiols lower boiling points and less solubility in water than alcohols of a similar molecular mass.

Phosphate group:

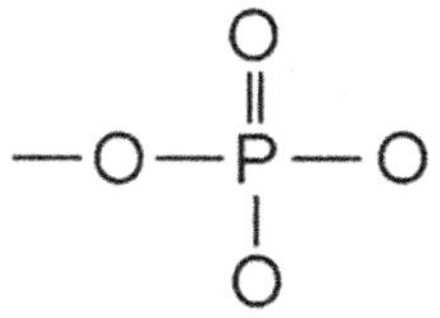

A phosphate ion can be attached to a hydrocarbon chain. This can make a molecule that is ideal for energy transfer reactions (such as those that occur with ATP) because of its symmetry and rotating double bond.

```
        O
        ‖
 —O—P—O
        |
        O
```

In biological systems, phosphates are most commonly found in the form of adenosine phosphates, (AMP, ADP and ATP) and in DNA and RNA.

SECTION VII PRACTICE QUESTIONS

CHAPTER 15

1. **What is the formula for phosphoric acid?**

   (A) $H_2PO_3(aq)$
   (B) $H_2PO_4(aq)$
   (C) $H_3PO_3(aq)$
   (D) $H_3PO_4(aq)$
   (E) $H_3P_2O_4(aq)$

   The correct answer is D.

2. **What is the name of Na_2CO_3 according to the IUPAC nomenclature?**

   (A) sodium carbonate
   (B) sodium tricarbonate
   (C) sodium carbon trioxide
   (D) disodium tricarbonate
   (E) disodium carbonate

   The correct answer is A, sodium carbonate.

Questions 3-5 refer to the following:

(A) Alkenes
(B) Benzene
(C) Pentane
(D) Amino-acid
(E) None of the above

3. **may have cis-trans isomers**

   Answer: A some alkenes have cis-trans isomers

4. **has double bond resonance**

   Answer: B – Benzene is an example of double bond resonance

5. **is found in proteins**

   Answer: D – amino acids are found in proteins.

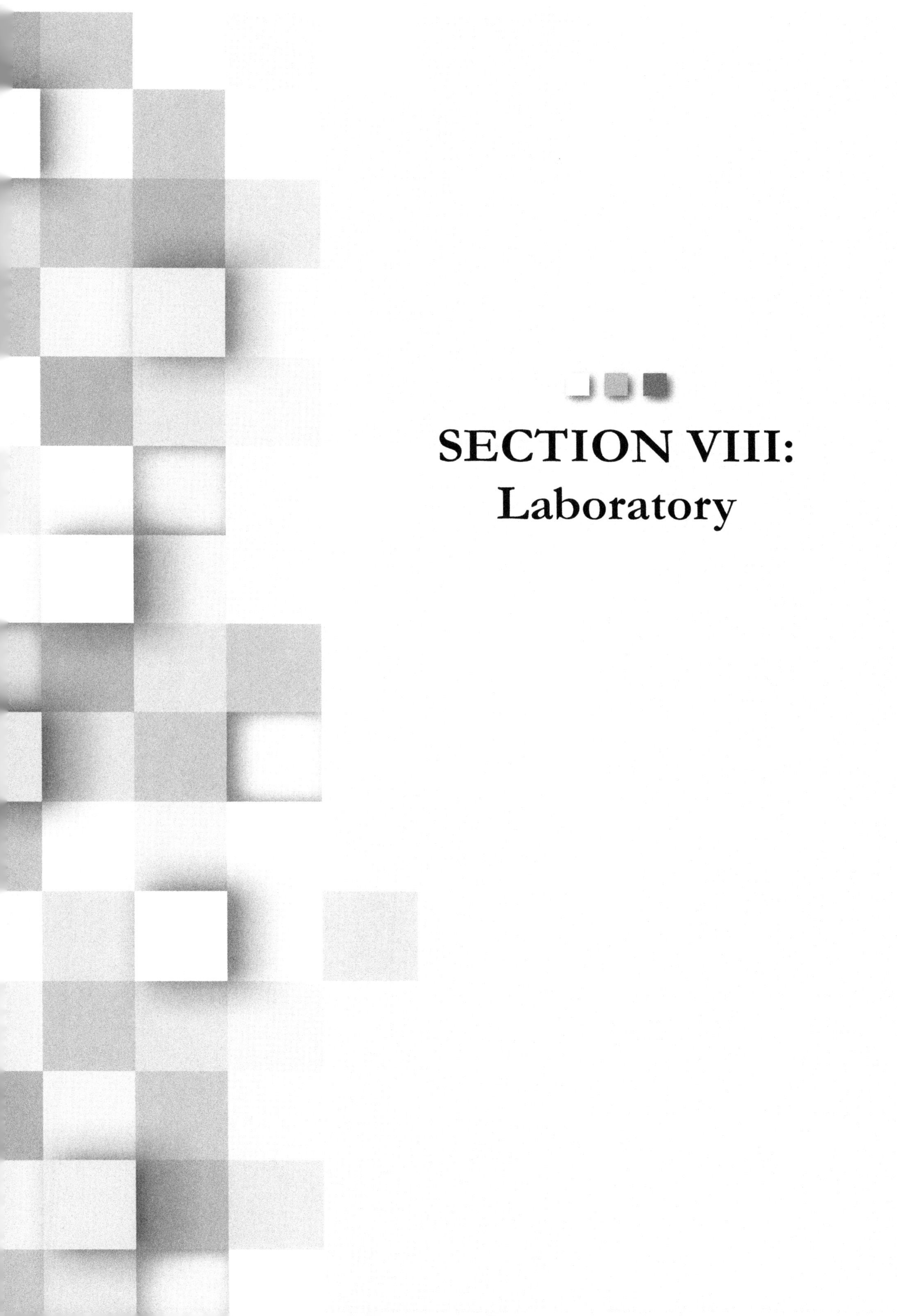

SECTION VIII: Laboratory

Chapter 16: Laboratory

16.1 Knowledge of Laboratory Equipment

Choose the correct laboratory equipment for a particular procedure.

The descriptions and diagrams in this subtitle are included to help you **identify** the techniques. They are **not meant as a guide to perform the techniques** in the lab.

Handling liquids

A **beaker** (below left) is a cylindrical cup with a notch at the top. Beakers are often used for making solutions. An **Erlenmeyer** flask (below center) is a conical flask. A liquid in an Erlenmeyer flask will evaporate more slowly than when it is in a beaker and it is easier to swirl about. A **round-bottom flask** (below-right) is also called a Florence flask. It is designed for uniform heating, but it requires a stand to keep it upright.

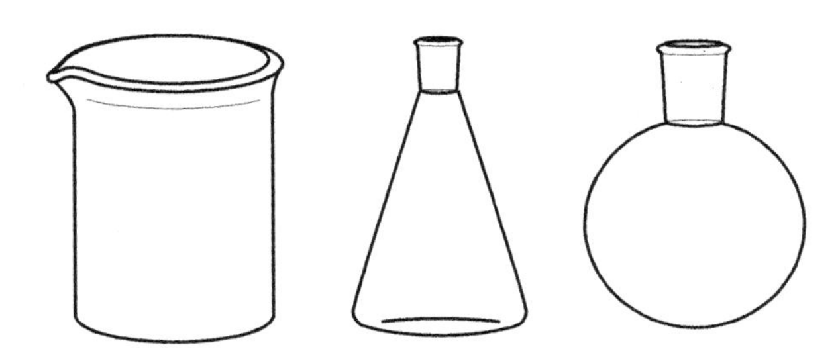

A **test tube** has a rounded bottom and is designed to hold and to heat small volumes of liquid. A **Pasteur pipet** is a long, thin glass tube with a capillary tip and a latex suction bulb, and it is used for transferring small amounts of liquid from one container to another.

A **crucible** is a cup-shaped container made of porcelain or metal for holding chemical compounds when heating them to very high temperatures. A **watch glass** is a concave circular piece of glass that is usually used as a surface for evaporating a liquid and observing precipitates or crystallization. A **Dewar flask** is a double walled vacuum flask with a metallic coating to provide good thermal insulation and short term storage of liquid nitrogen.

Fitting and cleaning glassware

If a thermometer, glass tube, or funnel must be threaded through a stopper or a piece of tubing and it won't fit, either make the hole larger or use a smaller piece of glass. Use soapy water or glycerol to lubricate the glass before inserting it. Hold the glass piece as close as possible to the stopper during insertion. It's also good practice to wrap a towel around the glass and the stopper during this procedure. **Never apply undue pressure**.

Glassware sometimes contains **tapered ground-glass joints** to allow direct glass-to-glass connections. A thin layer of joint **grease** must be applied when assembling an apparatus with ground-glass joints. Too much grease will contaminate the experiment, and too little will cause the components to become tightly locked together. Disassemble the glassware with a **twisting** motion immediately after the experiment is over.

Cleaning glassware becomes more difficult with time, so it should be cleaned soon after the experiment is completed. Wipe off any lubricant with paper towel moistened in a solvent like hexane before washing the glassware. Use a brush with lab soap and water. Acetone may be used to dissolve most organic residues. Spent solvents should be transferred to a waste container for proper disposal.

Heating

A **hot plate** (shown below) is used to heat Erlenmeyer flasks, beakers and other containers with a flat bottom. Hot plates often have a built-in **magnetic stirrer**. A **heating mantle** has a hemispherical cavity that is used to heat round-bottom flasks. A **Bunsen burner** is designed to burn natural gas. Bunsen burners are useful for heating high-boiling point liquids, water, or solutions of non-flammable materials. They are also used for bending glass tubing. Smooth boiling is achieved by adding **boiling stones** to a liquid.

Boiling and melting point determination

The **boiling point** is determined by heating a liquid along with a boiling stone in a clamped test tube with a clamped thermometer positioned just above the liquid surface and away from the tube walls. The thermometer measures the temperature of the vapor above the liquid, which is at the same temperature as the liquid while boiling. The constant highest constant-value temperature reading after boiling is achieved is the boiling point.

The **melting point** is determined by placing a pulverized solid in a capillary tube and using a rubber band to fasten the capillary to a thermometer so that the sample is at the level of the thermometer bulb. The thermometer and sample are inserted into a **Thiele tube** filled with mineral or silicon oil. The Thiele tube has a sidearm that is heated with a Bunsen burner to create a flow of hot oil. This flow maintains an even temperature during heating. The melting point is read when the sample turns into a liquid. Many electric melting point devices are also available that heat the sample more slowly to give more accurate results. These are also safer to use than a Thiele tube with a Bunsen burner.

Centrifugation

A **centrifuge** separates two immiscible phases by spinning the mixture (placed in a **centrifuge tube**) at high speeds. A **microfuge** or microcentrifuge is a small centrifuge. The weight of material placed in a centrifuge must be **balanced**, so if one sample is placed in a centrifuge, a tube with roughly an equal mass of water should be placed opposite the sample.

Filtration

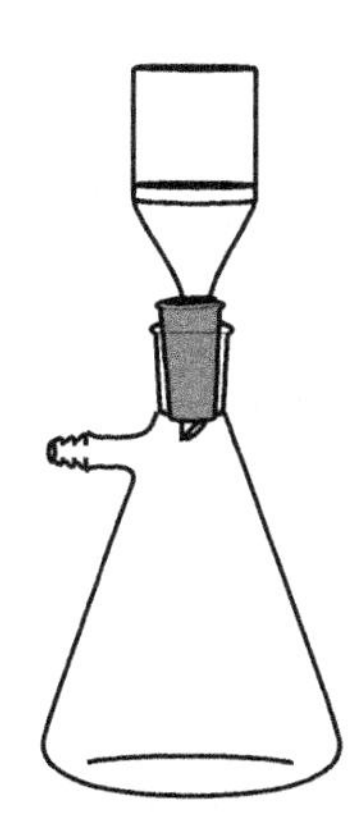

The goal of **gravity filtration** is to remove solids from a liquid and obtain a liquid without solid particulates. Filter paper is folded, placed in a funnel on top of a flask, and wetted with the solvent to seal it to the funnel. Next the mixture is poured through, and the solid-free liquid is collected from the flask.

The goal of **vacuum filtration** is usually to remove liquids from a solid to obtain a solid that is dry. An **aspirator** or a **vacuum pump** is used to provide suction though a rubber tube to a **filter trap**. The trap is attached to a **filter flask** (shown to the right) by a second rubber tube. The filter flask is an Erlenmeyer flask with a thick wall and a hose barb for the vacuum tube. Filter flasks are used to filter material using a **Büchner funnel** (shown to the right) or a smaller **Hirsch funnel**. These porcelain or plastic funnels hold a circular piece of filter paper. A single-hole rubber stopper supports the funnel in the flask while maintaining suction.

Mixing

Heterogeneous reaction mixtures in flasks are often mixed by **swirling**. To use a magnetic stirrer, a bar magnet coated with Teflon called a flea or a **stir bar** is placed into the container, and the container is placed on the stirrer. The container should be moved and the stir speed adjusted for smooth mixing. Mechanical stirring paddles, agitators, vortexers, or rockers are also used for mixing.

Decanting

When a coarse solid has settled at the bottom of a flask of liquid, **decanting** the solution simply means pouring out the liquid and leaving the solid behind.

Extraction

Compounds in solution are often separated based on their **solubility differences**. During **liquid-liquid extraction** (also called **solvent extraction**), a second solvent (immiscible with the first) is added to the solution in a **separatory funnel** (shown at right). Usually one solvent is nonpolar and the other is a polar solvent like water. The two solvents are immiscible and separate from each other after the mixture is shaken to allow solute exchange. One layer contains the compound of interest, and the other contains impurities to be discarded. The solutions in the two layers are separated from each other by draining liquid through the stopcock.

Titration

Titration is described in **16.7 Interpretation of graphical data**

Distillation

Liquids in solution are often separated based on their **boiling point differences**. During simple **distillation**, the solution is placed in a round-bottom flask called the **distillation flask** or **still pot**, and boiling stones are added. The apparatus shown below is assembled (note that clamps and stands are not shown), and the still pot is heated using a heating mantle.

While boiling, hot vapor escapes through the **distillation head**, enters the **condenser**, and is cooled and condensed back to a liquid. The vapor loses its heat to water flowing through the outside of the condenser. The condensate or **distillate** falls into the **receiving flask**. The apparatus is open to the atmosphere through a vent above the receiving flask. The distillate contains a higher concentration of the liquid with the lower boiling point. The less volatile liquid reaches a high concentration in the still pot. Head temperature is monitored during the process.

Distillation may also be used to remove a solid from a pure liquid by boiling and condensing the liquid into the
receiving flask.

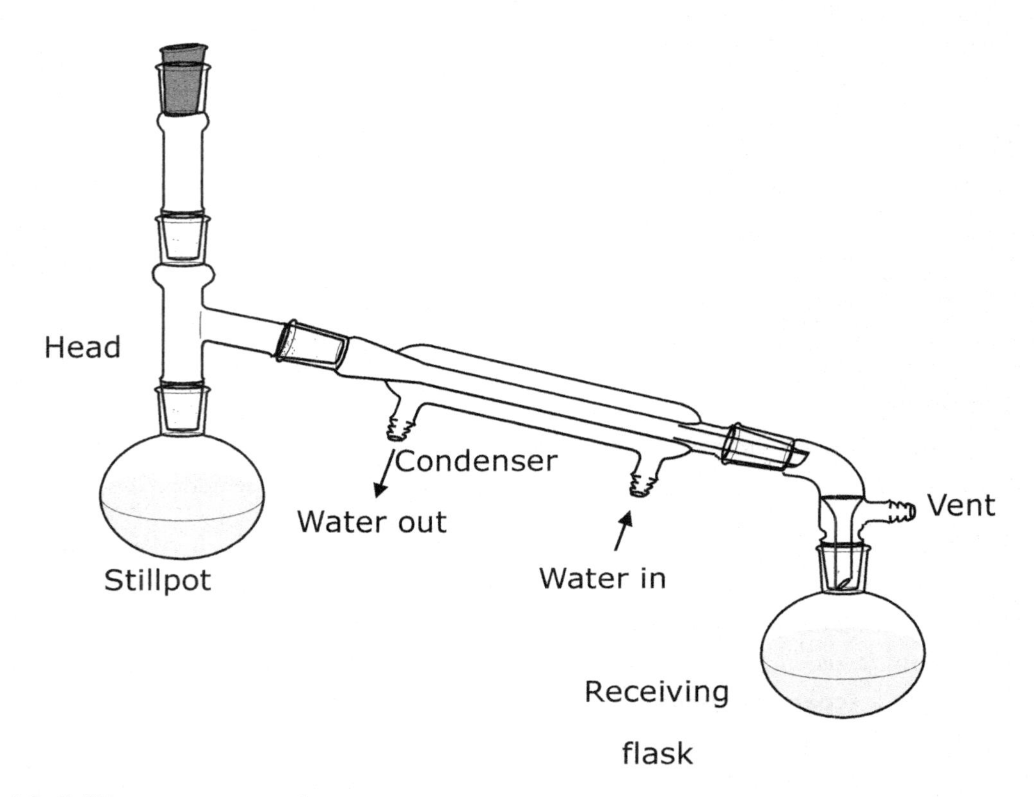

16.2 Measurements

Convert between dimensional units for 1-, 2-, and 3-dimensional measurements.

Dimensional analysis is a structured way to convert units. It involves a conversion factor that allows the units to be cancelled out when multiplied or divided. A conversion factor is the same measurement written as an equivalency between two different units such as 1 meter = 100 centimeters.

One Dimension Unit Conversions

These are the steps to converting one dimension measurements.

1. Write the term to be converted, (both number and unit)

   6.0 cm = ? m

2. Write the conversion formula(s) needed for the involved units.

   100 cm = 1 m

3. Make a fraction of the conversion formula, such that:
   a. if the unit in step 1 is in the numerator, that same unit in this step must be in the denominator

      $$\frac{1\text{ m}}{100\text{ cm}}$$

   b. if the unit in step 1 is in the denominator, that same unit in this step must be in the numerator

      $$\frac{100\text{ cm}}{1\text{ m}}$$

   Since both the numerator and denominator are equal in value, the fraction equals 1.
4. Multiply the term in step 1 by the correct fraction in step 3. Since the fraction equals 1, you can multiply by it without changing the size of the original term.
5. Cancel units : 6.0 ~~cm~~ × (1 m / 100 ~~cm~~)
6. Perform the indicated calculation, rounding the answer to the correct number of significant figures.

   6.0 × 1 m / 100 = 0. 060 m (or 6.0×10^{-2} m)

Two or Three Dimension Unit Conversions

The process is nearly the same for two and three dimension conversions.

Example: How many cm^3 is 1 m^3?

Solution: Remember, 100 cm = 1 m and that 1 m^3 is really 1m x 1m x 1m. Substituting in the 100 cm for every meter the problem can be rewritten as

100 cm × 100 cm × 100 cm or $1m^3$ = 1,000,000 cm^3

1 m^3 × 1,000,000 $cm^3/1m^3$ = 1,000,000 cm^3 (or 1.0×10^6 cm)

Example: Convert 4.17 kg/m^2 to g/cm^2

Solution: First to convert from kg to g use 1000 g = 1.00 kg as the conversion factor.

4.17 ~~kg~~/m^2 × 1000 g/1.00 ~~kg~~ = 4170g /m^2

Then use 1.00 m = 100 cm to convert the denominator. Remember that m^2 is m × m and replacing m with 100 cm, the denominator becomes 100 cm × 100 cm. or 10 000 cm^2

The conversion factor for the denominator becomes $1.00 m^2 = 10\,000 cm^2$ (cancel m)

$$4170 g/\cancel{m^2} \times 1.00 \cancel{m^2}/10{,}000 cm^2 = 0.417 g/cm^2$$

Dimensional analysis can also be used to help solve mathematical problems.

Example: The density of gold is 19.3 g/cm^3. How many grams of gold would be found in 55 cm^3?

Solution: Using dimensional analysis, some unit must be made to cancel. The answer needs to be in grams, so the cm^3 needs to be canceled out. Multiply or divide the units so that the cm^3 cancel. In this case the units part of the problem works out as follows:

$$g/\cancel{cm^3} \times \cancel{cm^3} = g$$

Now adding in the values from the problem, we get:

$$\mathbf{19.3 \, g/\cancel{cm^3} \times 55 \, \cancel{cm^3} = 1060 \, g \text{ of gold}}$$

Analyze the dimensional units of a mathematical formula.

Units are a part of every measurement. Without the units, the numbers could mean many things. For example, the distance 10 could mean 10 cm, 10 m, or 10 km. The units will even help solve mathematical problems.

Example: The density of gold is 19.3 g/cm^3. How many grams of gold would be found in 55 cm^3?

Solution: Using dimensional analysis, some unit must cancel. The answer needs to be in grams, so the cm^3 needs to cancel out. Multiply or divide the units so that the cm^3 cancel.

In this case:

$$g/cm^3 \times cm^3 = g \text{ (or density unit} \times \text{volume} = \text{mass)}$$

This will cancel out the cm^3, and then we perform the calculations to obtain the answer.

$$19.3 \, g/\cancel{cm^3} \times 55 \, \cancel{cm^3} = 1060 \, g \text{ of gold.}$$

Identify prefixes (e.g., kilo-, milli-, nano-) used in scientific measurements.

SI is an abbreviation of the French *Système International d'Unités* or the **International System of Units**. It is the most widely used system of units in the world and is the system used in science. The use of many SI units in the United States is increasing outside of science and technology. There are two types of SI units, **base units** and **derived units**. The base units are:

Quantity	Unit name	Symbol
Length	meter	m
Mass	kilogram	kg
Amount of substance	mole	mol
Time	second	s
Temperature	kelvin	K
Electric current	ampere	A
Luminous intensity	candela	cd

Amperes and candelas are rarely used in chemistry. The name "kilogram" occurs for the SI base unit of mass for historical reasons. Derived units are formed from the kilogram, but appropriate decimal prefixes are attached to the word "gram." See **13.1 Factors affecting reaction rates** for reaction rate constant units.

Derived units measure a quantity that may be expressed in terms of other units. The derived units important for chemistry are:

Derived quantity	Unit name	Expression in terms of other units	Symbol
Area	square meter	m^2	
Volume	cubic meter	m^3	
	liter	$dm^3 = 10^{-3}\ m^3$	L or l
Mass	unified atomic mass unit	$(6.022 \times 10^{23})^{-1}$ g	u or Da
Time	minute	60 s	min
	hour	60 min = 3600 s	h
	day	24 h = 86400 s	d
Speed	meter per second	m/s	
Acceleration	meter per second squared	m/s^2	
Temperature*	degree Celsius	K = 273.15	°C
Mass density	gram per liter	$g/L = 1\ kg/m^3$	
Amount-of-substance concentration (molarity)	molar	mol/L	M
Molality‡	molal	mol/kg	*m*
Chemical reaction rate	molar per second†	M/s = mol/(L•s)	
Force	newton	$m•kg/s^2$	N
Pressure	pascal	$N/m^2 = kg/(m•s^2)$	Pa
	standard atmosphere§	101325 Pa	atm
Energy, Work, Heat	joule	$N•m = m^3•Pa = m^2•kg/s^2$	J
	nutritional calorie§	4184 J	Cal
Heat (molar)	joule per mole	J/mol	
Heat capacity, entropy	joule per kelvin	J/K	
Heat capacity (molar), entropy (molar)	joule per mole kelvin	J/(mol•K)	
Specific heat	joule per kilogram kelvin	J/(kg•K)	
Power	watt	J/s	W
Electric charge	coulomb	s•A	C
Electric potential, electromotive force	volt	W/A	V
Viscosity	pascal second	Pa•s	
Surface tension	newton per meter	N/m	

*Temperature differences in Kelvin are the same as those differences in degrees Celsius. To obtain degrees Celsius from Kelvin, subtract 273.15.

‡Molality, *m*, is often considered obsolete. Differentiate *m* and meters (m) by context.

§These are commonly used non-SI units.

Decimal multiples of SI units are named by attaching a **prefix** directly before the unit and a symbol prefix directly before the unit symbol. SI prefixes range from 10^{-24} to 10^{24}. The most common prefixes you are likely to encounter in chemistry are shown below:

Factor	Prefix	Symbol	Factor	Prefix	Symbol
10^9	*giga–*	G	10^{-1}	*deci–*	d
10^6	*mega–*	M	10^{-2}	*centi–*	c
10^3	*kilo–*	k	10^{-3}	*milli–*	m
10^2	*hecto–*	h	10^{-6}	*micro–*	µ
10^1	*deca–*	da	10^{-9}	*nano–*	n
			10^{-12}	*pico–*	p

Example: The distance, 0.0000004355 meters is equal to both 4.355×10^{-7}m and 435.5×10^{-9}m. This length is also equal to 435.5 nm (nanometers).

Example: Find a unit to express the volume of a cubic crystal that is 0.2 mm on each side so that the number before the unit is between 1 and 1000.

Solution: Volume is equal to length x width x height, so this volume is $(0.0002\ \text{m})^3$ or $8 \times 10^{-12}\ \text{m}^3$. Conversions of volumes and areas using powers of units of length must take the power into account. Therefore:

$$1\ \text{m}^3 = 10^3\ \text{dm}^3 = 10^6\ \text{cm}^3 = 10^9\ \text{mm}^3 = 10^{18}\ \mu\text{m}^3$$

The length 0.0002 m is $2 \times 10^2\ \mu$m, so the volume is also $8 \times 10^6\ \mu\text{m}^3$. This volume could also be expressed as $8 \times 10^{-3}\ \text{mm}^3$, but none of these numbers are between 1 and 1000.

Expressing volume in liters is helpful in cases like these. There is no power on the unit of liters, therefore:

$$1\ \text{L} = 10^3\ \text{mL} = 10^6\ \mu\text{L} = 10^9\ \text{nL}$$

Converting cubic meters to liters (canceling the meters) gives:

$$8 \times 10^{-12}\ \text{m}^3 \times \frac{10^3\ \text{L}}{1\ \text{m}^3} = 8 \times 10^{-9}\ \text{L}$$

The crystal's volume is 8 nanoliters (8 nL).

Example: Determine the ideal gas constant, *R*, in L•atm/(mol•K) from its SI value of 8.3144 J/(mol•K).

Solution: One joule is identical to one m^3•Pa (see the table on the previous page).

$$8.3144\ \frac{m^3 \bullet Pa}{mol \bullet K} \times \frac{1000\ L}{1\ m^3} \times \frac{1\ atm}{101325\ Pa} = 0.082057\ \frac{L \bullet atm}{mol \bullet K}$$

Distinguish between accuracy and precision and between systematic and random error.

A measurement is **precise** when individual measurements of the same quantity agree with one another. A measurement is **accurate** when they agree with the true value of the quantity being measured. An **accurate** measurement is **valid**. We get the right answer. A **precise** measurement is **reproducible**, meaning that we get a similar answer each time. These terms are related to **sources of error** in a measurement.

Precise measurements are near the **arithmetic mean** of the values. The arithmetic mean is the sum of the measurements divided by the number of measurements. The **arithmetic mean** is commonly called the **average**. It is the **best estimate** of the quantity based on the measurements taken.

Random error results from **limitations in equipment or techniques**. **Random error decreases precision**. Remember that all measurements reported to a proper number of significant digits contain an imprecise final digit to reflect random error.

Systematic error results from **imperfect equipment or technique**. **Systematic error decreases accuracy**. Instead of a random error with random fluctuations, there is a biased result that, on average, is too large or small.

Example: An environmental engineering company creates a solution of 5.00 ng/L of a toxin and distributes it to four toxicology labs to test their protocols. Each lab tests the material 5 times. Their results are charted as points on the number lines below. Interpret these data in terms of precision, accuracy, and type of error.

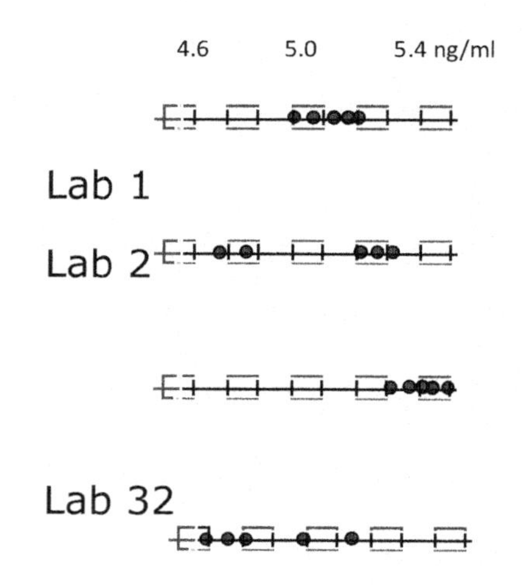

Solution: Results from lab 1 are both accurate and precise when compared to results from the other labs. Results from lab 2 are less precise than those from lab 1. Lab 2 seems to use a protocol that contains a greater random error. However, the mean result from lab 2 is still close to the known value. Lab 3 returned results that were about as precise as lab 1 but inaccurate compared to labs 1 and 2. Lab 3 most likely uses a protocol that yields a systematic error. The data from lab 4 is both imprecise and inaccurate. Systematic and random errors are larger than in lab 1.

Apply the correct number of significant figures in measurements or calculations.

Significant figures or **significant digits** are the digits indicating the **precision of a measurement**. There is uncertainty **only** in the last digit.

Example: You measure an object with a ruler marked in millimeters. The reading on the ruler is found to be about 2/3 of the way between 12 and 13 mm. What value should be recorded for its length?

Solution: Recording 13 mm does not give all the information that you found. Recording 12 2/3 t mm implies that an exact ratio was determined. Recording 12.666 mm gives more information than you found. A value of 12.7 mm or 12.6 mm should be recorded because there is uncertainty only in the last digit.

There are five rules for determining the **number of significant digits** in a quantity.

1. All nonzero digits are significant and all zeros between nonzero digits are significant.
   Example: 4.521 kJ and 7002 u both have four significant digits.

2. Zeros to the left of the first nonzero digit are not significant.
   Example: 0.0002 m contains one significant digit.

3. Zeros to the right of a nonzero digit and the decimal point are significant figures.
   Example: 32.500 g contains five significant digits.

4. The significance of numbers ending in zeros that are not to the right of the decimal point can be unclear, so **this situation should be avoided** by using scientific notation or a different decimal prefix. Sometimes a decimal point is used as a placeholder to indicate the units-digit is significant. A word like "thousand" or "million" may be used in informal contexts to indicate the remaining digits are not significant.
   Example: 12,000 Pa would be considered to have five significant digits by many scientists, but in the context, "The pressure rose from 11,000 Pa to 12,000 Pa," it almost certainly only has two. "12 thousand pascal" only has two significant figures, but 12000. Pa has five because of the decimal point. The value should be represented as 1.2×10^4 Pa (or 1.2000×10^4 Pa). The best alternative would be to use 12 kPa for two significant digits or 12.000 kPa for five significant digits.

5. Exact numbers have no uncertainty and contain an infinite number of significant digits. These relationships are **definitions**. They are not measurements.
   Example: There are exactly 1,000 L in one cubic meter.

There are four rules for **rounding off significant digits**:

1. If the rightmost digit to be removed is a four or less, then round down. The last remaining digit stays as it was.
   Example: Round 43.4 g to 2 significant figures. Answer: 43 g
2. If the rightmost digit to be removed is a six or more, then round up. The last remaining digit increases by one.
   Example: Round 6.772 to 2 significant figures. Answer: 6.8 g
3. If the rightmost digit to be removed is a five that is followed by nonzero digits, then round up. The last remaining digit increases by one.
   Example: Round 18.502 to 2 significant figures. Answer 19 g
4. If the rightmost digit to be removed is a five followed by nothing or by only zeros, force the last remaining digit to be even. If it is odd then round up by increasing it by one. If it is even (including zero) then it stays as it was.
   Examples: Round 18.50 g and 19.5 g to 2 significant figures.
   Answers: 18.50 g rounds off to 18 g and 19.5 g rounds off to 20 g

There are three rules for **calculating with significant digits**.

1. For multiplication or division, the result has the same number of significant digits as the term with the least number of significant digits.
   Example: What is the volume of a compartment in the shape of a rectangular prism 1.2 cm long, 2.4 cm high and 0.9 cm deep?
   Solution: Volume=length × height × width

Volume $= 1.2\text{ cm} \times 2.4\text{ cm} \times 0.9\text{ cm} = 2.592\text{ cm}$ (as read on a calculator)
Round to one digit because 0.9 cm has only one significant digit.
Volume $= 3\text{ cm}^3$

2. For addition or subtraction, the result has the same number of digits after the decimal point as the term with the least number of digits after the decimal point.
   Example: Volumes of 250.0 mL, 26 μL, and 4.73 mL are added to a flask. What is the total volume in the flask?
   Solution: Only identical units may be added to each other, so 26 μL is first converted to 0.026 mL.

Volume $= 250.0\text{ mL} + 0.026\text{ mL} + 4.73\text{ mL} = 254.756\text{ mL}$ (calculator value)
Round to one digit after the decimal because 250.0 mL has only one digit after the decimal. Volume $= 254.8$ mL.

3. For multi-step calculations, maintain all significant digits when using a calculator or computer and round off the final value to the appropriate number of significant digits after the calculation. When calculating by hand or when writing down an intermediate value in a multi-step calculation, maintain the first insignificant digit.

16.3 Procedures

Identify appropriate chemistry laboratory procedures for the safe storage, use, and disposal of materials and equipment.

Disclaimer: The information presented is intended as a starting point for identification purposes only and should not be regarded as a comprehensive guide for recognizing hazardous substances and reactions. It is the responsibility of the readers of this book to obtain the required information about chemical hazards in their laboratory.

Chemical purchase, use, and disposal

1. Inventory all chemicals on hand at least annually. Keep the list up-to-date as chemicals are consumed and replacement chemicals are received.
2. If possible, limit the purchase of chemicals to quantities that will be consumed within one year and that are packaged in small containers suitable for direct use in the lab without transfer to other containers.
3. Label all chemicals to be stored with date of receipt or preparation and have labels initialed by the person responsible.
4. Generally, bottles of chemicals should not remain:
   - Unused on shelves in the lab for more than one week. Move these chemicals to the storeroom or main stockroom.
   - Unused in the storeroom near the lab for more than one month. Move these chemicals to the main stockroom.
   - Check shelf life of chemicals. Properly dispose of any outdated chemicals.
5. Ensure that the disposal procedures for waste chemicals conform to environmental protection requirements.
6. Do not purchase or store large quantities of flammable liquids. Fire department officials can recommend the maximum quantities that may be kept on hand.
7. Never open a chemical container until you understand the label and the relevant portions of the MSDS.

Chemical storage plan for laboratories

1. Chemicals should be stored according to hazard class (ex. flammables, oxidizers, health hazards/toxins, corrosives, etc.).
2. Store chemicals away from direct sunlight or localized heat.
3. All chemical containers should be properly labeled, dated upon receipt, and dated upon opening.

4. Store hazardous chemicals below shoulder height of the shortest person working in the lab.
5. Shelves should be painted or covered with chemical-resistant paint or chemical-resistant coating.
6. Shelves should be secure and strong enough to hold chemicals being stored on them. Do not overload shelves.
7. Personnel should be aware of the hazards associated with all hazardous materials.
8. Separate solids from liquids.

Chemicals that require extreme care or should not be in schools

Materials that react with air or water

The risks of using these materials is considered to exceed their educational utility:

- Picric acid (must be kept wet, explosive when dry)
- Sodium metal (reacts with water, ignites in dry air)
- Phosphorus (white form reacts with air. Red form becomes white upon heating)

Extremely corrosive materials

These materials should not be in high school laboratories.

- Hydrofluoric acid (even dilute solutions cause adverse internal effects).
- Perchloric acid (also causes explosive products)
- Bromine

Highly toxic materials

These materials should not be in high school laboratories.

- Carbon disulfide (also explosive)
- Cyanide compounds
- Benzene and toluene
- Mercury and mercury compounds
- Cadmium, chromium, and arsenic compounds
- Carbon tetrachloride and chloroform

Recognize the impact of reagent concentration on hazard level

Recognize that a more dilute acid, base, oxidizer, or reducer presents a lower hazard level and requires a lower level of protection in the laboratory.

Organic peroxides and peroxide-forming materials

Organic peroxides have the general structure: R–O–O–R', where R and/or R' are organic substituents. These compounds are **very unstable** and may self-react in a **violent explosion** when triggered by heat, impact, friction, light, or vibration. **Benzoyl peroxide** and other organic peroxides should not be present in high school chemistry labs.

Peroxide-forming compounds are chemicals with the potential to form organic peroxides when they react with oxygen in the air (as the chemical is concentrated) and/

or become vaporized either by **erroneous boiling** or over time by **evaporation from a poorly sealed container**.

A peroxide-forming compound should be treated as a **potential explosive** and the local fire chief should be called if it:

- Is discolored
- Contains layering
- Contains crystals

Peroxide crystals may **form within the threads of the cap**, and the act of removing the cap may cause a fatal detonation. A bomb squad may be required.

Many high-risk peroxide-formers were used in high school chemistry in the past. In general, the risks of using the following materials are now thought to exceed their educational utility:

- Nearly all ethers including diisopropyl ether (the highest risk peroxide-forming compound), ethyl ether, and methyl ether.
- Potassium metal (also reacts with water)
- Tetrohydrofuran (THF)
- Cyclohexene and cyclohexanol
- Dioxanes

Several lower-risk peroxide-forming compounds are still commonly used in high schools. These materials **should never be concentrated by boiling (i.e. distilled)**:

- Isopropanol, 2-butanol, other secondary alcohols
- Ethylene glycol
- Acetaldehyde

http://www.govlink.org/hazwaste/publications/highrisktable.pdf

Below are examples of chemical groups that can be used to categorize storage. Use these groups as examples when separating chemicals for compatibility. Please note: reactive chemicals must be more closely analyzed since they have a greater potential for violent reactions. Contact Laboratory Safety if you have any questions concerning chemical storage.

Acids

1. Make sure that all acids are stored by compatibility (ex. separate inorganics from organics).
2. Store concentrated acids on lower shelves in chemical-resistant trays or in a corrosives cabinet. This will temporarily contain spills or leaks and protect shelving from residue.
3. Separate acids from incompatible materials such as bases, active metals (ex. sodium, magnesium, potassium) and from chemicals which can generate toxic gases when combined (ex. sodium cyanide and iron sulfide).

Bases

1. Store bases away from acids.
2. Store concentrated bases on lower shelves in chemical-resistant trays or in a corrosives cabinet. This will temporarily contain spills or leaks and protect shelving from residue.

Flammables

1. Approved flammable storage cabinets should be used for flammable liquid storage.
2. You may store 20 gallons of flammable liquids per 100 sq. ft. in a properly fire separated lab. The maximum allowable quantity for flammable liquid storage in any size lab is not to exceed 120 gallons.
3. You may store up to 10 gallons of flammable liquids outside of approved flammable storage cabinets.
4. An additional 25 gallons may be stored outside of an approved storage cabinet if it is stored in approved safety cans not to exceed 2 gallons in size.
5. Use only explosion-proof or intrinsically safe refrigerators and freezers for storing flammable liquids.

Peroxide-forming chemicals

1. Peroxide-forming chemicals should be stored in airtight containers in a dark, cool, and dry place.
2. Unstable chemicals such as peroxide-formers must always be labeled with date received, date opened, and disposal/expiration date.
3. Peroxide-forming chemicals should be properly disposed of before the date of expected peroxide formation (typically 6-12 months after opening).
4. Suspicion of peroxide contamination should be immediately investigated. Contact Laboratory Safety for procedures.

Water-reactive chemicals

1. Water-reactive chemicals should be stored in a cool, dry place.
2. Do not store water-reactive chemicals under sinks or near water baths.
3. Class D fire extinguishers for the specific water-reactive chemical being stored should be made available.

Oxidizers

1. Make sure that all oxidizers are stored by compatibility.
2. Store oxidizers away from flammables, combustibles, and reducing agents.

Toxins

1. Toxic compounds should be stored according to the nature of the chemical, with appropriate security employed when necessary.
2. A "Poison Control Network" telephone number should be posted in the laboratory where toxins are stored. Color coded labeling systems that may be found in your lab are shown below.

Hazard	Color Code
Flammables	Red
Health Hazards/Toxins	Blue
Reactives/Oxidizers	Yellow
Contact Hazards	White
General Storage	Gray, Green, Orange

Please Note: Chemicals with labels that are colored and striped may react with other chemicals in the same hazard class. See MSDS for more information. Chemical containers which are not color-coded should have hazard information on the label. Read the label carefully and store accordingly.

http://www.firn.edu/doe/curriculum/sc3/sc3dgc.html

Disposal of chemical waste

Schools are regulated by the Environmental Protection Agency, as well as state and local agencies, when it comes to disposing of chemical waste. Check with your state science supervisor, local college or university environmental health and safety specialists, and the Laboratory Safety Workshop for advice on the disposal of chemical waste. The American Chemical Society publishes an excellent guidebook, *Laboratory Waste Management, A Guidebook* (1994).

The following are basic guidelines for disposing of chemical waste.

You may dispose of hazardous waste as outlined below. It is the responsibility of the generator to ensure hazardous waste does not end up in ground water, soil or the atmosphere through improper disposal.

Sanitary Sewer - Some chemicals (acids or bases) may be neutralized and disposed to the sanitary sewer. This disposal option must be approved by the local waste water treatment authority prior to disposal. This may not be an option for some small communities that do not have sufficient treatment capacity at the

Identify emergency procedures and safety equipment needed in the science laboratory and classroom.

Disclaimer: The information presented below is intended as a starting point for identification purposes only and should not be regarded as a comprehensive guide for safety procedures in the laboratory. It is the responsibility of the readers of this book to consult with professional advisers about safety procedures in their laboratory.

Fire extinguisher

Fire extinguishers are rated for the type of fires they will extinguish. Chemical laboratories should have a combination ABC extinguisher along with a type D fire extinguisher. If a type D extinguisher is not available, a bucket of dry sand will do. Make sure you are trained to use the type of extinguisher available in your setting.

1. **Class A** fires are ordinary materials like burning paper, lumber, cardboard, plastics etc.
2. **Class B** fires involve flammable or combustible liquids such as gasoline, kerosene, and common organic solvents used in the laboratory.
3. **Class C** fires involve energized electrical equipment, such as appliances, switches, panel boxes, power tools, hot plates and stirrers. Water is usually a dangerous extinguishing medium for class C fires because of the risk of electrical shock unless a specialized water mist extinguisher is used.
4. **Class D** fires involve combustible metals, such as magnesium, titanium, potassium and sodium as well as pyrophoric organometallic reagents such as alkyllithiums, Grignards and diethylzinc. These materials burn at high temperatures and will react violently with water, air, and/or other chemicals. Handle with care!!
5. **Class K** fires are kitchen fires. This class was added to the NFPA portable extinguishers Standard 10 in 1998. Kitchen extinguishers installed before June 30, 1998 are "grandfathered" into the standard.

Some fires may be a combination of these! Your fire extinguishers should have ABC ratings on them. These ratings are determined under ANSI/UL Standard 711 and look something like "3-A:40-B:C." Higher numbers mean more firefighting power. In this example, the extinguisher has a good firefighting capacity for Class A, B and C fires. NFPA has a brief description of UL 711 if you want to know more.

Fire blanket

A fire blanket can be used to smother a fire. However, use caution when using a fire blanket on a clothing fire. Some fabrics are polymers that melt onto the skin. Stop, Drop and Roll is the best method for extinguishing clothing on fire.

Safety shower

Use a safety shower in the event of a chemical spill or fire. Pull the overhead handle and remove clothing that may be contaminated with chemicals, to allow the skin to be rinsed.

Eye protection

Everyone present must wear eye protection whenever anyone in the laboratory is performing any of the following activities:

1. Handling hazardous chemicals
2. Handling laboratory glassware
3. Using an open flame.

Safety glasses do not offer protection from splashing liquids. Safety glasses appear similar to ordinary glasses and may be used in an environment that only requires protection from **flying fragments**. Safety glasses with side-shields offer additional protection from **flying fragments approaching from the side**.

Safety goggles offer protection from both flying fragments and splashing liquids. **Only safety goggles** are suitable for eye protection where **hazardous chemicals** are used and handled. Safety goggles with no ventilation (type G) or with indirect ventilation (type H) are both acceptable. Goggles should be marked "Z87" to show they meet federal standards.

Eyewash

In the event of an eye injury or chemical splash, use the eyewash immediately. Help the injured person by holding their eyelids open while rinsing. Rinse copiously and have the eyes checked by a physician afterwards.

Skin protection

Wear **gloves** made of a material known to resist penetration by the chemical being handled. Check gloves for holes and the absence of interior contamination. Wash hands and arms and clean under fingernails after working in a laboratory.

Wear a lab coat or apron. Wear footwear that completely covers the feet.

Ventilation - Using a Fume Hood

A fume hood carries away vapors from reagents or reactions you may be working with. Using a fume hood correctly will reduce your personal exposure to potentially harmful fumes or vapors. When using a fume hood, keep the following in mind.

1. Place equipment or reactions as far back in the hood as is practical. This will improve the efficiency of fume collection and removal.
2. Turn on the light inside the hood using the switch on the outside panel, near the electrical outlets.
3. The glass sash of the hood is a safety shield. The sash will fall automatically to the appropriate height for efficient operation and should not be raised above this level, except to move equipment in and out of the hood. Keep the sash between your body and the inside of the hood. If the height of the automatic stop is too high to protect your face and body, lower the sash below this point. Do not stick your head inside a hood or climb inside a hood.

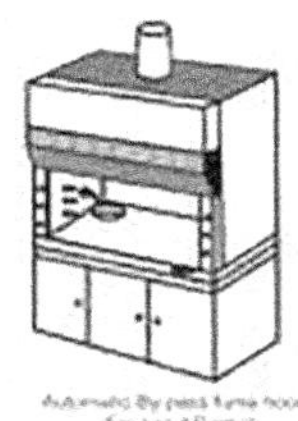

4. Wipe up all spills immediately. Clean the glass of your hood, if a splash occurs.
5. When you are finished using a hood, lower the sash to the level marked by the sticker on the side.

Work habits

1. Never work alone in a laboratory or storage area.
2. Never eat, drink, smoke, apply cosmetics, chew gum or tobacco, or store food or beverages in a laboratory environment or storage area.
3. Keep containers closed when they are not in use.
4. Never pipet by mouth.
5. Restrain loose clothing and long hair and remove dangling jewelry.
6. Tape all Dewar flasks with fabric-based tape.
7. Check all glassware before use. Discard if chips or star cracks are present.
8. Never leave heat sources unattended.
9. Do not store chemicals and/or apparatus on the lab bench or on the floor or aisles of the lab or storage room.
10. Keep lab shelves organized.
11. Never place a chemical, not even water, near the edges of a lab bench.
12. Use a fume hood that is known to be in operating condition when working with toxic, flammable, and/or volatile substances.
13. Never put your head inside a fume hood.
14. Never store anything in a fume hood.
15. Obtain, read, and be sure you understand the MSDS (see below) for each chemical that is to be used before allowing students to begin an experiment.
16. Analyze new lab procedures and student-designed lab procedures in advance to identify any hazardous aspects. Minimize and/or eliminate these components before proceeding. Ask yourself these questions:
    - What are the hazards?
    - What are the worst possible things that could go wrong?
    - How will I deal with them?
    - What are the prudent practices, protective facilities and equipment necessary to minimize the risk of exposure to the hazards?
        - Analyze close calls and accidents to eliminate their causes and prevent them from occurring again.
        - Identify which chemicals may be disposed of in the drain by consulting the MSDS or the supplier. Clear one chemical down the drain by flushing with water before introducing the next chemical.
        - Preplan for emergencies.
    - Keep the fire department informed of your chemical inventory and its location.
    - Consult with a local physician about toxins used in the lab and ensure that your area is prepared in advance to treat victims of toxic exposure.
    - Identify devices that should be shut off if possible in an emergency.
    - Inform your students of the designated escape route and alternate route.

Substitutions

- When feasible, substitute less hazardous chemicals for chemicals with greater hazards in experiments.
- Dilute substances when possible instead of using concentrated solutions.
- Use lesser quantities instead of greater quantities in experiments when possible.
- Use films, videotapes, computer displays, and other methods rather than experiments involving hazardous substances.

Label information

Chemical labels contain safety information in four parts:

1. There will be a signal word. From most to least potentially dangerous, this word will be "Danger!" "Warning!" or "Caution."
2. Statements of hazard (e.g., "Flammable," "May Cause Irritation") follow the signal word. Target organs may be specified.
3. Precautionary measures are listed such as "Keep away from ignition sources" or "Use only with adequate ventilation."
4. First aid information is usually included, such as whether to induce vomiting and how to induce vomiting if the chemical is ingested.

Chemical hazard pictorial

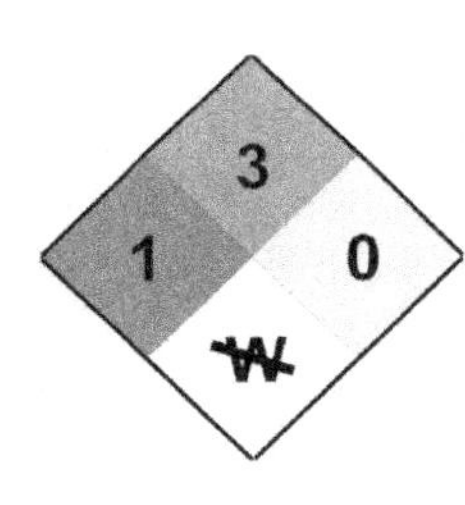

Several different pictorials are used on labels to indicate the level of a chemical hazard. The most common is the **"fire diamond" NFPA (National Fire Prevention Association) pictorial** shown at left. A zero indicates a minimal hazard and a four indicates a severe risk. Special information includes the ability of a chemical to react with water, **OX** for an oxidizer, **COR ACID** for a corrosive acid, and **COR ALK** for a corrosive base. The "Health" hazard level is for **acute toxicity only**.

Pictorials are designed for **quick reference in emergency situations**, but they are also useful as minimal summaries of safety information for a chemical. They are not required on chemicals you purchase, so it's a good idea to add a label pictorial to every chemical you receive if one is not already present. **The entrance to areas where chemicals are stored should carry a fire diamond label** to represent the materials present.

Procedures for flammable materials: minimize fire risk

The vapors of a flammable liquid **or solid** may travel across the room to an ignition source and cause a fire or explosion.

- Store flammables in an approved safety cabinet for flammable liquids. Store in safety cans if possible.
- Minimize volumes and concentrations used in an experiment with flammables.
- Minimize the time containers are open.
- Minimize ignition sources in the laboratory.
- Ensure that there is good air movement in the laboratory before the experiment.

- Check the fire extinguishers and be certain that you know how to use them.
- Tell the students that the "Stop, Drop, and Roll" technique is best for a clothing fire outside the lab, but in the lab they should walk calmly to the safety shower and use it. Practice this procedure with students in drills.
- A fire blanket should not be used for clothing fires because clothes often contain polymers that melt onto the skin. Pressing these fabrics into the skin with a blanket increases burn damage.
- If a demonstration of an exploding gas or vapor is performed, it should be done behind a safety shield using glass vessels taped with fabric tape.

Procedures for corrosive materials: minimize risk of contact

Corrosive materials **destroy or permanently change living tissue** through chemical action. **Irritants** cause inflammation due to an immune response but not through chemical action. The effect is usually reversible but can be severe and long lasting. **Sensitizers** are irritants that cause no symptoms after the first exposure but may cause irritation during a later exposure to the same or a different chemical.

- Always store corrosives below eye level.
- Only diluted corrosives should be used in pre-high school laboratories and their use at full-strength in high school should be limited.
- Dilute corrosive materials by slowly and carefully adding them to water. Adding water to a concentrated acid or base can cause rapid boiling and splashing.
- Wear goggles and face shield when handling hazardous corrosives. The face, ears, and neck should be protected.
- Wear gloves known to be impervious to the chemical. Wear sleeve gauntlets and a lab apron made of impervious material if splashing is likely.
- Always wash your hands after handling corrosives.
- Splashes on skin should be flushed with flowing water for 15 minutes while a doctor is called.
- If a corrosive material is splashed on their clothing:
    - First use the safety shower with clothing on.
    - Remove all clothing while under the safety shower including shoes and socks. This is no time for modesty.
    - Stay under the shower for 15 minutes while a doctor is called.
- Splashes in eyes should be dealt with as follows:
    - Go to the eyewash fountain within 30 seconds.
    - Have someone else hold their eyelids open with thumb and forefinger and use the eyewash.
    - Continuously move the eyeballs during 15 minutes of rinsing to cleanse the optic nerve at the back of the eye. A doctor should be called.

Procedures for toxic materials: minimize exposure

Toxic effects are either **chronic** or **acute**. Chronic effects are seen after repeated exposures or after one long exposure. Acute effects occur within a few hours at most.

- Use the smallest amount needed at the lowest concentration for the shortest period of time possible. Weigh the risks against the educational benefits.
- Be aware of the five different routes of exposure
    - Inhalation – the ability to smell a toxin is not a proper indication of unsafe exposure. Work in the fume hood when using toxins. Minimize dusts and mists by cleaning often, cleaning spills rapidly, and maintaining good ventilation in the lab.
    - Absorption through intact skin – always wear impervious gloves if the MSDS indicates this route of exposure.
    - Ingestion
    - Absorption through other body orifices such as ear canal and eye socket.
    - Injection by a cut from broken contaminated glassware or other sharp equipment.
- Be aware of the first symptoms of overexposure described by the MSDS. Often these are headache, nausea, and dizziness. Get to fresh air and do not return until the symptoms have passed. If the symptom returns when you come back into the lab, contact a physician and have the space tested.
- Be aware of whether vomiting should be induced in case of ingestion
- Be aware of the recommended procedure in case of unconsciousness.

Procedures for reactive materials to minimize incompatibility

Many chemicals are **self-reactive**. For example, they explode when dried out or when disturbed under certain conditions or they react with components of air. These materials generally **should not be allowed into the high school**. Other precautions must be taken to minimize reactions between **incompatible pairs**.

- Store fuels and oxidizers separately.
- Store reducing agents and oxidizing agents separately.
- Store acids and bases separately.
- Store chemicals that react with fire-fighting materials (i.e., water or carbon dioxide) under conditions that minimize the possibility of a reaction if a fire is being fought in the storage area.
- MSDSs list other incompatible pairs.
- Never store chemicals in alphabetical order by name.
- When incompatible pairs must be supplied to students, do so under direct supervision with very dilute solutions and/or small quantities.

Material Safety Data Sheet (MSDS) information

Many chemicals have a mixture of toxic, corrosive, flammable, and reactive risks. The **Material Safety Data Sheet** or **MSDS** for a chemical contains detailed safety information beyond that presented on the label. This includes acute and chronic health effects, first aid and firefighting measures, what to do in case of a spill, and ecological and disposal considerations.

The MSDS will state whether the chemical is a known or suspected carcinogen, mutagen, or teratogen. A **carcinogen** is a compound that causes cancer (malignant tumors). A **mutagen** alters DNA with the potential effects of causing cancer or birth defects in children not yet conceived. A **teratogen** produces birth defects by damaging the fetus during fetal development.

There are many parts of an MSDS that are not written for the layperson. Their level of detail and technical content may be difficult for many people outside the fields of toxicology and industrial safety to understand. In a high school chemistry lab, the value of an MSDS is in the words and not in the numerical data it contains, but some knowledge of the numbers is useful for comparing the dangers of one chemical to another. Numerical results of animal toxicity studies are often presented in the form of **LD_{50} values**. These represent the **dose required to kill 50% of animals** tested.

Exposure limits via inhalation may be presented in three ways:

1. PEL (Permissible Exposure Limit) or TLV-TWA (Threshold Limit Value-Time Weighted Average). This is the maximum permitted concentration of the airborne chemical in volume parts per million (ppm) for a **worker exposed 8 hours daily**.
2. TLV-STEL (Threshold Limit Value-Short Term Exposure Limit). This is the maximum concentration permitted for a 15-minute exposure period.
3. TLV-C (Threshold Limit Value-Ceiling). This is the concentration that should never be exceeded at any moment.

http://hazard.com/msds/index.php contains a large database of MSDSs. http://www.ilpi.com/msds/ref/demystify.html contains a useful "MSDS demystifier." Cut and paste an MSDS into the web page, and hypertext links will appear to a glossary of terms.

Facilities and equipment

- Use separate labeled containers for general trash, broken glass, and for each type of hazardous chemical waste—ignitable, corrosive, reactive, and toxic.
- Keep the floor area around safety showers, eyewash fountains, and fire extinguishers clear of all obstructions.
- Never block escape routes.
- Never prop open a fire door.
- Provide safety guards for all moving belts and pulleys.
- Instruct everyone in the lab on the proper use of the safety shower and eyewash fountain (see corrosive materials above). Most portable eyewash devices cannot maintain the required flow for 15 minutes. A permanent eyewash fountain is preferred.
- If contamination is suspected in the breathing air, arrange for sampling to take place.
- Regularly inspect fire blankets, if present, for rips and holes. Maintain a record of inspection.
- Regularly check safety showers and eyewash fountains for proper rate of flow. Maintain a record of inspection.
- Keep up-to-date emergency phone numbers posted next to the telephone.

- Place fire extinguishers near an escape route.
- Regularly maintain fire extinguishers and maintain a record of inspection.
- Arrange with the local fire department for training of teachers and administrators in the proper use of extinguishers.
- Regularly check fume hoods for proper airflow. Ensure that fume hood exhaust is not drawn back into the intake for general building ventilation.
- Secure compressed gas cylinders at all times and transport them only while secured on a hand truck.
- Restrict the use and handling of compressed gas to those who have received formal training.
- Install chemical storage shelves with lips. Never use stacked boxes for storage instead of shelves.
- Only use an explosion-proof refrigerator for chemical storage.
- Have appropriate equipment and materials available in advance for spill control and cleanup. Consult the MSDS for each chemical to determine what is required. Replace these materials when they become outdated.
- Provide an appropriate supply of first aid equipment and instruction on its proper use.

Additional comments: teach safety to students

- Weigh the risks and benefits inherent in lab work, inform students of the hazards and precautions involved in their assignment, and involve students in discussions about safety before every assignment.
- If an incident happens, it can be used to improve lab safety via student participation. Ask the student involved. The student's own words about what occurred should be included in the report.
- Safety information supplied by the manufacturer on a chemical container should be seen by students who actually use the chemical. If you distribute chemicals in smaller containers to be used by students, copy the hazard and precautionary information from the original label onto the labels for the students' containers. Students interested in graphic design may be able to help you perform

16.4 Safety

Safety is an important issue in every lab. There are a few safety rules which you must remember.

Goggle policy

There is only one policy on goggles that needs to be enforced. "Any time chemicals, glassware, or heat is used, you must wear your laboratory goggles. No exceptions!" If you don't wear your goggles then you must leave the lab.

Chemical terms

Many chemistry experiments include safety instructions terms like corrosive, flammable, oxidizer, reducer, etc. Make sure that you understand the meaning of these terms

16.5 Calculations

Calculations based on chemical equations

A chemical equation can be interpreted in several ways. For example, consider the balanced equation for the combustion of ethanol, C_2H_5OH, the alcohol that is blended with gasoline in the fuel known as gasohol.

$$C_2H_5OH + 3O_2 \longrightarrow 2CO_2 + 3\,H_2O$$

On a molecular submicroscopic level, we can view this as a reaction between individual molecules.

$$1 \text{ molecule } C_2H_5OH + 3 \text{ molecules } O_2 \longrightarrow 2 \text{ molecules } CO_2 + 3 \text{ molecules } H_2O$$

But we can just as easily scale this up to lab-sized amounts by applying the same "mole reasoning ". You have learned that the ratio by which the atoms of the elements are combined in compounds is exactly the same as the ratio by which moles of the atoms are combined; the atom ratios and the mole ratios are identical.

For a chemical reaction, we can make a similar statement. The ratios by which the molecules react or are formed are exactly the same as the ratios by which moles of these substances react or are formed. This means that for the combustion of ethanol, we can also write

$$1 \text{ mol } C_2H_5OH + 3 \text{ mol } O_2 \longrightarrow 2 \text{ mol } CO_2 + 3 \text{ mol } H_2O$$

We could increase the ratio of this equation by any number. For example, if we burn 2 moles of ethanol then we have

$$2 \text{ mol } C_2H_5OH + 6 \text{ mol } O_2 \longrightarrow 4 \text{ mol } CO_2 + 6 \text{ mol } H_2O$$

In fact, we can use any amounts we want, but we will always find three times as many moles of O_2 are consumed as moles of C_2H_5OH, and for every mole of C_2H_5OH consumed, 2 moles of CO_2 and 3 moles of H_2O will be formed. This is an important point in getting the information for calculations of a chemical equation because the coefficients in a chemical equation provide the ratios by which moles of one substance react with or from moles of another.

For example, the equation for the combustion of C_2H_5OH gives six chemical equivalences that we can use to form conversion factors for calculations:

$$1 \text{ mol } C_2H_5OH \longleftrightarrow 3 \text{ mol } O_2$$

$$1 \text{ mol } C_2H_5OH \longleftrightarrow 2 \text{ mol } CO_2$$

$$1 \text{ mol } C_2H_5OH \longleftrightarrow 3 \text{ mol } H_2O$$

$3 \text{ mol } O_2 \longleftrightarrow 2 \text{ mol } CO_2$

$3 \text{ mol } O_2 \longleftrightarrow 3 \text{ mol } H_2O$

$2 \text{ mol } CO_2 \longleftrightarrow 3 \text{ mol } H_2O$

Other calculations were covered in Chapters 10 and 11.

16.6 Data Analysis

Differentiate between the uses of qualitative and quantitative data.

Qualitative data refers to observations that are descriptive in nature. For example, "the sample was yellow," is descriptive and thus an example of qualitative data. Quantitative data is any observation collected that is numerical in nature. "The sample had a mass of 1.15 grams," is quantitative data. Both types of data have their place in scientific research.

Qualitative Data

The purpose of qualitative data analysis is a complete and detailed, though subjective, description. Qualitative research involves collecting, analyzing, and interpreting data by observing what the subject does.

Qualitative research is much more subjective than quantitative research and uses very different methods of collecting information. This type of data generates rich, detailed and valid observations that contribute to in-depth understanding of the experiment. For example, color change is a type of qualitative observation that is useful in chemistry. However, this is a subjective type of data, since two scientists may view the color change in different ways. Qualitative analysis allows for fine distinctions to be made because it is not necessary to force the data into a finite number of categories.

The main disadvantage of qualitative approaches to data analysis is that their findings cannot be extended to other experiments with the same degree of certainty that quantitative analyses can. This is because the findings of the research are not tested to discover whether they are either statistically significant or due to chance.

Quantitative Data

Analysis of quantitative data involves taking objective measurements using standard instruments during an experiment and analyzing the data in an attempt to explain the experiment. These findings, then, can be repeated by other scientists and direct comparisons can be made between two data sets as long as valid scientific techniques have been used.

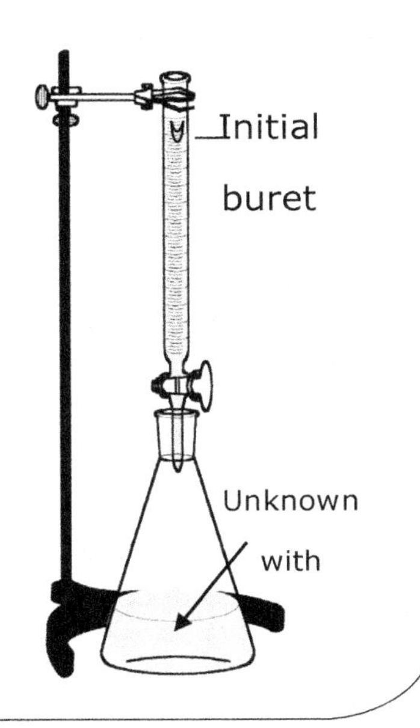

Most observations made during a scientific experiment will provide quantitative data. Any data point that involves a number obtained by an instrument is a piece of quantitative data. This includes temperature, pressure, mass, time, and many others.

Quantitative research generates reliable, generalizable data and is well-suited to establishing cause-and-effect relationships, such as the ideal gas law. For example, we can observe that when the temperature of an ideal gas increases by a specific amount, then the pressure will also increase by a specific, measurable amount (assuming the volume remains constant).

The decision of whether to choose a quantitative or a qualitative design is a philosophical question. Which methods to choose will depend on the nature of the experiment and the type of information needed.

It is important to keep in mind that these are two different types of data, not necessarily polar opposites. Elements of both designs can be used together in scientific observation. Combining qualitative and quantitative research is not uncommon.

It is advantageous for scientists to combine both types of research data for several reasons:

1. Research development (one approach is used to inform the other)
2. Increased validity (confirmation of results by means of different data sources)
3. Complementary (adding information such as words to numbers and vice versa)
4. Creating new ideas (emerging fresh perspectives)

Quantitative data involves a number, and **qualitative** data does not. Qualitative data may be a description such as a slight/moderate/intense color change or a weak/strong/explosive reaction or it may simply be the absence/presence of an event.

Basically, quantitative research is objective; qualitative is subjective. Quantitative research seeks explanatory laws; qualitative research aims at in-depth description. Quantitative research measures what it assumes to be a static reality in hopes of developing universal laws and is well suited to establishing cause-and-effect relationships. Quantitative data are measurements that are numerical in nature. "The sample had a mass of 1.15 grams" is a quantitative data point. Qualitative research is an exploration of what is assumed to be a dynamic reality. It does not claim that what is discovered in the process is universal, and thus necessarily replicable.

A **lab notebook** is used as the **record of lab work and data collection as it occurs**. Researchers often use lab notebooks to document hypotheses and data analysis. A good lab notebook should allow another scientist to follow the same steps. Electronic lab notebooks are growing in popularity.

16.7 Interpretation of graphical data

Interpret graphical and numerical titration data.

Standard titration

In a typical acid-base **titration, an acid-base indicator** (such as phenolphthalein) or a **pH meter** is used to monitor the course of a **neutralization reaction. This was discussed in detail in Chapter 7.3.**

Interpreting titration curves

A **titration curve** is a plot of a solution's **pH charted against the volume of an added acid or base**. Titration curves are obtained if a pH meter is used to monitor the titration, instead of an indicator. At the equivalence point, the titration curve is nearly vertical. This is the point where the most rapid change in pH occurs. In addition to determining the equivalence point, the **shape of titration curves** may be interpreted to determine **acid/base strength and the presence of a polyprotic acid.**

The pH at the equivalence point of a titration is the **pH of the salt solution obtained when the amount of acid is equal to amount of base**. For a strong acid and a strong base, the equivalence point occurs at the neutral pH of 7. For example, an equimolar solution of HCl and NaOH will contain NaCl(*aq*) at its equivalence point.

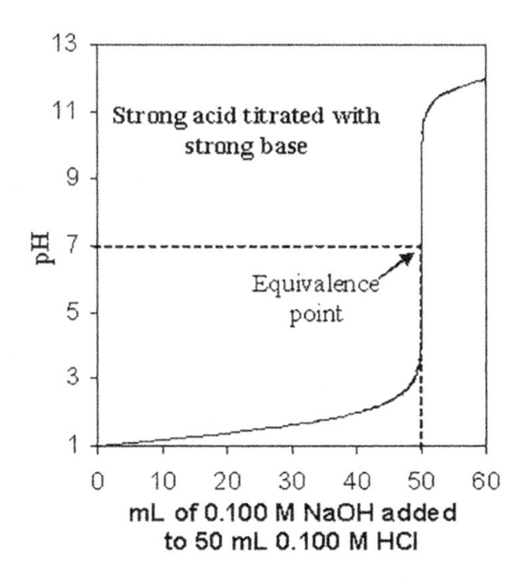

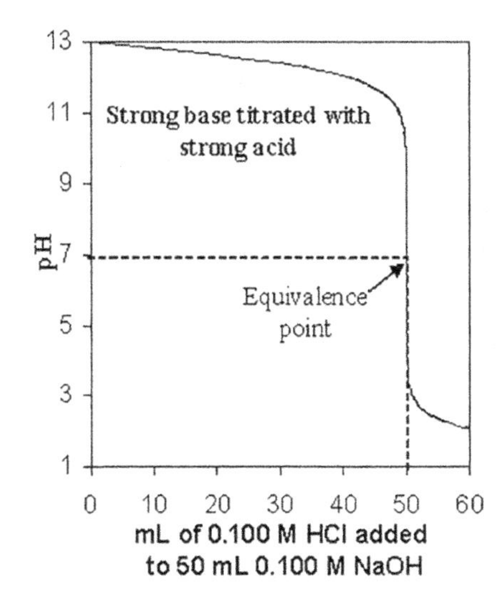

The salt solution at **the equivalence point of a titration involving a weak acid or base will not be at neutral pH**. For example, an equimolar solution of NaOH and hypochlorous acid HClO at the equivalence point of a titration will be a base because it is indistinguishable from a solution of sodium hypochlorite. A pure solution of NaClO(*aq*) will be a base because the ClO^- ion is the conjugate base of HClO, and it consumes H^+(*aq*) in the reaction:

$$ClO^- + H^+ \longrightarrow HClO$$

In a similar fashion, an equimolar solution of HCl and NH_3 will be an acid because a solution of NH_4Cl(*aq*) is an acid. It generates H^+(*aq*) in the reaction:

$$NH_4^+ \longrightarrow NH_3 + H^+$$

Contrast the following **titration curves for a weak acid or base** with those for a strong acid and strong base on the preceding page:

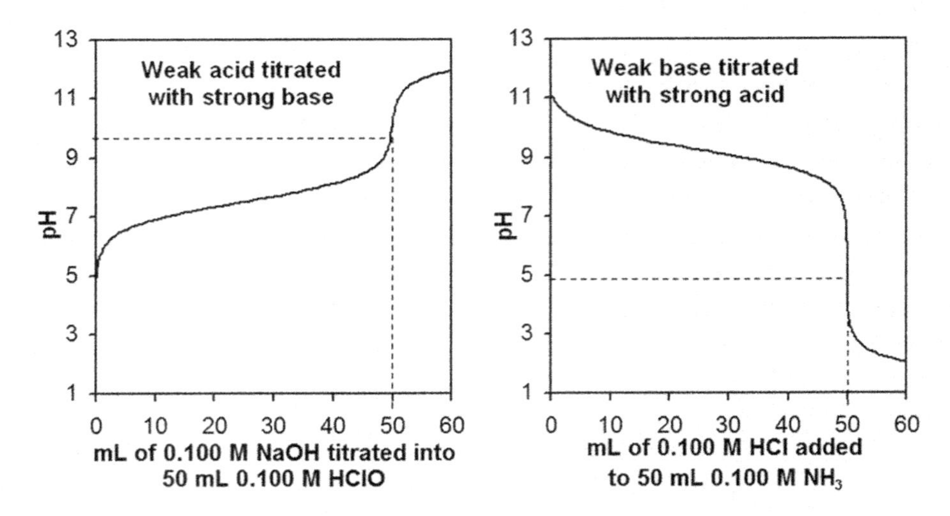

Titration of a polyprotic acid results in **multiple equivalence points** and a curve with more "bumps" as shown below for sulfurous acid and the carbonate ion.

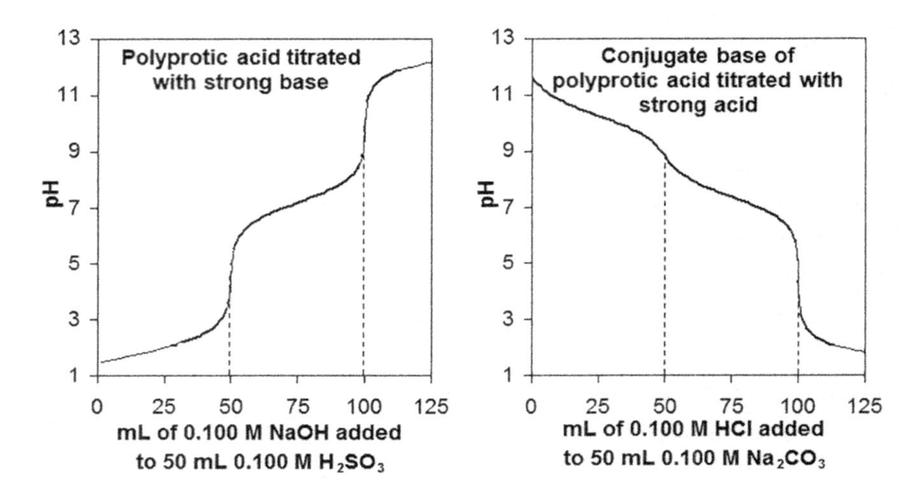

Evaluate, interpret, and predict from empirical data.

Galileo studied the behavior of falling bodies in the 1590's. Almost a century later, Newton followed up on his work and established calculus and physics which have governed mechanics for centuries. From that point on, science was on a new pathway. No longer would it merely be the philosophy of the ancient Greeks. It would be a qualitative and quantitative discourse.

Lavoisier made careful measurements in his conservation of mass experiments. Since then, data collection has become a central part of all science investigations. Data collected, however, takes varied forms depending of the scientific field and complexity of the inquiry.

Scientific data are initially organized into tables, spreadsheets, or databases. However, trends or patterns in data can be difficult to identify using tables of numbers. Therefore, data are more commonly compiled into graphs. Graphs help scientists visualize and interpret the variation in data. There are many types of graphs, and the nature of the data will determine which is best to use. Bar graphs, pie charts and line graphs are just a few methods used to pictorially represent numerical data.

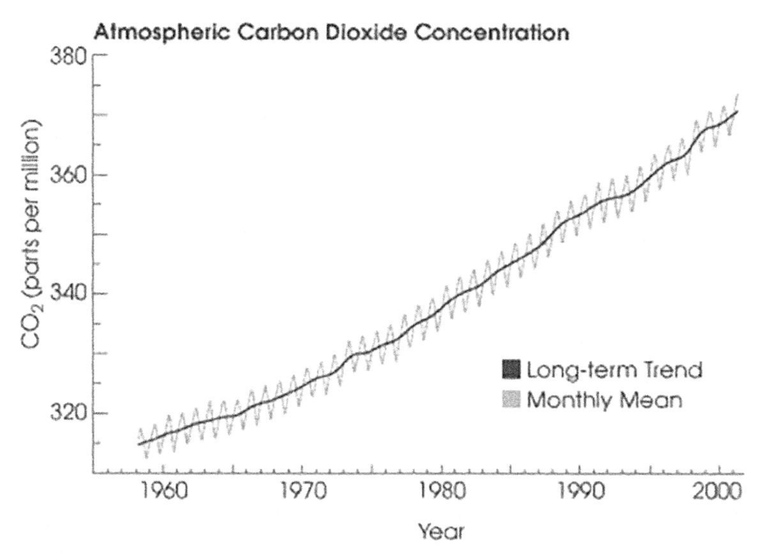

Atmospheric CO_2 measured at Mauna Loa
This is a famous graph called the Keeling Curve (courtesy of NASA).

Interpretation of graphical data shows that on the x-axis is the variable of time in units of years and the y-axis represents the variable of CO_2 concentration in units of parts per million (ppm). The best fit line (solid dark line) shows the trend in CO_2 concentrations during the time period. The steady upward-sloping line indicates a trend of increasing CO_2 concentrations during the time period. However, the light blue line, which indicates monthly mean CO_2, levels shows a periodic trend in CO_2 levels during the year. This periodic trend is accounted for by the changes in the seasons. In the spring and summer, deciduous trees and plants undergo increased photosynthesis and remove more CO_2 from the atmosphere in the Northern Hemisphere than in the fall and winter when they have no leaves.

Interpret graphical data.

The interpretation of data and the construction and interpretation of graphs are central practices in science. Graphs are effective visual tools which relay information quickly and reveal trends easily. While there are several different types of graphical displays, extracting information from them can be described in three basic steps.

1. **Describe the graph:** What does the title say? What is displayed on the x-axis and y-axis, including the units?
   - Determine the set-up of the graph.

- Make sure the units used are understood. For example, $g \cdot cm^{-3}$ means g/cm^3
- Notice any symbols used and check for a legend or explanation.

2. **Describe the data**: Identify the range of data. Are patterns reflected in the data?
3. **Interpret the data:** How do patterns seen in the graph relate to other variables? What conclusions can be drawn from the patterns?

16.8 Drawing conclusions from observations and data

Identify the characteristics and components of scientific inquiry.

Modern science began around the late 16th century with a new way of thinking about the world. Few scientists will disagree with Carl Sagan's assertion that "science is a way of thinking much more than it is a body of knowledge" (Broca's Brain, 1979). Science is a process of inquiry and investigation. It is a way of thinking and acting, not just a body of knowledge to be acquired by memorizing facts and principles. This way of thinking, the scientific method, is based on the idea that scientists begin their investigations with observations. From these observations they develop a hypothesis, which is further developed into a prediction. The hypothesis is challenged through experimentation and further observations and is refined as necessary. Science has progressed in its understanding of nature through careful observation, a lively imagination, and increasingly sophisticated instrumentation. Science is distinguished from other fields of study in that it provides guidelines or methods for conducting research, and the research findings must be reproducible by other scientists in order for those findings to be valid.

It is important to recognize that scientific practice is not always this systematic. Discoveries have been made that are serendipitous and others have been predicted based on theory rather than observation of phenomena. Einstein's theory of relativity was developed not from the observation of data but with a kind of mathematical puzzle. Only later were experiments able to be conducted that validated his theory.

The Scientific Method is a logical set of steps that a scientist goes through to solve a problem. The main purpose of using the Scientific Method is to eliminate, as much as possible, preconceived ideas, prejudices and biases by presenting an objective way to study possible answers to a question. Only by designing a way to study one variable at a time can each possible answer be ruled out or accepted for further study.

While an inquiry may start at any point in this method and may not involve all of the steps, the overall approach can be described as follows:

Making Observations

Scientific questions frequently result from observations of events in nature or events observed in the laboratory. An observation is not just a look at what happens. It also includes measurements and careful records of the event. Records could include photos, drawings, or written descriptions. The observations and data collection may provide answers, or they may lead to one or more questions. In chemistry, observations almost always deal with the behavior of matter.

Having arrived at a question, a scientist usually researches the scientific literature to see what is known about the question. Perhaps the question has already been answered, or another experimenter has found part of the solution. The scientist then may want to test or reproduce the answer found in the literature. Or, the research might lead to a new question.

Sometimes the same observations are made over and over again and are always the same. For example, one can observe that daylight lasts longer in summer than in winter. This observation never varies. Such observations are called **laws of nature**. For example, one of the most important laws in chemistry was discovered in the late 1700s. Chemists observed that no mass was ever lost or gained in chemical reactions. This law became known as the law of conservation of mass. Explaining this law was a major topic of chemistry in the early 19th century.

Developing a Hypothesis

If the question has not yet been answered, the scientist may prepare for an experiment by making a hypothesis. A hypothesis is a statement of a possible answer to the question. It is a tentative explanation for a set of facts and can be tested by experiments. Although hypotheses are usually based on observations, they may also be based on a sudden idea, intuition, or mathematical theory.

Conducting an Experiment

An experiment tests the hypothesis to determine whether it may be a correct answer to the question or a solution to the problem. Some experiments may test the effect of one thing on another under controlled conditions. Such experiments have two variables. The experimenter controls one variable, called the **independent variable**. The other variable, the **dependent variable**, shows the result of changing the independent variable.

For example, suppose a researcher wanted to test the effect of Vitamin A on the ability of rats to see in dim light. The independent variable would be the dose of Vitamin A added to the rats' diet. The dependent variable would be the intensity of light to which the rats respond. All other factors, such as time, temperature, age, water and other nutrients given to the rats are held constant.

Chemists sometimes do short experiments "just to see what happens" or to see what a certain reaction produces. Often, these are not formal experiments. Rather, they are ways of making additional observations about the behavior of matter.

In most experiments scientists collect **quantitative data**, which are data that can be measured with instruments. They also collect **qualitative data**, descriptive information from observations other than measurements. (16.7 Data Analysis)

Interpreting data and analyzing observations are an important part of the scientific method. If data are not organized in a logical manner, incorrect conclusions can be drawn. Also, other scientists may not be able to follow or reproduce the results. (See 16.8 Interpretation of graphical data) By placing data into charts and graphs, the scientist may see patterns or lack thereof. The scientist will also be able to understand if the experiment truly tested the hypothesis. Induction is drawing conclusions based on facts or observations. Deduction is drawing conclusions based on generalizations.

Drawing a Conclusion

Finally, a scientist must draw conclusions from the experiment. A conclusion must address the hypothesis on which the experiment was based. The conclusions state whether or not the data support the hypothesis. If not, the conclusion should state what the experiment did show. If the hypothesis is not supported, the scientist uses the observations from the experiment to make a new or revised hypothesis. Then, new experiments are planned.

Effective written communication is necessary to present the research to a teacher or to a scientific journal. Effective oral communication is needed to present the research to a group whether that group is a class or other scientists. Students must recognize that, in this age of communication, those who cannot communicate effectively will be left behind. Accordingly, the evaluation system of the use of the scientific method should make provision for communication skills and activities.

Defending results is as important as conducting an experiment. One can honestly defend one's own results only if the results are reliable, and experiments must be well-controlled and repeated at least twice to be considered reliable. It must be emphasized to the students that *honesty and integrity are the foundation for any type of investigation.*

Steps of the Scientific Method

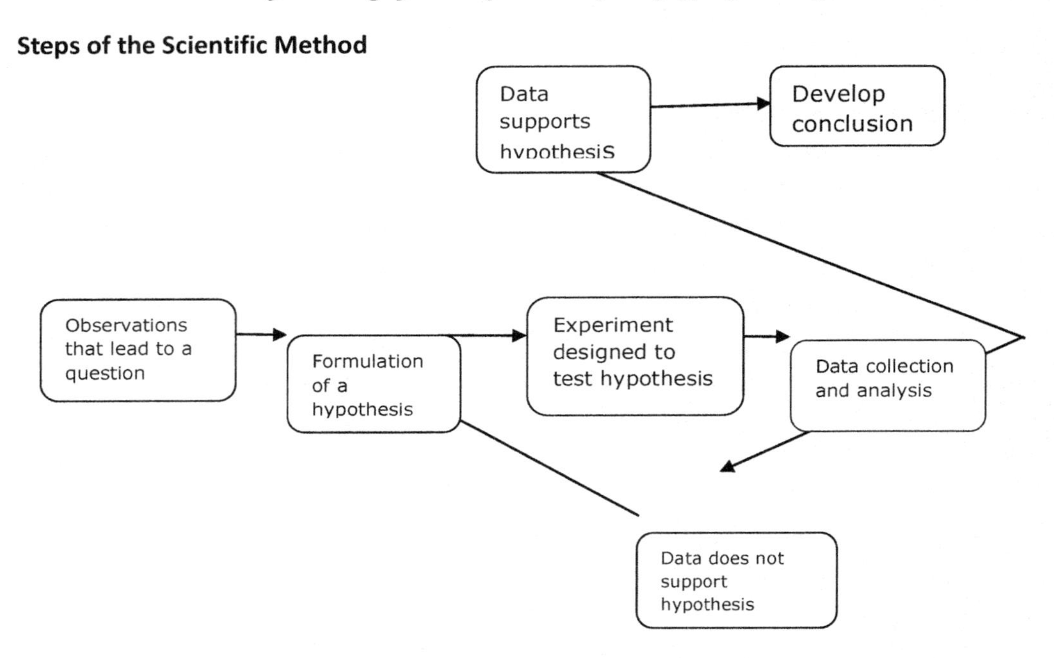

Collect quantitative data, data that can be measured with instruments and qualitative data, descriptive information from observations.

Are your notes complete and clear? Did you graph your data to look for a trend or pattern?

Did you look at the statistics? Percentage average comparison?

Interpreting data and analyzing observations are important. If data are not organized in a logical manner, wrong conclusions can be drawn and other scientists may not be able to follow your work.

Do the data indicate that your hypothesis is supported? Or do they refute your hypothesis?

If your hypothesis was refuted:

Was it a flaw in your experimental design?

Was it an incorrect hypothesis?

Or do the data support another hypothesis?

Developing a Theory

When a hypothesis survives many experimental tests to determine its validity, the hypothesis may be developed into a **theory**. A theory explains a body of facts and laws that are based on the facts. A theory also reliably predicts the outcome of related events in nature. For example, the law of conservation of matter and many other experimental observations led to a theory proposed early in the 19th century. This theory explained the conservation law by proposing that all matter is made up of atoms which are never created or destroyed in chemical reactions, only rearranged. This atomic theory also successfully predicted the behavior of matter in chemical reactions that had not been studied at the time. As a result, the atomic theory has stood for 200 years with only minor modifications.

A theory also serves as a scientific **model**. A model can be a physical model made of wood or plastic, a computer program that simulates events in nature, or simply a mental picture of an idea. A model illustrates a theory and explains nature. For example, in your chemistry course, you will develop a mental (and possibly a physical) model of the atom and its behavior. Outside of science, the word theory is often used to describe an unproven notion. In science, theory means much more. It is a thoroughly tested explanation of things and events observed in nature.

A theory can never be proven true, but it can be proven untrue.

SAT Chemistry Practice Test 1

Questions

Part I: Classification Questions

Questions 1-4 refer to the following:

(A) Temperature
(B) Energy
(C) Entropy
(D) Gibbs free energy

1. **Is negative for a spontaneous reaction**
2. **Is the average kinetic energy of the system**
3. **Is always conserved**
4. **Is a measure of the disorder of a system**

Questions 5-8 refer to the following

(A) London Forces
(B) H-bonds
(C) Covalent Bonds
(D) Ionic Bonds and Covalent Bonds

5. **These bonds are important in the structure of DNA and proteins.**
6. **The bond type found in H_2**
7. **The bond type found in $Ca_3(PO_3)_2$**
8. **Increase with the size of the molecule for long chain alkanes**

Part II: Relationship Questions

Determine if each statement is TRUE or FALSE.

If BOTH statements are true – decide if the first statement is TRUE BECAUSE of the second statement.

9. In the reaction of hydrogen with oxygen

$$2\,H_2 + O_2 \longrightarrow 2\,H_2O$$

22.4 L of oxygen will react exactly with 44.8 L of hydrogen to produce water

BECAUSE

At STP (standard temperature and pressure) the molar volume of any gas, the volume that one mole of any gas occupies, is 28.4 L.

10. In the Combustion reaction of propane:

$$C_3H_8 + 5\,O_2 \longrightarrow 3CO_2 + 4H_2O$$

The limiting reagent of the reaction of one mole of propane and 4 moles of oxygen to produce carbon dioxide and water is oxygen.

BECAUSE

The limiting reagent of a reaction is the reactant that runs out first. This reactant determines the amount of products formed, and any other reactants remain unconverted to product and are called excess reagents.

11. In ethylene and other alkene hydrocarbons, one s orbital and two p orbitals of carbon atoms combine together to form three sp^2 hybrid orbitals?

Each carbon in ethylene has four bonds including three sigma bonds, two between a carbon and two hydrogens and one between carbon and another carbon and one pi bond between the two carbons

12. A balloon filled with 10 liters of hydrogen gas at room temperature at the seaside was sent into the sky to a height where the pressure is 0.25 atm. Knowing that the maximum volume capacity of this balloon before exploding is 40 liters, we know the balloon was blown out at this altitude.

Since pressure of a gas is inversely proportional to its volume, as the balloon moves further away from the ground level into the sky, the air pressure decreases, and the volume of that gas will increase.

Part II. 5-Choice Completion Questions

13. Question 1- What is the electron configuration of phosphorus ion, P^{5+}?

(A) $1s^2\ 2s^2\ 2p^6\ 3s^2\ 3p^6$
(B) $1s^2\ 2s^2\ 2p^6\ 3s^2\ 3p^3$
(C) $1s^2\ 2s^2\ 2p^6$
(D) $1s^2\ 2s^2\ 2p^6\ 3s^1$
(E) $1s^2\ 2s^2\ 2p^6\ 3s^2\ 3p^4$

14. Question 2- Which of the following configurations represents the Cobalt (Co) atom?

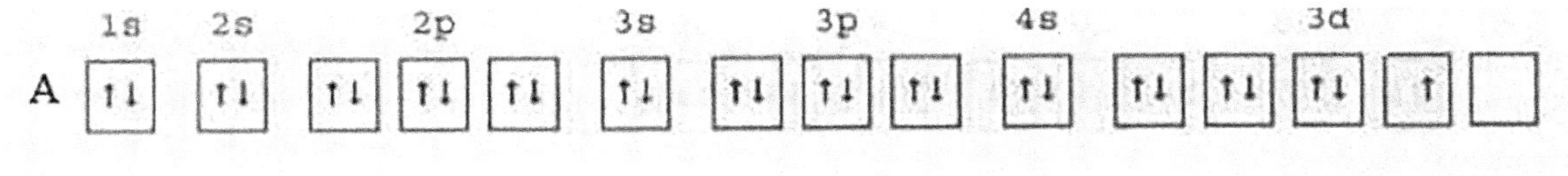

1s 2s 2p 3s 3p 4s 3d
B ↑↓ ↑↓ ↑↓ ↑↓ ↑↓ ↑↓ ↑↓ ↑↓ ↑↓ ↑↓ ↑↓ ↑↓ ↑ ↑ ↑

1s 2s 2p 3s 3p 4s 3d
C ↑↓ ↑↓ ↑↓ ↑↓ ↑↓ ↑↓ ↑↓ ↑↓ ↑↓ ↑↓ ↑↓ ↑ ↑ ↑ ↑↓

1s 2s 2p 3s 3p 4s 3d
D ↑↓ ↑↓ ↑↓ ↑↓ ↑↓ ↑↓ ↑↓ ↑↓ ↑↓ ↑↓ ↑↓ ↑ ↑↓ ↑↓

1s 2s 2p 3s 3p 4s 3d
E ↑↓ ↑↓ ↑↓ ↑↓ ↑↓ ↑↓ ↑↓ ↑↓ ↑↓ ↑↓ ↑↓ ↑↓ ↑ ↑↓

15. Which of the following Lewis dot structures for CH_3COOH is correct?

A
H :O:
H:C::C:O:H
H

B
H :O:
H:C::C:O:H
H

C
H :O:
H:C :C:O:H
H

D
H :O:
H:C::C::O:H
H

E
H :O:
H:C::C:O:H
H

16. 100 μL of SO_2 contains 2.2×10^{17} molecules. According to Avogadro's hypothesis, how many molecules are in 200 μL of methane gas, CH_4, at STP?

(A) 2.2×10^{17}
(B) $.44 \times 10^{17}$
(C) 4.4×10^{17}
(D) 0.22×10^{17}
(E) 22×10^{17}

17. Which one of the following atoms is the most chemically reactive atom?

(A) Ga
(B) Na
(C) Al
(D) Kr
(E) Mg

18. What is the shape of an NH_3 molecule? Use the VSEPR model.

(A) Trigonal planar
(B) Tetrahedral
(C) Trigonal bipyramidal
(D) Trigonal pyramidal
(E) Octahedral

19. Which of the following compounds is the most soluble in water?

(A) CO_2
(B) CH_2ClCH_3
(C) CH_2CH_2OH
(D) SiO_4
(E) $CH_3CH_2\ CH_2CH_2CH_2CH_3$

20. What is the number of neutrons in lead (Pb)?

(A) 132
(B) 88
(C) 123
(D) 130
(E) 125

21. How many mL of water is needed to add to 150 mL of 6M HNO_3 to prepare a solution of 0.6M HNO_3?

(A) 350
(B) 2050
(C) 1350
(D) 1200
(E) 120

22. Consider the unbalanced chemical equation: __CO (g) + __O_2 (g) ⟶ __CO_2 (g).

When this equation is completely balanced using the smallest whole numbers, what is the sum of the coefficients?

(A) 4
(B) 3
(C) 6
(D) 5
(E) 7

23. Which of the following solutions is the most concentrated?

(A) 2 moles of potassium hydroxide is dissolved in 10 liters of water
(B) 4 moles of potassium hydroxide is dissolved in 100 liters of water
(C) 2 moles of potassium hydroxide is dissolved in 20 liters of water
(D) 6 moles of potassium hydroxide is dissolved in 50 liters of water
(E) 2 moles of potassium hydroxide is dissolved in 40 liters of water

24. At STP, how many molecules of CO_2 are present in 224 liters of CO_2?

(A) 12.04×10^{23}
(B) 60.2×10^{23}
(C) 0.602×10^{23}
(D) 1.204×10^{23}
(E) 6.02×10^{23}

25. A gas in a locked piston is heated up. Which of the following statements about gas properties is correct in this regard?

(A) gas pressure increases
(B) average speed of molecules increases
(C) kinetic energy increases
(D) molecular collisions with container walls per second decreases
(E) A, B, and C

26. How many grams of H_2SO_4 are needed to prepare one liter of a 0.02 M solution of this acid?

(A) 1.96g
(B) 3.92g
(C) 19.6g
(D) 098g
(E) 9.8g

27. Which of the following compounds have the highest percentage of hydrogen (ratio of the mass of hydrogen to the mass of the compound)?

(A) CH_4
(B) H_2O_2
(C) C_2H_6
(D) H_2O
(E) NH_3

28. Which of the following reactions are oxidation-reduction reactions?

I. $H_2SO_4 + 2KOH \longrightarrow K_2SO_4 + 2H_2O$
II. $Ag\ NO_3 + NaCl \longrightarrow AgCl + NaNO_3$
III. $S + O_2 \longrightarrow SO_2$
IV. $2NaOH + Pb(NO_3)_2 \longrightarrow 2NaNO_3 + Pb(OH)_2$

(A) reaction II
(B) reaction III
(C) reaction IV
(D) both I and IV
(E) both II and III

29. The solubility product constants (K_{sp}) of the following five hydroxides are as follows:

I. $Ca(OH)_2$ is 5.5×10^{-6}
II. $Mn(OH)_2$ is 1.2×10^{-11}
III. $Zn(OH)_2$ is 4.5×10^{-17}
IV. $Al(OH)_3$ is 2×10^{-33}
V. $Ni(OH)_2$ is 1.6×10^{-14}

Which of the following statements are correct from this data?

(A) $Zn(OH)_2$ is the most soluble of these compounds
(B) $Al(OH)_3$ is the least soluble of these compounds
(C) $Ca(OH)_2$ is the most soluble of these compounds
(D) $Al(OH)_3$ has almost zero solubility in water and is a precipitate
(E) B, C, and D

30. Methane gas is combusted to produce CO_2 and water. How many liters of methane must be combusted with 0.2 moles of O_2?

$$CH_4 + 2O_2 \longrightarrow CO_2 + 2H_2O$$

(A) 4.48 L
(B) 2.24 L
(C) 22.4 L
(D) 44.8 L
(E) 0.448 L

31. What is the pH and pOH of a 0.00001M solution of HCl?

(A) pH= 4, pOH=10
(B) pH =3, pOH= 9
(C) pH=5, pOH =9
(D) pH= 5, pOH=13
(E) pH =1, pOH=11

32. Consider the following electrochemical overall reaction.

$$Ni + 2MnO_2 + 2NH_4^+(aq) \longrightarrow Ni_2^+ + 2\ Mn\ (OH)_2 + 2\ NH_3$$

What phenomenon occurs during this reaction?

(A) Ni is oxidized and its oxidation number increases
(B) MnO_2 is oxidized and its oxidation number decreases
(C) Ni is reduced and its oxidation number decreases
(D) MnO_2 is reduced and its oxidation number increases
(E) NH_4^+ is oxidized and its oxidation number increases

33. Butane, the volatile liquid fuel in disposable cigarette lighters, has the formula C_4H_{10}. How many atoms of carbon are present in 2.24 liters of butane in the gas form? (Avogadro's number is 6.02×10^{23})

(A) 2.4×10^{23}
(B) 6.02×10^{23}
(C) 0.602×10^{23}
(D) 4.8×10^{22}
(E) 0.48×10^{23}

Practice Test One

34. Consider the following four reactions:

I. $S + H_2 \longrightarrow SH_2$
II. $S + O_2 \longrightarrow SO_2$
III. $S + Na \longrightarrow SNa_2$
IV. $S + F_2 \longrightarrow SF_2$

If 3.2 grams of sulfur in four separate experiments are reacted with, respectively, 0.1 g hydrogen, 6.4 g oxygen, 0.23 g sodium, and 0.38 g fluorine, in which reaction is sulfur the limiting reagent?

(A) I
(B) III
(C) IV
(D) II
(E) both III and IV

35. Write the equilibrium expression Keq for the reaction

$$2NO_{2(g)} + O_{2(g)} \rightleftarrows 2NO_{2(g)}$$

(A) $4[NO_2]^2 / [NO]^2[O_2]$
(B) $4[NO_2]^2 / 4 [NO]^2[O_2]$
(C) $2[NO_2]^2 / 2 [NO]^2[O_2]$
(D) $[NO_2]^2 / [NO]^2[O_2]$
(E) $[NO_2]^2 / 4 [NO]^2[O_2]$

36. Which statements about reaction rates are not true?

(A) Most reaction mechanisms are multi-step processes involving reaction intermediates
(B) Intermediates are chemicals that are formed during one elementary step and consumed during another
(C) If one elementary reaction is the slowest reaction, it determines the overall reaction rate.
(D) This slowest reaction in the series is called the rate-limiting step or rate determining step.
(E) Many reactions have two rate-determining steps

37. 5 kilograms of crushed ice are added to a container of freezing water at 0° C. The ice water is then heated while a thermometer is placed in the container. The heat is applied until the water in the container boils for 10 minutes. Then, we start to cool the container down very quickly, to the freezing point of the water, and we stop recording the temperature. Which of the following types of graphs is obtained by plotting temperature versus time for this experiment?

(A) Temperature / Time

(B) Temperature / Time

(C)

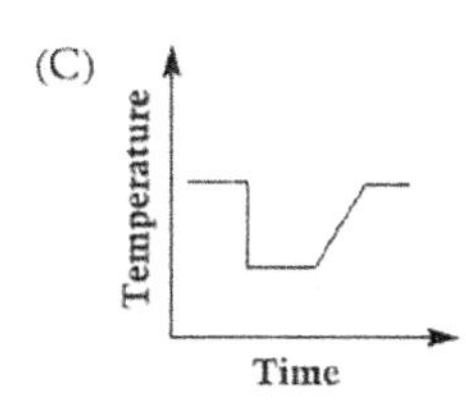

(D) Temperature / Time

(E)

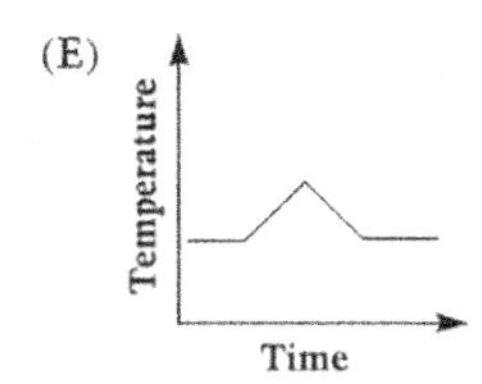

38. When dissolving $AlCl_3$ in water, the water becomes substantially warmer. Now if we gradually heat up the container, what happens to the solubility of $AlCl_3$?

(A) It will increase
(B) It will decrease
(C) There will be no change in temperature
(D) Cannot be determined with the information given
(E) It will increase and the solution will start to boil

39. Which is the correct formula for 3-methyl-4-methyl- 2-pentene?

(A)

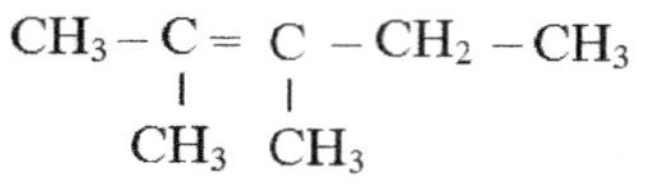

$$\begin{array}{ccccccccc} CH_3 & - & C & = & C & - & CH_2 & - & CH_3 \\ & & | & & | & & & & \\ & & CH_3 & & CH_3 & & & & \end{array}$$

(B)

$$\begin{array}{ccccccccc} & & H & & & & & & \\ & & | & & & & & & \\ CH_3 & - & C & - & C & = & C & - & CH_3 \\ & & | & & | & & | & & \\ & & CH_3 & & CH_3 & & H & & \end{array}$$

(C)

$$\begin{array}{ccccccccc} & & & & H & & & & \\ & & & & | & & & & \\ CH_3 & - & C & - & C & = & C & - & CH_3 \\ & & | & & & & | & & \\ & & CH_3 & & & & CH_3 & & \end{array}$$

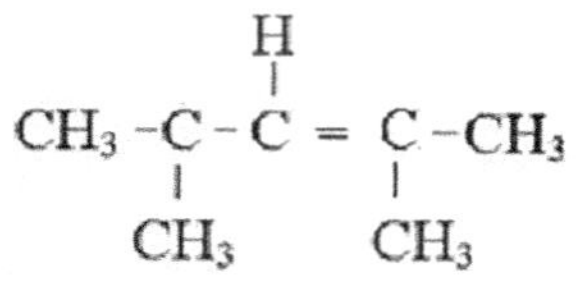

(D)

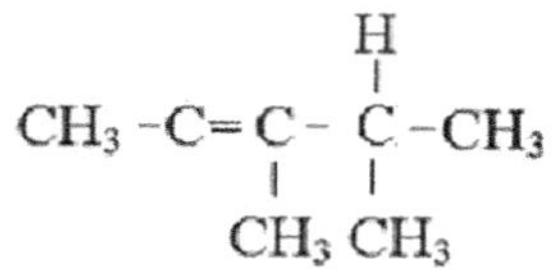

$$\begin{array}{ccccccccc} & & & & & & H & & \\ & & & & & & | & & \\ CH_3 & - & C & = & C & - & C & - & CH_3 \\ & & & & | & & | & & \\ & & & & CH_3 & & CH_3 & & \end{array}$$

(E)

$$\begin{array}{ccccccccc} CH_2 & = & C & - & C & - & CH_2 & - & CH_3 \\ & & | & & | & & & & \\ & & CH_3 & & CH_3 & & & & \end{array}$$

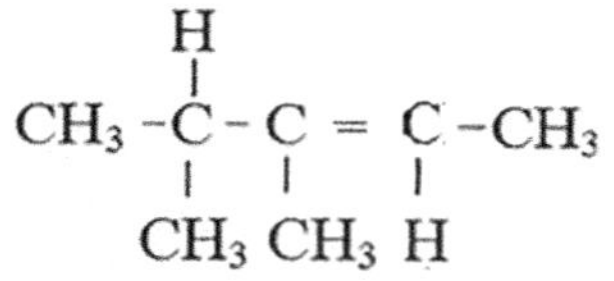

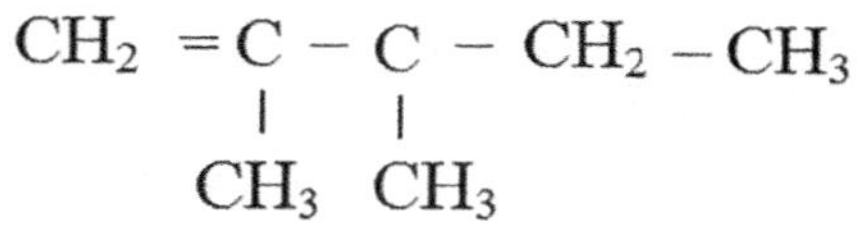

40. Which of the following statements regarding a chemical storage plan for laboratories is incorrect?

(A) All chemical containers should be properly labeled, dated upon receipt, and dated upon opening
(B) Although it is preferable to separate corrosives from flammables, in some conditions they can be stored together
(C) Shelves should be secure and strong enough to hold chemicals being stored on them. Shelves should not be overloaded.
(D) Personnel should be aware of the hazards associated with all hazardous materials
(E) The important chemical containers should be properly labeled, dated upon receipt, and dated upon opening

41. Which of the following statements is incorrect?

(A) A catalyst is a material that increases the rate of a chemical reaction without changing itself permanently in the process.
(B) The weight of the catalyst remains unchanged in the reaction it catalyzes.
(C) Catalysts reduce the activation energy of a reaction
(D) A catalyst reduces the heat of reaction.
(E) A catalyst increases the rate of both the forward and reverse reactions by lowering the activation energy for the reaction.

42. Based on trends in the periodic table, which of the following properties would you expect to be greater for Ba than for Be?

I. Atomic radius
II. Atomic number
III. Ionization energy
IV. Electronegativity
V. Nuclear charge

(A) I, II, and V
(B) Only I
(C) II and V
(D) III, IV
(E) III, V

43. Consider two elements, designated L and M, in the Periodic Table. Electron affinity and ionization energy is greater for L than for M, and atomic radius and metallic character are greater for M than L. Locate the approximate location of L and M in the table (in a period or in a column) with respect to each other.

(A) M is to the right of L in a period
(B) M is higher than L in a column
(C) L is to the right of M in a period
(D) Both A and B
(E) Both B and C

44. Which method would be most useful for separating a specific protein from a mixture of proteins?

(A) Extraction
(B) Gas chromatography
(C) Distillation
(D) Liquid chromatography
(E) Crystallization

45. Which separation method is best for separation of an aqueous solution containing methanol and ethanol?

(A) Extraction
(B) Chromatography
(C) Distillation
(D) Vaporization
(E) Both A and C

46. A 2.0 g coin of silver alloy was dissolved in nitric acid to give a solution of $AgNO_3$. This solution was in turn reacted with enough sodium chloride solution to give 1.435 g of white precipitate, AgCl. What is the percentage of silver in the coin?

$$AgNO_{3\,(l)} + NaCl_{(l)} \longrightarrow AgCl_{(s)} + NaNO_{3\,(l)}$$

(A) 68%
(B) 10.8%
(C) 5.4%
(D) 54%
(E) 35%

47. Which of the following group elements are more electronegative?

(A) Ar, Kr, Ne
(B) Ba, Sr, Ca
(C) As, P, N
(D) Se, S, O
(E) Ga, Al, B

48. In each set of the following elements the size of atom increases from left to right except in?

(A) Li, Na, K
(B) As, Ge, Ga
(C) Br, Se, As
(D) F, Cl, Br
(E) Al, Si, P

49. Which of the following substance (s) are conductors of electricity?

(A) P
(B) He
(C) Ni
(D) Cl
(E) Br

50. Rank the following bonds from least to most polar:

C-H, C-Cl, H-H, C-F

(A) C-H < H-H < C-F < C-Cl
(B) H-H < C-H < C-F < C-Cl
(C) C-F < C-Cl < C-H < H-H
(D) H-H < C-H < C-Cl < C-F
(E) H-H < C-CL < C-H < C-F

51. Mercury is the only liquid metal and one of the heaviest elements with a density of 13.546 g/mL. What is the mass of 200 mL of mercury (almost the volume of a cup of coffee)?

(A) 857.8g
(B) 2709.2g
(C) 1709.6g
(D) 270.92g
(E) 5418.4g

52. 250 grams of a mixture of sugar water (dissolved sugar) and gravel (insoluble) were filtered by suction filtration. 75 grams of gravel was left in the Büchner funnel. What is the weight percentage of sugar and gravel in the mixture?

(A) 60% sugar , 40% gravel
(B) 55% sugar, 45% gravel
(C) 75% sugar, 25% gravel
(D) 80%sugar, 20% gravel
(E) 70% sugar , 30% gravel

53. Which statement about acids and bases is not true?

(A) A Lewis base transfers an electron pair to a Lewis acid
(B) All strong acids are completely ionized
(C) All Brønsted bases use OH– as a proton acceptor.
(D) All Arrhenius acids form H+ ions in water.
(E) All Brønsted acids and bases exist in conjugate pairs with and without a proton.

54. Calcium phosphate, $Ca_3(PO_4)_2$ is the main constituent of human bone. What is the weight percentage of phosphorous, P, in this compound?

(A) 30%
(B) 10%
(C) 20%
(D) 40%
(E) 25%

55. There are 30.1 x 10^{23} neon atoms (Ne) in a sealed neon lamp. How many moles of neon are present in the lamp?

(A) 15
(B) 25
(C) 5
(D) 12
(E) 8

56. In the following phase diagram, sublimation occurs as P is decreased from A to B at constant T and condensation occurs as T is decreased from D to C at constant P. Which of the following statements are TRUE?

(A) A is liquid and B is solid
(B) C is gas and D is solid
(C) A is solid and B is gas
(D) D is liquid and C is gas
(E) B is liquid and C is gas.

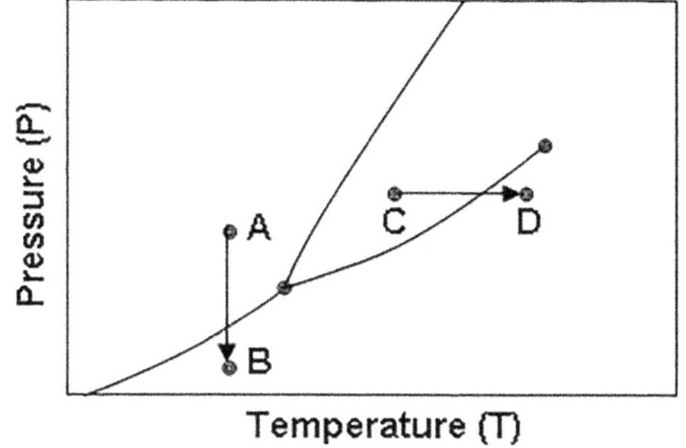

57. A solid which has a very high melting point, is very hard, and does not conduct electricity is probably :

(A) an ionic crystal
(B) a covalent network solid
(C) a metallic solid
(D) a polar solid
(E) a crystalline solid

58. A change in the disorder of a system is expressed as

(A) Enthalpy
(B) Entropy
(C) Kinetic Energy
(D) Specific Heat
(E) Heat of formations

59. Which of the following molecules are tetrahedral?

(A) H_2O
(B) CO_2
(C) CCl_4
(D) BeF_2
(E) none of the above

60. Which of the following sets of elements is in order of decreasing metallic character?

(A) Al > Ga > In > Ti
(B) In >Ga > Al >Ti
(C) Ti >Ga > In > Al
(D) Ti > In > Ga >Al
(E) None of the above

61. When comparing ethanol, acetone and ethylene glycol, which would have the lowest boiling point?

(A) Ethanol because it has an OH group
(B) Acetone because it has weak intermolecular forces
(C) Ethylene glycol because it has the most lone pairs
(D) All three have similar BP
(E) It cannot be determined without additional data.

62. For a titration of a strong acid and a weak base, a group of students need an indicator in the pH range 3 -4. Which indicator should they choose?

(A) Methyl Orange
(B) Bromthymol Blue
(C) Litmus
(D) Phenolphthalein
(E) any of the above

63. Chlorine has 7 valance electrons. An atom that would have similar chemical properties would be atomic number:

(A) 7
(B) 9
(C) 15
(D) 19
(E) 14

64. If a student needs to make a 1 L solution that is 6 M into a 3 M solution, how much water must be added ?

(A) 2 L
(B) 1 L
(C) 0.5 L
(D) 3 L
(E) 0.75 L

65. The cathode is the electrode:

(A) That is always +
(B) That is always -
(C) Where reduction takes place
(D) where ions are formed
(E) where the main reaction occurs.

66. Which of the following are true about an exothermic reaction?

(A) The reaction will not occur
(B) The reaction will occur and give off energy
(C) The reaction requires heat
(D) The reaction will require an outside current.
(E) The reverse reaction will occur

67. The reaction Zn + S ⟶ ZnS is an example of:

(A) a synthesis reaction
(B) a decomposition reaction
(C) a combustion reaction
(D) a redox reaction
(E) a single replacement reaction

68. The following statements are correct steps for drawing Lewis dot structure models except:

(A) Determine the number of valence shell electrons for each atom
(B) If the compound is an anion, the charge of the ion is added to the total electron count
(C) If the compound is a cation, subtract the charge of the ion.
(D) every atom has an octet of electrons except hydrogen and oxygen atoms with two electrons
(E) the remaining electrons are placed around the atoms as unshared pairs

69. Which of the following are in order of increasing boiling points?

(A) I, Br, Cl, F
(B) Cl,F, Br, I
(C) Br, I, Fl, Cl
(D) F, Cl, I, Br
(E) None of the above.

70. What is the partial pressure of O_2 in a container of H_20, O_2 and N_2 at 200 KPa and 100° C? The vapor pressures of H_2O and N_2 at 100° C are, respectively, 20 and 30 Kpa.

(A) 50 KPa
(B) 175 KPa
(C) 75KPa
(D) 125 Kpa
(E) 150 Kpa

71. 250 mL of a 0.5 M solution of HCl is neutralized completely with a solution of 0.025 M KOH. What is the volume of the KOH solution consumed in this titration reaction?

(A) 1,250 mL
(B) 2,500 mL
(C) 800 mL
(D) 5,000 mL
(E) 500 mL

72. In the oxidation-reduction reaction

$$8\ HNO_3 + Cu_2 \longrightarrow 2\ Cu\ (NO_3)_2 + 4\ NO_2 + 4\ H_2O$$

Which of the following statements is correct?

(A) N is reduced and its oxidation number increases
(B) N is reduced and its oxidation number decreases
(C) Cu is oxidized and its oxidation number decreases
(D) Cu is oxidized and its oxidation number increases
(E) Both B and D

73. The combustion reaction of pentane (airplane fuel, saturated hydrocarbon) produces carbon dioxide and water.

$$C_5H_{12}\ (l) + O_2\ (g) \longrightarrow CO_2\ (g) + H_2O\ (g)$$

Suppose this reaction is balanced with the lowest integer (whole-number) coefficient for each reactant and product. What is the summation of all of the coefficients in the reaction?

(A) 20
(B) 14
(C) 15
(D) 12
(E) 10

74. Which of the following expresses standard atmospheric pressure?

(A) 29.9 in Hg
(B) 76.0 cm Hg
(C) 760 torr
(D) 14.7 psi
(E) all of the above

75. A sample of ozone gas is at 0° C. If both the volume and pressure double, what is the new Kelvin temperature?

(A) 68 K
(B) 135 K
(C) 273 K
(D) 546 K
(E) 1092 K

76. A chemical reaction that absorbs heat from the surroundings is said to be ________ and has a _______ value of Δ H.

(A) endothermic, positive
(B) endothermic, negative
(C) exothermic, negative
(D) exothermic, positive
(E) none of the above

77. Which of the following formulas is an empirical formula?

(A) C_2O_2
(B) C_2H_6O
(C) $C_6H_{12}O_2$
(D) N_2O_4
(E) $Na_2Cl_2O_2$

78. The amount of heat required to raise 1 g of water by one degree Celsius is called:

(A) The heat of formation
(B) Specific heat of water
(C) 1 calorie
(D) delta-H
(E) delta-S

79. In an aqueous solution, the solubility of $Ce_2(SO_4)_3$ can be expressed as:

$$Ce_2(SO_4)_3 \longrightarrow 2Ce^{3+} + 3\,SO_4^{2-} + \text{Heat}$$

According to Le Châtelier's principle, the addition of heat will:

(A) allow $Ce_2(SO_4)_3$ to dissolve completely
(B) make the solute less soluble
(C) make the solute more soluble
(D) cause the ions to react with the H_2O in the solution
(E) none of the above

80. One mole of each of the following four salts is dissolved in a 2L beaker. The heat of solution of each of these salts is:

ΔH KCl = 17.2 KJ/mol
ΔH KBr = 19.9 KJ/mol
ΔH LiI = –59 KJ/mol
ΔH $LiNO_3$ = –1.3 KJ/mol

Which of the following statements are correct?

(A) The overall heat of solution is –2.32 KJ/mol and the beaker warms up.
(B) The overall heat of solution is –2.32 KJ/mol and the beaker cools down.
(C) The overall heat of solution is – 23.2 KJ/mol and the beaker cools down.
(D) There is no change in the temperature of the beaker.
(E) The overall heat of solution is – 23.2 KJ/mol and the beaker warms up.

81. Pentane, butane, hexanol, and hexane are all liquids at room temperature. Rank them in order of decreasing viscosity.

(A) n-pentane > n-butane > 1-hexanol > n-hexane
(B) 1-hexanol > n-pentane > n-butane > > n-hexane
(C) 1-hexanol > n- hexane > n-pentane > n-butane
(D) n-hexane > 1-hexanol > n-pentane > n-butane
(E) n-butane > 1-hexanol > n-pentane > n-hexane

82. What is the molar mass of aspirin, $C_9H_8O_4$?

(A) 116.0 g/mol
(B) 180.0 g/mol
(C) 188.0 g/mol
(D) 244.0 g/mol
(E) 132.0 g/mol

83. What term refers to an insoluble substance produced by a reaction in aqueous solution?

(A) colloid
(B) precipitate
(C) product
(D) salt
(E) none of the above

84. Which energy sublevel immediately follows the 4s sublevel according to increasing energy?

(A) 3p
(B) 3d
(C) 4p
(D) 4d
(E) 5s

85. Which of the following best defines the term isotopes?

(A) same number of protons -- different number of electrons
(B) same number of protons -- different number of neutrons
(C) same number of neutrons -- different number of protons
(D) same number of neutrons -- different number of electrons
(E) same number of electrons -- different number of protons

SAT Chemistry Practice Test 1

Answer Key

Question Number	Correct Answer	Your Answer
1	D	
2	A	
3	B	
4	C	
5	B	
6	C	
7	D	
8	A	
9	T, F	
10	T, T	
11	T, T	
12	T, T	
13	C	
14	B	
15	C	
16	C	
17	B	
18	D	
19	C	
20	E	
21	C	
22	D	
23	A	
24	B	
25	E	
26	A	
27	A	
28	B	
29	E	
30	B	

Question Number	Correct Answer	Your Answer
31	C	
32	A	
33	A	
34	D	
35	D	
36	E	
37	D	
38	B	
39	D	
40	B	
41	D	
42	A	
43	C	
44	D	
45	C	
46	D	
47	D	
48	E	
49	C	
50	D	
51	B	
52	E	
53	C	
54	C	
55	C	
56	C	
57	B	
58	B	
59	C	
60	D	

Question Number	Correct Answer	Your Answer
61	B	
62	A	
63	B	
64	B	
65	C	
66	B	
67	A	
68	D	
69	A	
70	E	
71	D	
72	E	
73	A	
74	E	
75	E	
76	A	
77	B	
78	B	
79	B	
80	E	
81	C	
82	B	
83	B	
84	B	
85	B	

SAT Chemistry Practice Test 1

Explanations

Part I: Classification Questions

Questions 1-4 refer to the following:

(A) Temperature
(B) Energy
(C) Entropy
(D) Gibbs free energy

1. **Is negative for a spontaneous reaction**
2. **Is the average kinetic energy of the system**
3. **Is always conserved**
4. **Is a measure of the disorder of a system**

1. **The correct answer is D.**
   Gibbs Free energy or delta-G is always negative for a spontaneous reaction

2. **The correct answer is A.**
   Temperature is the average kinetic energy of the system

3. **The correct answer is B.**
   Energy is always conserved. It can change form, from potential energy to heat energy, for example, but within a system all forms of energy are conserved.

4. **The correct answer is C.**
   Entropy is the measure of the disorder of a system.

Questions 5-8 refer to the following

(E) London Forces
(F) H-bonds
(G) Covalent Bonds
(H) Ionic Bonds and Covalent Bonds

5. **These bonds are important in the structure of DNA and proteins.**
6. **The bond type found in H_2**
7. **The bond type found in $Ca_3(PO_3)_2$**
8. **Increase with the size of the molecule for long chain alkanes**

Practice Test One

5. **The correct answer is B.**
   These bonds are important in the structure of DNA and proteins.

6. **The correct answer is C.**
   The bond type found in H2

7. **The correct answer is D.**
   Both ionic and covalent bonds are found in $Ca_3 (PO_3)_2$. (Ionic bonds between the calcium and phosphate ions and covalent bonds between phosphorous and oxygen atoms.)

8. **The correct answer is A.**
   Increase with the size of the molecule for long chain alkanes

Part II: Relationship Questions

Determine if each statement is TRUE or FALSE.

If BOTH statements are true – decide if the first statement is TRUE BECAUSE of the second statement.

9. In the reaction of hydrogen with oxygen

$$2\,H_2 + O_2 \longrightarrow 2\,H_2O$$

22.4 L of oxygen will react exactly with 44.8 L of hydrogen to produce water

BECAUSE

At STP (standard temperature and pressure) the molar volume of any gas, the volume that one mole of any gas occupies, is 28.4 L.

10. In the Combustion reaction of propane:

$$C_3H_8 + 5\,O_2 \longrightarrow 3CO_2 + 4H_2O$$

The limiting reagent of the reaction of one mole of propane and 4 moles of oxygen to produce carbon dioxide and water is oxygen.

BECAUSE

The limiting reagent of a reaction is the reactant that runs out first. This reactant determines the amount of products formed, and any other reactants remain unconverted to product and are called excess reagents.

11. In ethylene and other alkene hydrocarbons, one s orbital and two p orbitals of carbon atoms combine together to form three sp^2 hybrid orbitals?

Each carbon in ethylene has four bonds including three sigma bonds, two between a carbon and two hydrogens and one between carbon and another carbon and one pi bond between the two carbons

12. A balloon filled with 10 liters of hydrogen gas at room temperature at the seaside was sent into the sky to a height where the pressure is 0.25 atm. Knowing that the maximum volume capacity of this balloon before exploding is 40 liters, we know the balloon was blown out at this altitude.

Since pressure of a gas is inversely proportional to its volume, as the balloon moves further away from the ground level into the sky, the air pressure decreases, and the volume of that gas will increase.

Explanations for Questions 9-12

9. True, False

The stoichiometry of this reaction indicates that two moles (2 x 22.4 L= 44.8 L) of hydrogen reacts with one mole (1 x 22.4 = 22.4 L) of oxygen to produce 2 moles of water. Therefore the statement I is correct

According to the gas law one mole of an ideal gas at STP occupies 22.4 L. Thus statement II is false because it states that the molar volume of gases is 28.2 L.

The stoichiometry of this reaction indicates that two moles (2 x 22.4 L = 44.8 L) of hydrogen reacts with one mole (1 x 22.4 = 22.4 L) of oxygen to produce 2 moles of water. Therefore the statement I is correct and the correct answer to this question is true, false.

10. True, True

The limiting reagent may be determined by dividing the number of moles of each reactant by its stoichiometric coefficient. This determines the moles of reactant if each reactant were limiting. The lowest result will indicate the actual limiting reagent. Here we have one mole of propane which is divided by the stoichiometric coefficient of propane, one (1). The number of oxygen molecules is four which is divided by the stoichiometric coefficient of oxygen, 5, which equals 4/5 = 0.8. Since 0.8 is smaller than 1, we conclude that oxygen is the limiting reagent. Therefore, the correct answer is true, true, correct explanation.

11. True, True

In alkene hydrocarbons, the carbon has a double bond. One s orbital and two p orbitals of the carbon atom combine to create three sp2 hybrids of trigonal planar orientation with one electron in each orbital. So statement I is correct. Carbon combines with two hydrogens and another carbon atom to form three sigma bonds by these three sp2 orbitals. The fourth bond of carbon will be a pi bond between the two carbons of ethylene molecule. So statement II is also correct. Therefore the correct answer is true, true, correct explanation.

12. True, True

At constant temperature according to the gas law P1 x V1 = P2 x V2

Since the pressure at the ground is 1 atmosphere and the volume of hydrogen is 10 liters as the balloon climbs up in the air the pressure of atmosphere decreases while the volume of the balloon increases because the inside and outside pressures have to be balanced by automatic equilibration. Thus when the volume of the balloon reaches 40 liters (equivalent to .25 atmospheres) it blows out. Therefore the correct answer is true, true, correct explanation.

Part II. 5-Choice Completion Questions

13. Question 1- What is the electron configuration of phosphorus ion, P^{5+}?

(A) $1s^2\ 2s^2\ 2p^6\ 3s^2\ 3p^6$
(B) $1s^2\ 2s^2\ 2p^6\ 3s^2\ 3p^3$
(C) $1s^2\ 2s^2\ 2p^6$
(D) $1s^2\ 2s^2\ 2p^6\ 3s^1$
(E) $1s^2\ 2s^2\ 2p^6\ 3s^2\ 3p^4$

The correct answer is C.

The number of electrons of phosphorus as a neutral atom is 15 electrons. Hence, its electronic configuration is: $1s^2\ 2s^2\ 2p^6\ 3s^2\ 3p^3$.

In the cation P^{5+}, five electrons are lost from the valance shell; three of these electrons are from the 3p orbitals and two are from the 3s orbitals. This will give an electronic configuration for P^{5+} as: $1s^2\ 2s^2\ 2p^6$, which fits with answer C.

14. Question 2- Which of the following configurations represents the Cobalt (Co) atom?

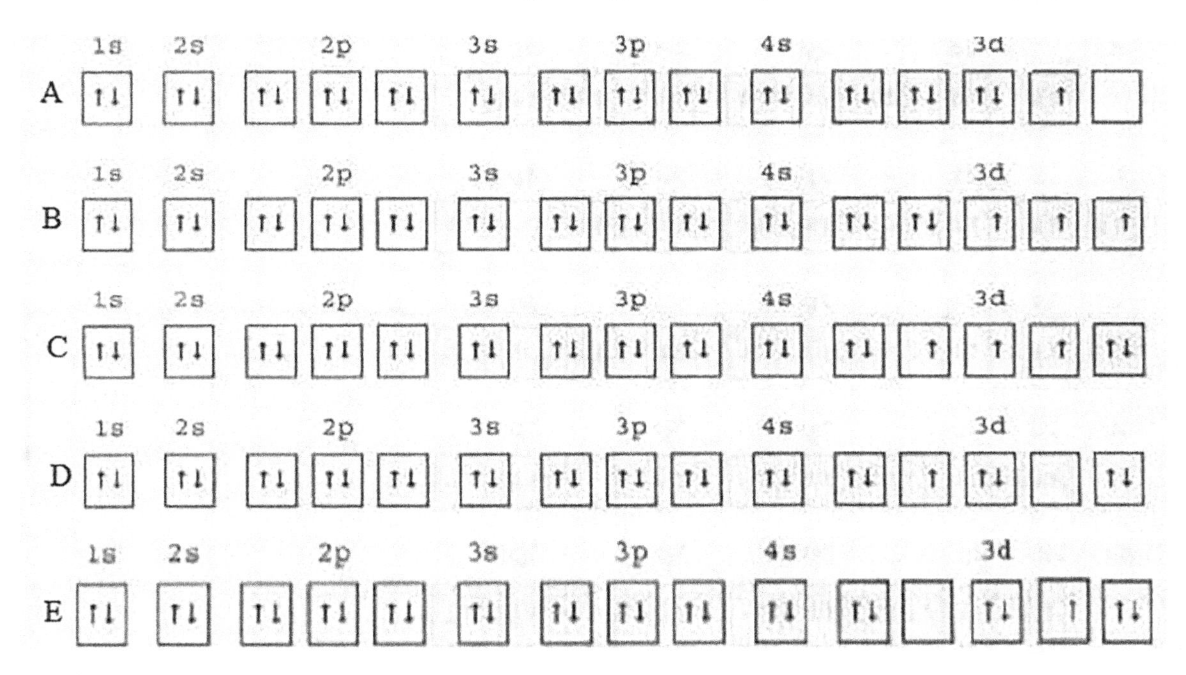

The correct answer is B.

From the periodic table, Co has 27 electrons. All of the proposed answers show 27 electrons, but only answer B does not violate any of the filling rules. In other words, in the 3d orbital, the 7 electrons must first fill in each orbital with 1 electron in each before completing with a second reverse spin electron.

15. Which of the following Lewis dot structures for CH_3COOH is correct?

```
A     H :O:
      ..  ::  ..
    H:C::C:O:H
      ..    ..
      H

B     H :O:
      ..  ::
    H:C::C:O:H
      ..
      H

C     H :O:
      ..  ::  ..
    H:C :C:O:H
      ..      ..
      H

D     H :O:
      ..  ..
    H:C::C::O:H
      ..    ..
      H

E     H :O:
      ..  ..
    H:C::C:O:H
      ..
      H
```

The correct answer is C.

In the Lewis structure of a molecule, each atom in the molecule must be surrounded by an octet of electrons (except hydrogen, which is surrounded by two electrons).

The only structure which has the correct octet electron dot configuration is C. The others are not correct and have more than 8 electrons around one or two atoms.

16. 100 μL of SO_2 contains 2.2×10^{17} molecules. According to Avogadro's hypothesis, how many molecules are in 200 μL of methane gas, CH_4, at STP?

(A) 2.2×10^{17}
(B) $.44 \times 10^{17}$
(C) 4.4×10^{17}
(D) 0.22×10^{17}
(E) 22×10^{17}

The correct answer is C.

Avogadro's hypothesis states that equal volumes of different gases at the same temperature and pressure contain equal numbers of molecules. Therefore when 100 μL of SO_2 contains 2.2×10^{17} molecules, 200 μL of methane contains double the amount of molecules, which is 4.4×10^{17}, answer C.

17. Which one of the following atoms is the most chemically reactive atom?

(A) Ga
(B) Na
(C) Al
(D) Kr
(E) Mg

The correct answer is B.

Sodium is the most electropositive of these atoms and is the most chemically reactive. Magnesium is just to the right of sodium in the periodic table. The chemical reactivity in a period is reduced from left to right and hence magnesium is less chemically reactive than sodium. Gallium and aluminum are in group thirteen of the elements (very stable) and the most stable of all is Kr, an inert noble gas. The inert gases are the least reactive elements because they are saturated by electrons and have neither electron affinity nor intention to lose electron.

18. What is the shape of an NH_3 molecule? Use the VSEPR model.

(A) Trigonal planar
(B) Tetrahedral
(C) Trigonal bipyramidal
(D) Trigonal pyramidal
(E) Octahedral

The correct answer is D.

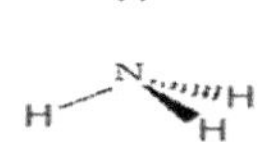

The Lewis structure for NH3 is given to the right. This structure contains 4 electron pairs around the central atom, so the geometric arrangement is tetrahedral. However, the shape of a molecule is given by its atom locations, and there are only three atoms so choices B and E are not correct. Four electrons pairs in single bonds with one unshared pair (3 single bonds and one lone pair) give a trigonal pyramidal shape. A tetrahedron is also a trigonal pyramid in shape, but in chemistry, it is reserved for the sort of tetrahedron with all sides equal that results when a central atom is bonded to four others, creating a tetrahedron with the central atom in the middle.

19. Which of the following compounds is the most soluble in water?

(A) CO_2
(B) CH_2ClCH_3
(C) CH_2CH_2OH
(D) SiO_4
(E) CH_3CH_2 $CH_2CH_2CH_2CH_3$

The correct answer is C.

Carbon dioxide (CO_2) is a linear and nonpolar gas which is not soluble in water at atmospheric pressure. n- Hexane is an alkane and is not soluble in water. SiO_4 is also a nonpolar and will not dissolve in water. CH_2ClCH_3 has one chlorine atom in the molecule and makes that part of the molecule partially negative and therefore somewhat polar, but not enough to make it very soluble in water. However, ethanol has a strong hydrogen bond with water, and is therefore quite soluble in water. It is the most water-soluble of all the choices.

20. What is the number of neutrons in lead (Pb)?

(A) 132
(B) 88
(C) 123
(D) 130
(E) 125

The correct answer is E.

The number of neutrons of an atom is the subtraction of atomic number (number of protons) from mass number. The mass number is 207 and atomic number is 82. E is the correct answer.

21. How many mL of water is needed to add to 150 mL of 6M HNO_3 to prepare a solution of 0.6M HNO_3?

(A) 350
(B) 2050
(C) 1350
(D) 1200
(E) 120

The correct answer is C.

The equation of volume-concentration relation of solutions at constant temperature is M1 x V1 = M2 x V2

6 x 150 = 0.6 x V2 $\longrightarrow$ V2 = 6 x 150/0.6 =1500 mL

But this is the volume of the final solution after adding water to the original solution. So for answering the question of this problem we have to subtract 150mL from the final volume.

The amount of water needed to be added to the concentrated HNO3 is 1500 – 150 = 1350 mL

22. Consider the unbalanced chemical equation: __CO (g) + __O_2 (g) $\longrightarrow$ __CO_2 (g).

When this equation is completely balanced using the smallest whole numbers, what is the sum of the coefficients?

(A) 4
(B) 3
(C) 6
(D) 5
(E) 7

The correct answer is D.

The balanced chemical equation is 2CO (g) + O_2 (g) $\longrightarrow$ 2CO_2 (g). Coefficients are the numbers in front of the element symbols. Add 2 + 1 + 2 to get 5.

23. Which of the following solutions is the most concentrated?

(A) 2 moles of potassium hydroxide is dissolved in 10 liters of water
(B) 4 moles of potassium hydroxide is dissolved in 100 liters of water
(C) 2 moles of potassium hydroxide is dissolved in 20 liters of water
(D) 6 moles of potassium hydroxide is dissolved in 50 liters of water
(E) 2 moles of potassium hydroxide is dissolved in 40 liters of water

The correct answer is A.

Molarity (moles of solute/liter of solution) for each proposition is (A) 2 mole / 10 liter or 0.2 M, (B) 4 moles / 100 liters or 0.04 M, (C) 2 moles / 20 liters or 0.1 M, (D) 6 moles / 50 liters or 0.12 M, (E) 2 moles / 40 liters or 0.05 M. 0.2 M is the most concentrated solution.

24. At STP, how many molecules of CO_2 are present in 224 liters of CO_2?

(A) 12.04×10^{23}
(B) 60.2×10^{23}
(C) 0.602×10^{23}
(D) 1.204×10^{23}
(E) 6.02×10^{23}

The correct answer is B.

Since at STP one mole of any gas (containing Avogadro's number of molecules of that gas) has the volume of 22.4 liters. Here we have 224 liters of CO_2 which is 224/22.4 = 10 moles. Thus:

The number of CO_2 molecules in 224 liters of that gas = $10 \times (6.02 \times 10^{23}) = 60.2 \times 10^{23}$

25. A gas in a locked piston is heated up. Which of the following statements about gas properties is correct in this regard?

(A) gas pressure increases
(B) average speed of molecules increases
(C) kinetic energy increases
(D) molecular collisions with container walls per second decreases
(E) A, B, and C

The correct answer is E.

Since the piston is locked the volume is constant and according to the gas law equation PV= nRT, pressure is proportional to temperature. Therefore pressure increases by heating up the gas. Also the velocity of molecules is related to temperature and so the average speed of molecules increases. The kinetic energy is also increased by temperature. Since the pressure is increased the molecular collisions with container walls per second increases, the opposite of answer D.

26. How many grams of H_2SO_4 are needed to prepare one liter of a 0.02 M solution of this acid?

(A) 1.96g
(B) 3.92g
(C) 19.6g
(D) 098g
(E) 9.8g

The correct answer is A.

The molecular weight of H_2SO_4 is 98 g/mol. In order to prepare a 0.02 M solution we have to first calculate the number of grams in .02 moles of H_2SO_4:

0.02 mol x 98 g/mol = 1.96 g. So, a solution of 1.96 g in one liter of the solution corresponds to Answer A.

27. Which of the following compounds have the highest percentage of hydrogen (ratio of the mass of hydrogen to the mass of the compound)?

(A) CH_4
(B) H_2O_2
(C) C_2H_6
(D) H_2O
(E) NH_3

The correct answer is A.

The weight percentage of hydrogen in each compound is calculated as follows:

CH_4 (mass of H /mass of CH_4) x 100 = 4/16 x 100 = 25%
H_2O_2 (mass of H /mass of H_2O_2) x 100 = 2/32 x 100 = 6%
C_2H_6 (mass of H /mass of C_2H_6) x 100 = 6/30 x 100 = 20%
H_2O (mass of H /mass of H_2O) x 100 = 2/18 x 100 = 11%
NH_3 (mass of H /mass of NH_3) x 100 = 3/17 x 100 = 17%

Therefore CH_4 has the highest percentage of hydrogen in the molecule

28. Which of the following reactions are oxidation-reduction reactions?

I. $H_2SO_4 + 2KOH \rightarrow K_2SO_4 + 2H_2O$
II. $Ag\,NO_3 + NaCl \rightarrow AgCl + NaNO_3$
III. $S + O_2 \rightarrow SO_2$
IV. $2NaOH + Pb(NO_3)_2 \rightarrow 2NaNO_3 + Pb(OH)_2$

(A) reaction II
(B) reaction III
(C) reaction IV
(D) both I and IV
(E) both II and III

The correct answer is B.

In oxidation-reduction reactions the oxidation states of oxidized and reduced atoms are changed. The oxidation states of the oxidized atoms are increased and the oxidation states of the reduced elements decrease. Only in the B choice do the oxidation states of S and O change. The oxidation state of S is changed from zero to plus 4 (oxidized) and the oxygen changes from zero to minus two (reduced). The other reactions are not redox reactions. They are substitution reactions.

29. The solubility product constants (K_{sp}) of the following five hydroxides are as follows:

I. $Ca(OH)_2$ is 5.5×10^{-6}
II. $Mn(OH)_2$ is 1.2×10^{-11}
III. $Zn(OH)_2$ is 4.5×10^{-17}
IV. $Al(OH)_3$ is 2×10^{-33}
V. $Ni(OH)_2$ is 1.6×10^{-14}

Which of the following statements are correct from this data?

(A) $Zn(OH)_2$ is the most soluble of these compounds
(B) $Al(OH)_3$ is the least soluble of these compounds
(C) $Ca(OH)_2$ is the most soluble of these compounds
(D) $Al(OH)_3$ has almost zero solubility in water and is a precipitate
(E) B, C, and D

The correct answer is E.

The solubility product of a compound is proportional to the solubility power of that compound in water. The smaller this coefficient is, the less soluble the compound is, and the more likely it is to precipitate in the solvent liquid. $Al(OH)_3$ has almost zero solubility in water and is considered a precipitate. So it is the least soluble compound of these five. $Ca(OH)_2$ has the largest solubility product and therefore is the most soluble of these five compounds.

30. Methane gas is combusted to produce CO_2 and water. How many liters of methane must be combusted with 0.2 moles of O_2?

$$CH_4 + 2O_2 \longrightarrow CO_2 + 2H_2O$$

(A) 4.48 L
(B) 2.24 L
(C) 22.4 L
(D) 44.8 L
(E) 0.448 L

The correct answer is B

According to the ideal gas law, the standard molar volume of any gas at STP is 22.4 L, i.e. one mole of an ideal gas at STP occupies 22.4 L. In this reaction, one mole of methane (equivalent to 22.4 L) reacts with 2 moles of oxygen. Thus for 0.2 moles of oxygen, 0.1 moles of methane will react. Since 1 mole of methane takes up 22.4 L, one tenth of one mole of methane will take up 2.24 L.

=> 2.24 L of methane are needed

Practice Test One

31. What is the pH and pOH of a 0.00001M solution of HCl?

(A) pH= 4, pOH=10
(B) pH =3, pOH= 9
(C) pH=5, pOH =9
(D) pH= 5, pOH=13
(E) pH =1, pOH=11

The correct answer is C

The concentration of HCl is 0.00001 = 1 /100,000 = 1×10^{-5}

HCL is a strong acid and dissociates completely or 100% to H+ and Cl–

$[H+]= 1 \times 10^{-5}$ pH = – log10 [H+] = – log10 [1 x 10–5] = (–1)(–5) log10 [1 x 10]= 5

pH + pOH = 14 pOH = 14 – pH = 14–5 = 9 so the correct answer is C

32. Consider the following electrochemical overall reaction.

$$Ni + 2MnO_2 + 2NH_4^+(aq) \longrightarrow Ni_2^+ + 2\ Mn\ (OH)_2 + 2\ NH_3$$

What phenomenon occurs during this reaction?

(A) Ni is oxidized and its oxidation number increases
(B) MnO_2 is oxidized and its oxidation number decreases
(C) Ni is reduced and its oxidation number decreases
(D) MnO_2 is reduced and its oxidation number increases
(E) NH_4^+ is oxidized and its oxidation number increases

The correct answer is A

First examination of the overall reaction requires writing down the two half electrochemical reactions.

At the anode: $Ni(s) \longrightarrow Ni_2^+(aq) + 2e^-$

At the cathode: $MnO_2(s) + NH_4^+(aq) + 2e^- \longrightarrow Ni\ (OH)_2(s) + NH_3(aq)$

By definition, oxidation occurs when electrons are lost and the ion becomes more positive with higher oxidation state. This fits with the proposition A

33. Butane, the volatile liquid fuel in disposable cigarette lighters, has the formula C_4H_{10}. How many atoms of carbon are present in 2.24 liters of butane in the gas form? (Avogadro's number is 6.02×10^{23})

(A) 2.4×10^{23}
(B) 6.02×10^{23}
(C) 0.602×10^{23}
(D) 4.8×10^{22}
(E) 0.48×10^{23}

The correct answer is A

Explanation: Each molecule of C_4H_{10} has four atoms of carbon and each mole of any gas at STP has the volume of 22.4 L. Therefore, 2.24 liters of butane gas includes 0.1 moles of butane molecules.

Since there are 4 atoms of carbon for every mole of butane, the number of carbon atoms in 2.24 liters of butane is:

4×0.1 mole x (6.02×10^{23}) atoms/mole = 2.4×10^{23}

34. Consider the following four reactions:

I. $S + H_2 \longrightarrow SH_2$
II. $S + O_2 \longrightarrow SO_2$
III. $S + Na \longrightarrow SNa_2$
IV. $S + F_2 \longrightarrow SF_2$

If 3.2 grams of sulfur in four separate experiments are reacted with, respectively, 0.1 g hydrogen, 6.4 g oxygen, 0.23 g sodium, and 0.38 g fluorine, in which reaction is sulfur the limiting reagent?

(A) I
(B) III
(C) IV
(D) II
(E) both III and IV

The correct answer is D

3.2 grams S is 0.1 mol. This is the amount that is reacting in the four different reactions. The limiting reagent between two substrates in a reaction is the reagent with the smaller number of moles. The number of moles of the four substrates in these reactions is: 0.05 mol H_2, 0.2 mol O_2, 0.01 mol Na, and 0.01 mol F_2. In these four reactions the only reaction that the no. of sulfur moles is smaller than the other substrate is the reaction II with O_2. Hence it is limiting reagent in that reaction.

35. Write the equilibrium expression Keq for the reaction

$$2NO_{2(g)} + O_{2(g)} \rightleftarrows 2NO_{2(g)}$$

(A) $4[NO_2]^2 / [NO]^2[O_2]$
(B) $4[NO_2]^2 / 4 [NO]^2[O_2]$
(C) $2[NO_2]^2 / 2 [NO]^2[O_2]$
(D) $[NO_2]^2 / [NO]^2[O_2]$
(E) $[NO_2]^2 / 4 [NO]^2[O_2$

The correct answer is D

In an equilibrium expression, product concentrations to the power of their stoichiometric coefficient are multiplied together in the numerator and reactant concentrations to the power of their stoichiometric coefficient are multiplied together in the denominator, eliminating choices A, B, C, and E. because in all these four choices we have extra figures of 4 and 2. D is correct.

36. Which statements about reaction rates are not true?

(A) Most reaction mechanisms are multi-step processes involving reaction intermediates
(B) Intermediates are chemicals that are formed during one elementary step and consumed during another
(C) If one elementary reaction is the slowest reaction, it determines the overall reaction rate.
(D) This slowest reaction in the series is called the rate-limiting step or rate determining step.
(E) Many reactions have two rate-determining steps

The correct answer is E

Explanation: Choice A is correct and fits with the mechanism of reactions. Choice B is also correct and is definition of intermediate as part of the mechanism. Choices C and D are correct and define the rate-determining step. Choice E is wrong. No reaction can have two rate-determining steps. So that is the correct answer.

37. 5 kilograms of crushed ice are added to a container of freezing water at 0° C. The ice water is then heated while a thermometer is placed in the container. The heat is applied until the water in the container boils for 10 minutes. Then, we start to cool the container down very quickly, to the freezing point of the water, and we stop recording the temperature. Which of the following types of graphs is obtained by plotting temperature versus time for this experiment?

(A)
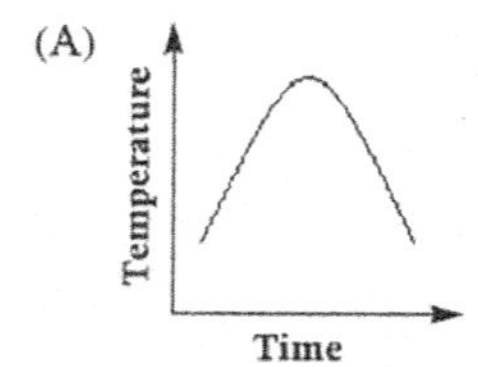

(B)
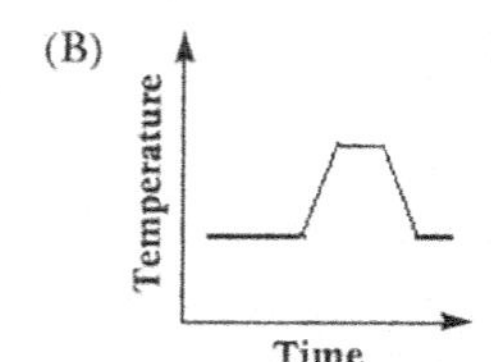

(C)
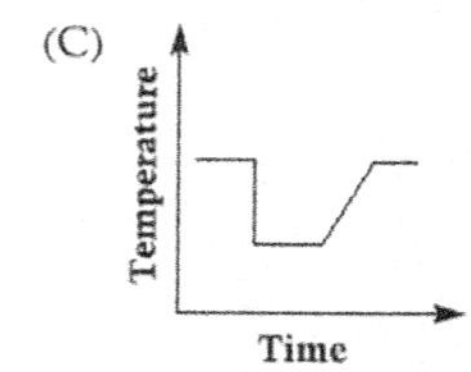

(D)
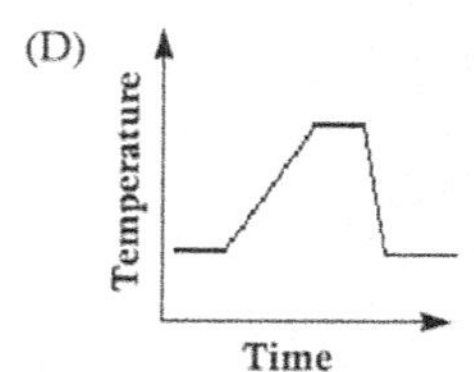

(E)
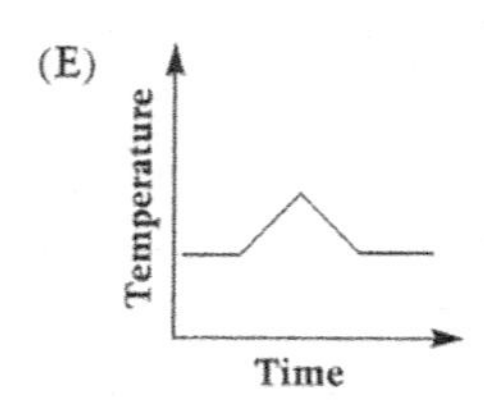

The correct answer is D

The first step of heating is heat of fusion to melt the crystal solid structure of ice (0°C) to liquid water of 0°C. At this period the graph is horizontal linear. Then with continuous heating, the temperature gradually rises up to the boiling point and the graph turns horizontal again, this time due to heat energy being used for vaporization. The temperature stays constant because the energy transferred to the water is consumed to evaporate water to steam. Now by cooling again very rapidly, the boiling water cools, and the temperature graph sharply descends to the freezing point again. While the water continues to lose heat, the graph is linear horizontal because water is turning to crystal ice solid.

38. When dissolving $AlCl_3$ in water, the water becomes substantially warmer. Now if we gradually heat up the container, what happens to the solubility of $AlCl_3$?

(A) It will increase
(B) It will decrease
(C) There will be no change in temperature
(D) Cannot be determined with the information given
(E) It will increase and the solution will start to boil

The correct answer is B

Explanation: By dissolving $AlCl_3$ the following equilibrium will take place:

$AlCl_3$ (s) $\longrightarrow$ Al^{3+}(l) + $3Cl^-$ (l)

The rise in water temperature indicates that the net solution process is exothermic (releases heat). A temperature increase supplying more heat will reverse the solubility reaction and the reaction will shift to the left according to Le Châtelier's principle. The solubility of $AlCl_3$ is reduced by increasing the temperature and less solid will dissolve. If we want more aluminum chloride to be dissolved we have to cool down the reaction container.

39. Which is the correct formula for 3-methyl-4-methyl- 2-pentene?

(A)

$CH_3 - C = C - CH_2 - CH_3$

$\quad\quad | \quad\;\; |$

$\quad\; CH_3 \; CH_3$

(B)

$\quad\quad\;\; H$

$\quad\quad\;\; |$

$CH_3 - C - C = C - CH_3$

$\quad\quad\;\; | \quad\; | \quad\;\; |$

$\quad\;\; CH_3 \; CH_3 \; H$

(C)

$\quad\quad\quad\quad H$

$\quad\quad\quad\quad |$

$CH_3 - C - C = C - CH_3$

$\quad\quad\;\; | \quad\quad\quad |$

$\quad\;\; CH_3 \quad\;\; CH_3$

(D)

$\quad\quad\quad\quad\quad\quad H$

$\quad\quad\quad\quad\quad\quad |$

$CH_3 - C = C - C - CH_3$

$\quad\quad\quad\quad\; | \quad\; |$

$\quad\quad\quad\; CH_3 \; CH_3$

(E)

$CH_2 = C - C - CH_2 - CH_3$

$\quad\quad\;\; | \quad\;\; |$

$\quad\;\; CH_3 \; CH_3$

The correct answer is D

To write the formula based on IUPAC nomenclature, write a five carbon chain

Draw a double bond on carbon-2 (between 2 and 3)

Draw a methyl on each of carbons 3 and 4

40. Which of the following statements regarding a chemical storage plan for laboratories is incorrect?

(A) All chemical containers should be properly labeled, dated upon receipt, and dated upon opening

(B) Although it is preferable to separate corrosives from flammables, in some conditions they can be stored together

(C) Shelves should be secure and strong enough to hold chemicals being stored on them. Shelves should not be overloaded.

(D) Personnel should be aware of the hazards associated with all hazardous materials

(E) The important chemical containers should be properly labeled, dated upon receipt, and dated upon opening

The correct answer is B.

Corrosives should be stored separately from flammables

41. Which of the following statements is incorrect?

(A) A catalyst is a material that increases the rate of a chemical reaction without changing itself permanently in the process.
(B) The weight of the catalyst remains unchanged in the reaction it catalyzes.
(C) Catalysts reduce the activation energy of a reaction
(D) A catalyst reduces the heat of reaction.
(E) A catalyst increases the rate of both the forward and reverse reactions by lowering the activation energy for the reaction.

The correct answer is D

Choice A is the definition of a catalyst with its main properties and it is correct. Choices B and C are two other important properties of catalysts and are true. Choice D is incorrect. Choice E is also correct.

42. Based on trends in the periodic table, which of the following properties would you expect to be greater for Ba than for Be?

I. Atomic radius
II. Atomic number
III. Ionization energy
IV. Electronegativity
V. Nuclear charge

(A) I, II, and V
(B) Only I
(C) II and V
(D) III, IV
(E) III, V

The correct answer is A

Ba is underneath Be in the alkali earth metal column (group 2) of the periodic table. There is a general trend for atomic radius to increase lower on the table for elements in the same group, so we select choice I. Atomic number (number of protons, number of electrons) naturally is greater as we go lower in the table and is greater in Ba than Be. So we select Choice 2 too. Ionization energy decreases for larger atoms further down the periodic table, so we do not choose III. Also Electronegativity decreases for larger atoms further down the periodic table, so we do not choose IV. Nuclear charge increases lower in the table. So we select choice V too. Therefore the correct answer is A.

43. Consider two elements, designated L and M, in the Periodic Table. Electron affinity and ionization energy is greater for L than for M, and atomic radius and metallic character are greater for M than L. Locate the approximate location of L and M in the table (in a period or in a column) with respect to each other.

(A) M is to the right of L in a period
(B) M is higher than L in a column
(C) L is to the right of M in a period
(D) Both A and B
(E) Both B and C

The correct answer is C

All of the properties could be due to L being to the right of M in the same row. M could also be below L in the periodic table, but the increased atomic radius and metallic character could also be due to M being to the left of L in the same row.

44. Which method would be most useful for separating a specific protein from a mixture of proteins?

(A) Extraction
(B) Gas chromatography
(C) Distillation
(D) Liquid chromatography
(E) Crystallization

The correct answer is D

Extraction is not a good method for separation of proteins because the solubility of proteins are very similar. Neither distillation nor crystallization are good methods to separate proteins, which are solid and large molecules. Gas chromatography is used for small molecules in the gas phase. Proteins are too large to exist in the gas phase. Liquid chromatography (Choice D) is used to separate large molecules.

45. Which separation method is best for separation of an aqueous solution containing methanol and ethanol?

(A) Extraction
(B) Chromatography
(C) Distillation
(D) Vaporization
(E) Both A and C

The correct answer is C

The best method for separation of these two compounds is distillation. Extraction is not a good method because the solubility of methanol and ethanol in organic solvents are close. Chromatography is not a good method because it is not usually used for separation of liquids and also for the high amount materials. Vaporization also is not a good method because it is a long process.

46. A 2.0 g coin of silver alloy was dissolved in nitric acid to give a solution of $AgNO_3$. This solution was in turn reacted with enough sodium chloride solution to give 1.435 g of white precipitate, AgCl. What is the percentage of silver in the coin?

$$AgNO_{3\,(l)} + NaCl_{(l)} \longrightarrow AgCl_{(s)} + NaNO_{3\,(l)}$$

(A) 68%
(B) 10.8%
(C) 5.4%
(D) 54%
(E) 35%

The correct answer is D

All of the silver in the coin is converted to silver chloride. Thus the stoichiometric equation of the reaction sequence is:

$Ag \longrightarrow AgNO_3 \longrightarrow AgCl$

Therefore one mole of Ag yields one mole of AgCl. The amount of Ag in the coin can be calculated:

1mole Ag = 108 g

1mole AgCl = 143.5g

108 g Ag = 1.435g AgCl/143.5 g/mol AgCl = 0.1 x 108 g Ag = 1.08g Ag in the coin

So the purity of silver in the coin is 1.08g Ag/2.0g coin x 100 = 54%

47. Which of the following group elements are more electronegative?

(A) Ar, Kr, Ne
(B) Ba, Sr, Ca
(C) As, P, N
(D) Se, S, O
(E) Ga, Al, B

The correct answer is D

In the Periodic Table the electronegativity of elements increases through a period from left to right. The A set of elements are all inert gases or rare gases with no electronegativity. B set of elements are metals of group ii of the periodic table with the lowest electronegativity in these sets. E set of elements are all in group iii of the periodic table with low electronegativity. The C set of elements are all in group v of the periodic table with higher electronegativity; but the D set of elements are all nonmetals in group vi of the table with highest electronegativity.

48. In each set of the following elements the size of atom increases from left to right except in?

(A) Li, Na, K
(B) As, Ge, Ga
(C) Br, Se, As
(D) F, Cl, Br
(E) Al, Si, P

The correct answer is E

The size of elements in the periodic table is increased from left to right in each period and from top to bottom in each column. In A and D the set of elements are located in columns so their sizes are increased from left to right. In B and C the set of elements are located in the periods with the order of right to left and since the size of atom is decreased in the periods from left to right; so in B and C the sizes of atoms increase too. But in E the size of atom must decrease in the set of Al, Si, P, while it is increased. Therefore it is the right answer.

49. Which of the following substance (s) are conductors of electricity?

(A) P
(B) He
(C) Ni
(D) Cl
(E) Br

The correct answer is C

Metals are the best conductors of electricity. Hence, Ni is a conductor of electricity. By contrast, He is an inert gas with no free electron to conduct electricity. Phosphorus is an amorphous nonmetal solid and is a nonconductor. Bromine is found as the molecular Bromine (Br_2) as liquid with no electrical conductivity. Chlorine is also a gas that does not conduct electricity.

50. Rank the following bonds from least to most polar:

C-H, C-Cl, H-H, C-F

(A) C-H < H-H < C-F < C-Cl
(B) H-H < C-H < C-F < C-Cl
(C) C-F < C-Cl < C-H < H-H
(D) H-H < C-H < C-Cl < C-F
(E) H-H < C-CL < C-H < C-F

The correct answer is D

Bonds between atoms of the same element are completely non-polar, so H-H is the least polar bond in the list, eliminating choices A and C. The C-H bond is considered to be non-polar even though the electrons of the bond are slightly unequally shared. C-Cl and C-F are both polar covalent bonds, but C-F is more strongly polar because F has a greater electronegativity.

51. Mercury is the only liquid metal and one of the heaviest elements with a density of 13.546 g/mL. What is the mass of 200 mL of mercury (almost the volume of a cup of coffee)?

(A) 857.8g
(B) 2709.2g
(C) 1709.6g
(D) 270.92g
(E) 5418.4g

The correct answer is B

Explanation: The equation of density is D = M/V -> M = DV
M = 13.546 g/mL x 200 mL = 2709.2g
So B is the correct answer

52. 250 grams of a mixture of sugar water (dissolved sugar) and gravel (insoluble) were filtered by suction filtration. 75 grams of gravel was left in the Büchner funnel. What is the weight percentage of sugar and gravel in the mixture?

(A) 60% sugar , 40% gravel
(B) 55% sugar, 45% gravel
(C) 75% sugar, 25% gravel
(D) 80%sugar, 20% gravel
(E) 70% sugar , 30% gravel

The correct answer is E

The weight of gravel in 250 grams of mixture is 75g
The weight of sugar in 250 grams of mixture is 250 – 75 = 175g
The weight percentage of sugar = 175gsugar/250g mixture x 100 = 70%
The weight percentage of gravel = 75gsugar/250g mixture x 100 = 30%

53. Which statement about acids and bases is not true?

(A) A Lewis base transfers an electron pair to a Lewis acid
(B) All strong acids are completely ionized
(C) All Brønsted bases use OH– as a proton acceptor.
(D) All Arrhenius acids form H+ ions in water.
(E) All Brønsted acids and bases exist in conjugate pairs with and without a proton.

The correct answer is C

Choice A is the definition of a Lewis acid, choice B is the definition of a strong acid, and choice D is the definition of an Arrhenius acid. By definition, all Arrhenius bases form OH– ions in water. Choice E is also correct and all Brønsted acids and bases exist in conjugate pairs with and without a proton. But not all Brønsted bases use OH– as a proton acceptor.

54. Calcium phosphate, $Ca_3(PO_4)_2$ is the main constituent of human bone. What is the weight percentage of phosphorous, P, in this compound?

(A) 30%
(B) 10%
(C) 20%
(D) 40%
(E) 25%

The correct answer is C

Explanation: The molar mass of P is 31 and the molar mass of Calcium Phosphate is 310. (The molar mass of Ca is 40, 40 x 3 = 120; the molar mass of P is 31, 31 x 2 = 62; the molar mass of oxygen is 16, 16 x 8 = 128. 120+62+128 = 310.) Thus, the P weight percentage in calcium phosphate is:

2 x 31/310 x 100 = 20%

55. There are 30.1 x 10^{23} neon atoms (Ne) in a sealed neon lamp. How many moles of neon are present in the lamp?

(A) 15
(B) 25
(C) 5
(D) 12
(E) 8

The correct answer is C

Explanation: This could be estimated by dividing 30.1x 10^{23} by 6.02 x 10^{23} to find the number of moles of neon gas (= 5 moles).

56. In the following phase diagram, sublimation occurs as P is decreased from A to B at constant T and condensation occurs as T is decreased from D to C at constant P. Which of the following statements are TRUE?

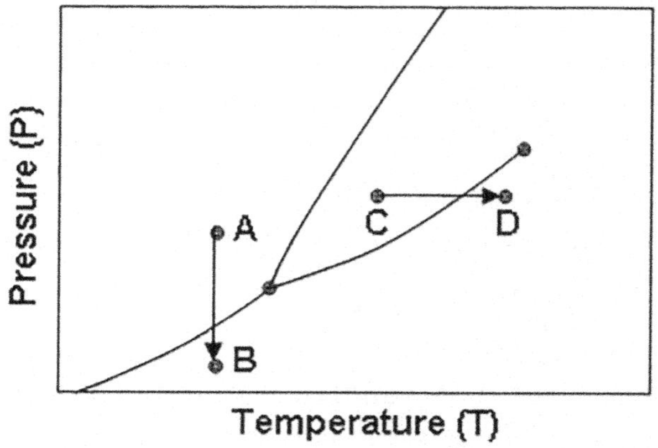

(A) A is liquid and B is solid
(B) C is gas and D is solid
(C) A is solid and B is gas
(D) D is liquid and C is gas
(E) B is liquid and C is gas.

The correct answer is C

Point A is located in the solid phase; point C is located in the liquid phase. Points B and D are located in the gas phase. The transition from solid to gas is sublimation and the transition from liquid to gas is vaporization and the reverse of it is condensation.

57. A solid which has a very high melting point, is very hard, and does not conduct electricity is probably :

(A) an ionic crystal
(B) a covalent network solid
(C) a metallic solid
(D) a polar solid
(E) a crystalline solid

The correct answer is B
A covalent network solid has all three of these properties.

58. A change in the disorder of a system is expressed as

(A) Enthalpy
(B) Entropy
(C) Kinetic Energy
(D) Specific Heat
(E) Heat of formations

The correct answer is B
Entropy is a measure of the disorder of a system.

59. Which of the following molecules are tetrahedral?

(A) H_2O
(B) CO_2
(C) CCl_4
(D) BeF_2
(E) none of the above

The correct answer is C
CCl_4 is a central carbon with 4 Cl molecules attached in a tetrahedral arrangement.

60. Which of the following sets of elements is in order of decreasing metallic character?

(A) Al > Ga > In > Ti
(B) In >Ga > Al >Ti
(C) Ti >Ga > In > Al
(D) Ti > In > Ga >Al
(E) None of the above

The correct answer is D

According to the periodic trend the metallic character decreases from left to right and also decreases from top to bottom. Titanium (Ti) is the most metallic by far; it is located far to the left of the other elements. Among these five choices the only choice that follows the period trend is D in which the metallic character decreases in the order Ti > In > Ga >Al. A and B choices are wrong because the trend in them is opposite to the correct answer. Choice C has a mixed trend and is not correct.

61. When comparing ethanol, acetone and ethylene glycol, which would have the lowest boiling point?

(A) Ethanol because it has an OH group
(B) Acetone because it has weak intermolecular forces
(C) Ethylene glycol because it has the most lone pairs
(D) All three have similar BP
(E) It cannot be determined without additional data.

The correct answer is B

Acetone is a gas at 56-57 °C
Ethanol boils at 78 °C
Ethylene glycol boils at 197 °C.

62. For a titration of a strong acid and a weak base, a group of students need an indicator in the pH range 3 -4. Which indicator should they choose?

(A) Methyl Orange
(B) Bromthymol Blue
(C) Litmus
(D) Phenolphthalein
(E) any of the above

The correct answer is A

Methyl Orange is an indicator used in the pH 3-4 range.
Bromthymol Blue is used in the pH 6-7.6 range
Litmus is used in the pH 4.5-8.3 range
Phenolphthalein in the pH 8.3-10.0 range

63. Chlorine has 7 valance electrons. An atom that would have similar chemical properties would be atomic number:

(A) 7
(B) 9
(C) 15
(D) 19
(E) 14

The correct answer is B

Chlorine is a halogen. Other halogens would have similar properties. Therefore F, atomic number 9 would be the most similar.

64. If a student needs to make a 1 L solution that is 6 M into a 3 M solution, how much water must be added ?

(A) 2 L
(B) 1 L
(C) 0.5 L
(D) 3 L
(E) 0.75 L

The correct answer is B
M1 V1 = M2 V2
(6M * 1L = 3M * ?L
? = 2L
The final volume must be two liters, therefore 1 L of water must be added.

65. The cathode is the electrode:

(A) That is always +
(B) That is always -
(C) Where reduction takes place
(D) where ions are formed
(E) where the main reaction occurs.

The correct answer is C
Reduction always takes place at the cathode.

66. Which of the following are true about an exothermic reaction?

(A) The reaction will not occur
(B) The reaction will occur and give off energy
(C) The reaction requires heat
(D) The reaction will require an outside current.
(E) The reverse reaction will occur

The correct answer is B
The reaction will occur and give off energy

67. The reaction $Zn + S \longrightarrow ZnS$ is an example of:

(A) a synthesis reaction
(B) a decomposition reaction
(C) a combustion reaction
(D) a redox reaction
(E) a single replacement reaction

The correct answer is A
This is a synthesis reaction. Two elements are combined to form a new compound.

68. The following statements are correct steps for drawing Lewis dot structure models except:

(A) Determine the number of valence shell electrons for each atom
(B) If the compound is an anion, the charge of the ion is added to the total electron count
(C) If the compound is a cation, subtract the charge of the ion.
(D) every atom has an octet of electrons except hydrogen and oxygen atoms with two electrons
(E) the remaining electrons are placed around the atoms as unshared pairs

The correct answer is D

All the statements regarding Lewis dot structure are correct except statement D, which is wrong because every atom has an octet of electrons in Lewis structures except the hydrogen atom, with two electrons. Hydrogen is the only exception; oxygen must also have a full octet.

69. Which of the following are in order of increasing boiling points?

(A) I, Br, Cl, F
(B) Cl,F, Br, I
(C) Br, I, Fl, Cl
(D) F, Cl, I, Br
(E) None of the above.

The correct answer is A

In halogens fluorine is a gas, chlorine is heavy gas moving on the ground, bromine is liquid and iodine is solid

70. What is the partial pressure of O_2 in a container of H_20, O_2 and N_2 at 200 KPa and 100° C? The vapor pressures of H_2O and N_2 at 100° C are, respectively, 20 and 30 Kpa.

(A) 50 KPa
(B) 175 KPa
(C) 75KPa
(D) 125 Kpa
(E) 150 Kpa

The correct answer is E

According to Dalton's law, $Ptotal = PH_2O + PO_2 + PN_2$ => $PO_2 = Ptotal - PH_2O - PN_2$

$PO_2 = 200 - 20 - 30 = 150$ Kpa

71. 250 mL of a 0.5 M solution of HCl is neutralized completely with a solution of 0.025 M KOH. What is the volume of the KOH solution consumed in this titration reaction?

(A) 1,250 mL
(B) 2,500 mL
(C) 800 mL
(D) 5,000 mL
(E) 500 mL

The correct answer is D

The titration equation of strong acid with strong base is Ma x Va = Mb x Vb

Since H2SO4 is strong acid and KOH is strong base we can calculate the volume of used KOH

0.5 x 250 = 0.025 x Vb

Therefore Vb = 0.5 x 250/0.025 = 5,000 mL

72. In the oxidation-reduction reaction

$$8\ HNO_3 + Cu_2 \longrightarrow 2\ Cu\ (NO_3)_2 + 4\ NO_2 + 4\ H_2O$$

Which of the following statements is correct?

(A) N is reduced and its oxidation number increases
(B) N is reduced and its oxidation number decreases
(C) Cu is oxidized and its oxidation number decreases
(D) Cu is oxidized and its oxidation number increases
(E) Both B and D

The correct answer is E

The oxidation number in HNO_3 is +5 and in NO_2 is +4. So, it is reduced and its oxidation number is decreased. On the other hand the oxidation number of Cu at the left side of the reaction is zero and in the right side is +2. So it is oxidized and its oxidation number is increased.

73. The combustion reaction of pentane (airplane fuel, saturated hydrocarbon) produces carbon dioxide and water.

$$C_5H_{12}\ (l) + O_2\ (g) \longrightarrow CO_2\ (g) + H_2O\ (g)$$

Suppose this reaction is balanced with the lowest integer (whole-number) coefficient for each reactant and product. What is the summation of all of the coefficients in the reaction?

(A) 20
(B) 14
(C) 15
(D) 12
(E) 10

The correct answer is A

On the left side there are 5 carbons and 12 hydrogens. Balancing on the right side of the reaction, we come up with 5 CO_2 and 6 H_2O Now everything is balanced except oxygen. There are 16 oxygen atoms on the right side but only 2 on the left. We give the coefficient of 8 to the reacting oxygen. Now the equation is balanced and the number of atoms is equal on both sides of the reaction.

$$C_5H_{12}\ (l) + 8\ O_2\ (g) \longrightarrow 5\ CO_2\ (g) + 6\ H_2O\ (g)$$

74. Which of the following expresses standard atmospheric pressure?

(A) 29.9 in Hg
(B) 76.0 cm Hg
(C) 760 torr
(D) 14.7 psi
(E) all of the above

The correct answer is E

All of these are correct.

75. A sample of ozone gas is at 0° C. If both the volume and pressure double, what is the new Kelvin temperature?

(A) 68 K
(B) 135 K
(C) 273 K
(D) 546 K
(E) 1092 K

The correct answer is E

The temperature 0 °C, is approximately equal to 273 K. Thinking of the ideal gas equation, PV = nRT, it becomes clear that $P_1 * V_1 / T_1 = P_2 * V_2 / T_2$. Rearranging, $T_2 = P_2 * V_2 * T_1 / P_1 * V_1$.

Since $P_2 = 2P_1$ and $V_2 = 2V_1$,
$T_2 = 2P_1 * 2V_1 * 273 \text{ K} / P_1 * V_1$
$T_2 = 2 * 2 * 273$
$T_2 = 1092$

76. A chemical reaction that absorbs heat from the surroundings is said to be ________ and has a ______ value of Δ H.

(A) endothermic, positive
(B) endothermic, negative
(C) exothermic, negative
(D) exothermic, positive
(E) none of the above

The correct answer is A

77. Which of the following formulas is an empirical formula?

(A) C_2O_2
(B) C_2H_6O
(C) $C_6H_{12}O_2$
(D) N_2O_4
(E) $Na_2Cl_2O_2$

The correct answer is B

An empirical formula is the simplest formula using the smallest set of integers to express the ratio of atoms present in a molecule. The formulas A, C, D, and E have the empirical formulas of CO, C_3H_6O, NO_2, and NaOCl. But C_2H_6O is an empirical formula itself and cannot make it simpler.

78. The amount of heat required to raise 1 g of water by one degree Celsius is called:

(A) The heat of formation
(B) Specific heat of water
(C) 1 calorie
(D) delta-H
(E) delta-S

The correct answer is B

79. In an aqueous solution, the solubility of $Ce_2(SO_4)_3$ can be expressed as:

$$Ce_2(SO_4)_3 \longrightarrow 2Ce^{3+} + 3\,SO_4^{2-} + \text{Heat}$$

According to Le Châtelier's principle, the addition of heat will:

(A) allow $Ce_2(SO_4)_3$ to dissolve completely
(B) make the solute less soluble
(C) make the solute more soluble
(D) cause the ions to react with the H_2O in the solution
(E) none of the above

The correct answer is B

For an exothermic solution process this principle states that solubility will decrease with increasing temperature

80. One mole of each of the following four salts is dissolved in a 2L beaker. The heat of solution of each of these salts is:

ΔH KCl = 17.2 KJ/mol
ΔH KBr = 19.9 KJ/mol
ΔH LiI = –59 KJ/mol
ΔH $LiNO_3$ = –1.3 KJ/mol

Which of the following statements are correct?

(A) The overall heat of solution is –2.32 KJ/mol and the beaker warms up.
(B) The overall heat of solution is –2.32 KJ/mol and the beaker cools down.
(C) The overall heat of solution is – 23.2 KJ/mol and the beaker cools down.
(D) There is no change in the temperature of the beaker.
(E) The overall heat of solution is – 23.2 KJ/mol and the beaker warms up.

The correct answer is E

Two of these four compounds (LiI and $LiNO_3$) have negative heats of solution and they are exothermic reactions in dissolving in water. The other two (KCl and KBr) have positive heats of solution and they are endothermic reactions in dissolving in water. When we dissolve all the four compounds in water the overall heat of solution would be:

Δ Overall = ΔH KCl + ΔH KBr + ΔH LiI + ΔH LiNO3 = 17.2=19.9 – 59 – 1.3= –23.2 KJ/mol

Therefore the overall ΔHsoln is negative. The beaker warms up.

81. Pentane, butane, hexanol, and hexane are all liquids at room temperature. Rank them in order of decreasing viscosity.

(A) n-pentane > n-butane > 1-hexanol > n-hexane
(B) 1-hexanol > n-pentane > n-butane > > n-hexane
(C) 1-hexanol > n- hexane > n-pentane > n-butane
(D) n-hexane > 1-hexanol > n-pentane > n-butane
(E) n-butane > 1-hexanol > n-pentane > n-hexane

The correct answer is C

Higher viscosities result from stronger intermolecular attractive forces. The molecules listed fall into two main groups; two of them are alkanes and three of them are alcohols. Alkanes are composed entirely of non-polar C-C and C-H bonds, resulting in no dipole interactions or hydrogen bonding. London dispersion forces increase with the size of the molecule, resulting in a stronger intermolecular force in n-hexane ($CH_3CH_2CH_2CH_2CH_2CH_3$) than n-pentane ($CH_3CH_2CH_2CH_2CH_3$)

On the other hand alcohols are polar, with intermolecular hydrogen-bond interaction which makes them more viscous than alkanes.

82. What is the molar mass of aspirin, $C_9H_8O_4$?

(A) 116.0 g/mol
(B) 180.0 g/mol
(C) 188.0 g/mol
(D) 244.0 g/mol
(E) 132.0 g/mol

The correct answer is B

C= 12* 9 = 108

O = 4 * 16 = 64

H = 1 * 8 = 8

83. What term refers to an insoluble substance produced by a reaction in aqueous solution?

(A) colloid
(B) precipitate
(C) product
(D) salt
(E) none of the above

The correct answer is B

A precipitate is what falls out of the solution when it cannot be dissolved.

84. Which energy sublevel immediately follows the 4s sublevel according to increasing energy?

(A) 3p
(B) 3d
(C) 4p
(D) 4d
(E) 5s

The correct answer is B

The energy levels of the orbitals are 4s < 3d < 4p < 5s

85. Which of the following best defines the term isotopes?

(A) same number of protons -- different number of electrons
(B) same number of protons -- different number of neutrons
(C) same number of neutrons -- different number of protons
(D) same number of neutrons -- different number of electrons
(E) same number of electrons -- different number of protons

The correct answer is B

By definition an isotope has same number of protons — different number of neutrons

SAT Chemistry Practice Test 2

Questions

1. A piston compresses a gas at constant temperature. Which gas properties increase?

I. Average speed of molecules
II. Pressure
III. Molecular collisions with container walls per second

(A) I and II
(B) I and III
(C) II and III
(D) I, II, and III
(E) None of the above

2. The temperature of a liquid is raised at atmospheric pressure. Which property of liquids increases?

(A) Critical pressure
(B) Vapor pressure
(C) Surface tension
(D) Viscosity
(E) Boiling Point

3. Potassium crystallizes with two atoms contained in each unit cell. What is the mass of potassium found in a lattice $1.0 \cdot 10^6$ unit cells wide, $2.0 \cdot 10^6$ unit cells high, and $5.0 \cdot 10^5$ unit cells deep?

(A) A. 85 ng
(B) B. 32.5 μg
(C) C. 64.9 μg
(D) D. 13 μg
(E) E. 130 μg

4. A gas is heated in a sealed container. Which of the following occur(s)?

(A) Gas pressure rises
(B) Gas density decreases
(C) The average distance between molecules increases
(D) The volume increases
(E) All of the above

5. **How many molecules are in 2.20 pg of a protein with a molecular weight of 150 kDa?**

(A) $8.83 \cdot 10^{9}$
(B) $1.82 \cdot 10^{9}$
(C) $8.83 \cdot 10^{6}$
(D) $1.82 \cdot 10^{6}$
(E) 8.83×10^{15}

6. **At STP, 20 μL of O2 contain $5.4 \cdot 10^{16}$ molecules. According to Avogadro's hypothesis, how many molecules are in 20 μL of Ne?**

(A) $5.4 \cdot 10^{15}$
(B) $1.0 \cdot 10^{16}$
(C) $2.7 \cdot 10^{16}$
(D) $5.4 \cdot 10^{16}$
(E) $1.3 \cdot 10^{6}$

7. **An ideal gas at 50.0° C and 3.00 atm is in a 300 cm^3 cylinder. The cylinder volume changes by moving a piston until the gas is at 50.0° C and 1.00 atm. What is the final volume?**

(A) 100 cm^3
(B) 450 cm^3
(C) 900 cm^3
(D) 1.20 dm^3
(E) 150.0 cm^3

8. **1-butanol, ethanol, methanol, and 1-propanol are all liquids at room temperature. Rank them in order of increasing boiling point.**

(A) 1-butanol < 1-propanol < ethanol < methanol
(B) methanol < ethanol < 1-propanol < 1-butanol
(C) methanol < ethanol < 1-butanol < 1-propanol
(D) 1-propanol < 1-butanol < ethanol < methanol
(E) ethanol < methanol < 1-butanol < 1-propanol

9. **One mole of an ideal gas at STP occupies 22.4 L. At what temperature will one mole of an ideal gas at 1 atm occupy 31.0 L?**

(A) 34.6° C
(B) 139° C
(C) 378° C
(D) 442° C
(E) 28 ° C

10. What pressure is exerted by a mixture of 2.7 g of H_2 and 59 g of Xe at STP on a 50. L container?

(A) 0.69 atm
(B) 0.76 atm
(C) 0.88 atm
(D) 0.97 atm
(E) 27.0 atm

11. The boiling point of water at sea level on the Kelvin scale is closest to:

(A) 112 K
(B) 212 K
(C) 273 K
(D) 373 K
(E) 298 K

12. Which phases may be present at the triple point of a substance?

I. Gas
II. Liquid
III. Solid
IV. Supercritical fluid

(A) I, II, and III
(B) I, II, and IV
(C) II, III, and IV
(D) I, II, III, and IV
(E) I, III, IV

13. Carbonated water is bottled at 25° C under pure CO_2 at 4.0 atm. Later the bottle is opened at 4° C under air at 1.0 atm that has a partial pressure of $3 \cdot 10^{-4}$ atm CO_2. Why do CO_2 bubbles form when the bottle is opened?

(A) CO_2 falls out of solution due to a drop in solubility at the lower total pressure.
(B) CO_2 falls out of solution due to an increase in solubility at the lower CO_2 pressure.
(C) CO_2 falls out of solution due to a drop in solubility at the lower temperature.
(D) CO_2 falls out of solution due to the decreased concentration of water.
(E) CO_2 is formed by the decomposition of carbonic acid.

14. Which statement about molecular structures is false?

(A) 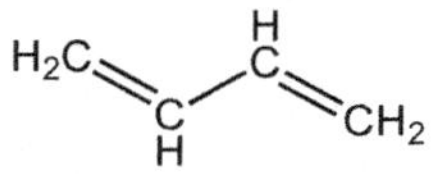 $H_2C{=}CH{-}CH{=}CH_2$ is a conjugated molecule.

(B) A bonding σ orbital connects two atoms in a double bond.

(C) A bonding π orbital connects two atoms via a single bond in a separate region from the straight line between them.

(D) The anion with resonance forms

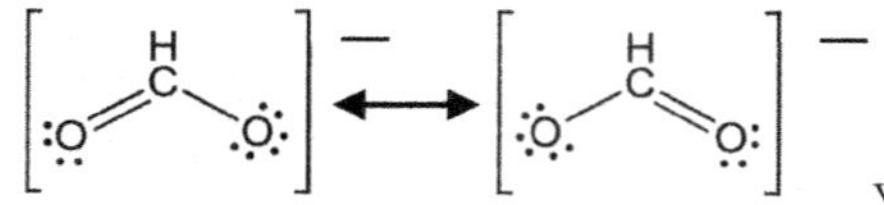

$[:O{=}CH{-}\ddot{O}:]^- \longleftrightarrow [:\ddot{O}{-}CH{=}O:]^-$ will always exist in one form or the other.

(E) They are all false

15. What is the chemical composition of magnesium nitrate?

(A) 11.1% Mg, 22.2% N, 66.7% O
(B) 50.1% Mg, 22.2% N, 33.0% O
(C) 16.4% Mg, 18.9% N, 64.7% O
(D) 20.9% Mg, 24.1% N, 55.0% O
(E) 28.2% Mg, 16.2% N, 55.7% O

16. How many neutrons are there in the radioactive isotope Cobalt-60 (atomic mass = 60)?

(A) 27
(B) 33
(C) 60
(D) 87
(E) 14

17. Select the list of atoms that is arranged in order of increasing size:

(A) Mg, Na, Si, Cl
(B) Si, Cl, Mg, Na
(C) Cl, Si, Mg, Na
(D) Na, Mg, Si, Cl
(E) Mg, Si, Cl, Na

18. Based on trends in the periodic table, which of the following properties would you expect to be greater for Rb than for K?

I. Density
II. Melting point
III. Ionization energy
IV. Oxidation number in a compound with chlorine

(A) I only
(B) I, II, and III
(C) II and III
(D) I, II, III, and IV
(E) None of the above

19. Why does $CaCl_2$ have a higher normal melting point than NH_3?

(A) London dispersion forces in $CaCl_2$ are stronger than covalent bonds in NH_3.
(B) Covalent bonds in NH_3 are stronger than dipole-dipole bonds in $CaCl_2$.
(C) Ionic bonds in $CaCl_2$ are stronger than London dispersion forces, the strongest intermolecular forces in NH_3.
(D) Ionic bonds in $CaCl_2$ are stronger than hydrogen bonds, the strongest intermolecular forces in NH_3.
(E) None of the above

20. Rank the following bonds from least to most polar:

C-H, C-Cl, H-H, C-F

(A) C-H < H-H < C-F < C-Cl
(B) H-H < C-H < C-F < C-Cl
(C) C-F < C-Cl < C-H < H-H
(D) H-H < C-H < C-Cl < C-F
(E) H-H < C-CL < C-H< C-F

21. In C_2H_2, each carbon atom contains the following valence orbitals:

(A) p only
(B) p and sp hybrids
(C) p and sp^2 hybrids
(D) sp^3 hybrids only
(E) s, p, and sp^2 hybrids

22. The boiling points of N_2, O_2 and Cl_2 are -196, -182, and -34 respectively. Cl_2 boils at much higher temperature than the others. This is explained by

(A) Single bonds are easier to break than double or triple bonds
(B) Cl_2 has a longer bond length than the others
(C) Cl_2 has greater London dispersion forces because it has more electrons.
(D) Hydrogen bonding is stronger in Cl_2
(E) None of the above

23. Which of the following statements regarding molecular geometries are true?

I. Double bonds are planar
II. A central atom with sp2 hybridization is trigonal
III. A central atom with sp2 hybridization is bipyramidal
IV. Single bonds are always planar

(A) I and II are true
(B) I and IV are true
(C) II and IV are true
(D) II and III are true
(E) I, II and IV are true

24. The electronic configuration or Iron (Fe) is

(A) $1s^2\ 2s^2\ 2p^6\ 3s^2\ 3p^6\ 4s^2\ 3d^6$
(B) $1s^2\ 2s^2\ 2p^6\ 3s^2\ 3p^6\ 3d^8$
(C) $1s^2\ 2s^2\ 2p^6\ 3s^2\ 3p^6\ 4s^2$
(D) $1s^2\ 2s^2\ 2p^6\ 3s^2\ 3p^6\ 4s^2\ 4p^6$
(E) $1s^2\ 2s^2\ 2p^6\ 3s^2\ 3p^6\ 4s^2\ 4p^3\ 4d^3$

25. Which of the following is a correct electron arrangement for oxygen?

(A) 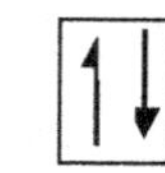 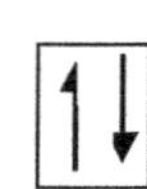 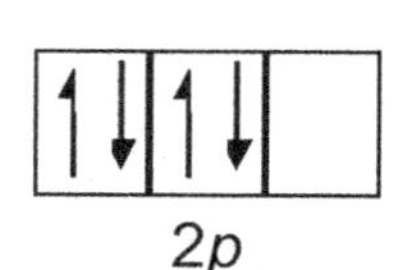
1s 2s 2p

(B) $1s^2 2s^2, 2s^2, 2p^2$
(C) 2, 2, 4
(D) 2, 2, 4, 1/2
(E) None of the above

26. **Which of the following is a proper Lewis dot structure of CHClO?**

(A) 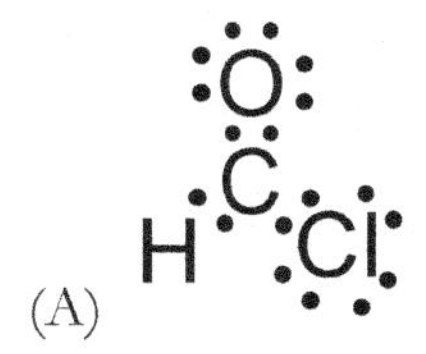

(B)

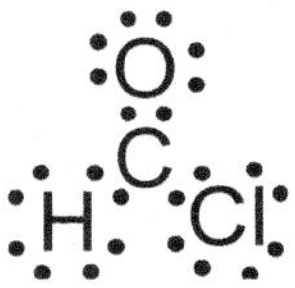

(C)

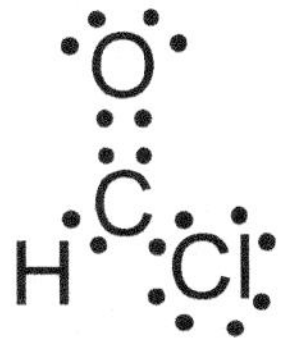

(D) 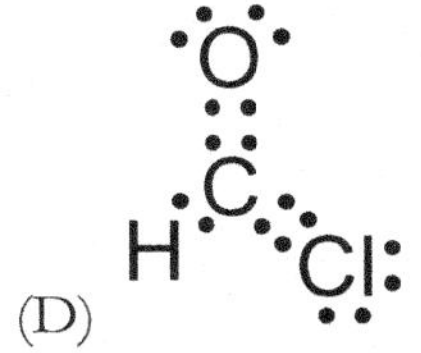

(E) None of the above

27. **Which intermolecular attraction explains the following trend in straight-chain alkanes?**

Condensed structural formula	Boiling point (° C)
CH_4	-161.5
CH_3CH_3	-88.6
$CH_3CH_2CH_3$	-42.1
$CH_3CH_2CH_2CH_3$	-0.5
$CH_3CH_2CH_2CH_2CH_3$	36.0
$CH_3CH_2CH_2CH_2CH_2CH_3$	68.7

(A) London dispersion forces
(B) hydrophobic interactions
(C) Dipole-dipole interactions
(D) Hydrogen bonding
(E) Ion-induced dipole interactions

28. **Match the theory with the scientist who first proposed it:**

I. Electrons, atoms, and all objects with momentum also exist as waves.
II. Electron density may be accurately described by a single mathematical equation.
III. There is an inherent indeterminacy in the position and momentum of particles.
IV. Radiant energy is transferred between particles in exact multiples of a discrete unit.

(A) I - de Broglie, II - Planck, III - Schrödinger, IV - Thomson
(B) I - Dalton, II - Bohr, III - Planck, IV - de Broglie
(C) I - Henry, II - Bohr, III - Heisenberg, IV - Schrödinger
(D) I - de Broglie, II - Schrödinger, III - Heisenberg, IV - Planck
(E) I - Schrödinger, II - de Broglie, III - Plank, IV - Heisenberg

29. **The terrestrial composition of an element is: 50% as a stable isotope with an atomic mass of 78 amu and 50% as a stable isotope with an atomic mass of 80 amu. Calculate the atomic mass of the element.**

(A) 79 amu
(B) 78 amu
(C) 77 amu
(D) 81 amu
(E) 80 amu

30. $^{3}_{1}H$ **decays with a half-life of 12 years. 3.0 g of pure** $^{3}_{1}H$ **were placed in a sealed container 24 years ago. How many grams of** $^{3}_{1}H$ **remain?**

(A) 0.38 g
(B) 0.75 g
(C) 1.5 g
(D) 0.125 g
(E) 3.0 g

31. **Moving down a column in the Periodic Table:**

I. the atomic radius increase
II. Ionization energy increase
III. Protons are added
IV. metallic characteristics increase

(A) I and III
(B) I and II only
(C) III only
(D) III and IV only
(E) None of the above

32. Moving from left to right on a Periodic Table (Li – to Ne) which of the following are true:

I. the atomic radius decreases
II. electrons are added
III. ionization energies increase
IV. electronegativity decreases

(A) I only
(B) I, II
(C) I, II, III
(D) I, II, III, IV
(E) None of the above

33. NH_4F is dissolved in water. Which of the following are conjugate acid/base pairs present in the solution?

I. NH_4^+/NH_4OH
II. HF/F^-
III. H_3O^+/H_2O
IV. H_2O/OH^-

(A) I only
(B) I, II, and III
(C) I, III, and IV
(D) II and IV
(E) II, III, and IV

34. Rank the following from lowest to highest pH. Assume a small volume for the component given in moles:

I. 0.01 mol HCl added to 1 L H_2O
II. 0.01 mol HI added to 1 L of an acetic acid/sodium acetate solution at pH 4.0
III. 0.01 mol NH_3 added to 1 L H_2O
IV. 0.1 mol HNO_3 added to 1 L of a 0.1 M $Ca(OH)_2$ solution

(A) I < II < III < IV
(B) I < II < IV < III
(C) II < I < III < IV
(D) II < I < IV < III
(E) IV < III < II < I

35. **Which statement about acids and bases is *not* true?**

(A) All strong acids ionize in water.
(B) All Lewis acids accept an electron pair.
(C) All Brønsted bases use OH^- as a proton acceptor
(D) All Arrhenius acids form H^+ ions in water.
(E) Water can act as either an acid or a base.

36. **What is the pH of a buffer solution made of 0.128 M sodium formate (NaCOOH) and 0.072 M formic acid (HCOOH)? The pK_a of formic acid is 3.75.**

(A) 2.0
(B) 3.0
(C) 3.75
(D) 4.0
(E) 5 .0

37. **A 100. L vessel of pure O_2 at 500. kPa and 20.° C is used for the combustion of butane:**

$$2\ C_4H_{10} + 13\ O_2 \longrightarrow 8\ CO_2 + 10\ H_20$$

Find the mass of butane that consumes all the O_2 in the vessel. Assume O_2 is an ideal gas and use a value of R = 8.314 J/(mol•K).

(A) 183 g
(B) 467 g
(C) 1.83 kg
(D) 2.6 kg
(E) 7.75 kg

38. **32.0 g of hydrogen and 32.0 grams of oxygen react to form water until the limiting reagent is consumed. What is present in the vessel after the reaction is complete?**

(A) 16.0 g O_2 and 48.0 g H_2O
(B) 24.0 g H_2 and 40.0 g H_2O
(C) 28.0 g H_2 and 36.0 g H_2O
(D) 28.0 g H_2 and 34.0 g H_2O
(E) 28.0 g H_2 and 16.0 g O_2

39. **Which reaction is *not* a redox process?**

(A) Combustion of octane: $2C_8H_{18} + 25O_2 \longrightarrow 16CO_2 + 18H_2O$
(B) Depletion of a lithium battery: $Li + MnO_2 \longrightarrow LiMnO_2$
(C) Corrosion of aluminum by acid: $2Al + 6HCl \longrightarrow 2AlCl_3 + 3H_2$
(D) Taking an antacid for heartburn:
$CaCO_3 + 2HCl \longrightarrow CaCl_2 + H_2CO_3 \longrightarrow CaCl_2 + CO_2 + H_2O$
(E) None of the above

Practice Test Two

40. Which of the following are true about Galvanic cells?

I. two half reactions take place in the two chambers
II. oxidation takes place at the anode
III. reduction takes place at the cathode
IV. if the salt bridge is removed the voltage drops to zero
V. La Châtelier's principle can be applied to the systematic

(A) I , II, III ARE TRUE
(B) ONLY II AND III ARE TRUE
(C) ONLY I AND IV ARE TRUE
(D) All are true
(E) none are true

41. Balance the equation for the neutralization reaction between phosphoric acid and calcium hydroxide by filling in the blank stoichiometric coefficients.

$$H_3PO_4 + \quad Ca(OH)_2 \longrightarrow Ca_3(PO_4)_2 + H_2O$$

(A) 4, 3, 1, 4
(B) 2, 3, 1, 8
(C) 2, 3, 1, 6
(D) 2, 1, 1, 2
(E) 4, 3, 1, 1

42. Write an equation showing the reaction between calcium nitrate and lithium sulfate in aqueous solution. Include all products.

(A) $Ca(NO_3)_2 + 2H_2O + Li_2SO_4 \longrightarrow CaSO_4\ (s) + 2\ LiNO_3 + 2OH- + H_2(g)$
(B) $Ca(NO_3)_2 + Li_2SO_4 \longrightarrow CaSO_4\ (s) + 2\ LiNO_3$
(C) $Ca(NO_3)_2$ and $Li_2SO_2 \longrightarrow CaSO_2\ (s) + 2\ LiNO_3$
(D) $Ca(NO_3)$ and $Li_2SO_4 \longrightarrow CaSO_4\ (s) + Li_2NO_3$
(E) None of the Above

43. Find the mass of CO_2 produced by the combustion of 15 kg of isopropyl alcohol in the reaction:

$$2\ C_3H_7OH\ (l) + 9\ O_2\ (g) \longrightarrow 6\ CO_2\ (g) + 8\ H_2O\ (g)$$

(A) 33 kg
(B) 44 kg
(C) 50 kg
(D) 60 kg
(E) 66 kg

Practice Test Two

44. What is the density of nitrogen gas at STP? Assume an ideal gas and a value of 0.08206 L•atm/(mol•K) for the gas constant.

(A) 0.62 g/L
(B) 1.14 g/L
(C) 1.25 g/L
(D) 2.03 g/L
(E) 3.38 g/L

45. Find the volume of methane that will produce 12 m^3 of hydrogen in the reaction:

$$CH_4(g) + H_2O(g) \rightarrow CO(g) + 3H_2(g)$$

Assume that the temperature and pressure remain constant.

(A) 4.0 m^3
(B) 32 m^3
(C) 36 m^3
(D) 64 m^3
(E) Cannot be determined

46. Household "chlorine bleach" is sodium hypochlorite. Which of the following best represents the production of sodium hypochlorite, sodium chloride, and water by bubbling chlorine gas through aqueous sodium hydroxide?

(A) $Cl + NaOH \longrightarrow NaCl + H_2O$
(B) $Cl_2 + 2NaOH \longrightarrow 2NaClO_2 + H_2O$
(C) $Cl + 2NaOH \longrightarrow Na^+ + NaOCl + H_2O$
(D) $Cl_2 + 2NaOH \longrightarrow NaCl + NaOCl + H_2O$
(E) None of the above.

47. For the reaction $M(s) + Cd^{2+} \longrightarrow M^+ + Cd(s)$ in an electrochemical cell at 25 °C Which of the following are true ?

(A) $M(s) \longrightarrow M^+ + e^-$ takes place at the anode
(B) As the reaction proceeds the $[M^+]$ decreases
(C) $Cd(2^+)$ looses electrons
(D) $Cd(2^+) \longrightarrow Cd(s)$ takes place at the anode
(E) There is not enough information to determine

48. Write the equilibrium expression K_{eq} for the reaction:

$$CO_2\ (g) + H_2\ (g) \longleftrightarrow CO\ (g) + H_2O\ (g)$$

(A) $K_{eq} = [CO_2]\ [H_2]^3\ /\ [H_2O]\ [CO]$
(B) $K_{eq} = [H_2O]\ [CO]\ /\ [CO_2]\ [H_2]$
(C) $K_{eq} = [CO_2]\ [H_2]^2\ /\ [H_2O]\ [CO]$
(D) $K_{eq} = [CO_2]\ [H_2]\ /\ [H_2O]\ [CO]$
(E) None of the above

49. The exothermic reaction $2NO\ (g) + Br_2\ (g) \longleftrightarrow 2NOBr\ (g)$ is at equilibrium. According to Le Châtelier's principle:

(A) Adding Br_2 will increase [NO].
(B) Adding [NO] will increase Br_2.
(C) An increase in container volume (with T constant) will increase [NOBr].
(D) An increase in pressure (with T constant) will increase [NOBr].
(E) An increase in temperature (with P constant) will increase [NOBr].

50. At a certain temperature, *T*, the equilibrium constant for the reaction $2NO\ (g) \longleftrightarrow N_2\ (g) + O_2\ (g)$ is $K_{eq} = 2 \times 10^3$. If a 1.0 L container at this temperature contains 90 mM N_2, 20 mM O_2, and 5 mM NO, what will occur?

(A) The reaction will make more N_2 and O_2.
(B) The reaction is at equilibrium.
(C) The reaction will make more NO.
(D) The temperature, *T*, is required to solve this problem.
(E) None of the above

51. $BaSO_4$ ($K_{sp} = 1 \times 10^{-10}$) is added to pure H_2O. How much is dissolved in 1 L of saturated solution?

(A) 2mg
(B) 10 μg
(C) 2 μg
(D) 100 pg
(E) Cannot determine from the information given

52. What are the pH and the pOH of 0.010 M HNO_3 (*aq*)?

(A) pH = 1.0, pOH = 9.0
(B) pH = 2.0, pOH = 12.0
(C) pH = 2.0, pOH = 8.0
(D) pH = 8.0, pOH = 6.0
(E) pH = 0.1, pOH = 0.9

Practice Test Two

53. Which statements about reaction rates are true?

I. Catalysts shift an equilibrium to favor product formation.
II. Catalysts increase the rate of forward and reverse reactions.
III. A greater temperature increases the chance that a molecular collision will overcome a reaction's activation energy.
IV. A catalytic converter contains a homogeneous catalyst.

(A) I and II
(B) II and III
(C) I, II, and III
(D) II, III, and IV
(E) I, III, and IV

54. Consider the reaction between iron and hydrogen chloride gas:

$$2\ Fe(s) + 6\ HCl(aq) = 2\ FeCl_3(aq) + 3\ H_2(g)$$

7 moles of iron and 10 moles of HCl react until the limiting reagent is consumed. Which statements are true?

I. HCl is the excess reagent
II. HCl is the limiting reagent
III. 7 moles of H_2 are produced
IV. 2 moles of the excess reagent remain

(A) I and III
(B) I and IV
(C) II and III
(D) II and IV
(E) I only

55. The reaction:

$$(CH_3)_3CBr(aq) + OH^-(aq) \longrightarrow (CH_3)_3COH(aq) + Br$$

occurs in three elementary steps:

$(CH_3)_3CBr \longrightarrow (CH_3)_3C^+ + Br^-$ is slow

$(CH_3)_3C^+ + H_2O \longrightarrow (CH_3)_3COH_2^+$ is fast

$(CH_3)_3COH_2^+ + OH^- \longrightarrow (CH_3)_3COH + H_2O$ is fast

What is the rate law for this reaction?

(A) Rate = $k[(CH_3)_3CBr]$
(B) Rate = $k[OH^-]$
(C) Rate = $k[(CH_3)_3CBr][OH^-]$
(D) Rate = $k[(CH_3)_3CBr]^2$
(E) Rate = k[(CH3)3 C Br]

56. Which statement about thermochemistry is true?

(A) Particles in a system move about less freely at high entropy.
(B) Water at 100° C has the same internal energy as water vapor at 100° C.
(C) A decrease in the order of a system corresponds to an increase in entropy.
(D) The Heat of Fusion is the energy needed to turn from a gas into a solid.
(E) At its sublimation temperature, dry ice has higher entropy than gaseous CO_2.

57. Which statement about reactions is true?

(A) All spontaneous reactions are exothermic and cause an increase in entropy.
(B) An endothermic reaction that increases the order of the system cannot be spontaneous.
(C) A reaction can be non-spontaneous in one direction and also non-spontaneous in the opposite direction.
(D) Melting snow is an exothermic process.
(E) Thermodynamic functions are dependent on the reaction pathway.

58. Given:

$E° = -2.37V$ for

Mg^{2+} (*aq*) $+2e^- \longrightarrow$ Mg (*s*)

and

$E°=0.80$ V for

Ag^+ (*aq*) + $e^- \longrightarrow$ Ag (*s*)

what is the standard potential of a voltaic cell composed of a piece of magnesium dipped in a 1 M Ag^+ solution and a piece of silver dipped in 1 M Mg^{2+}?

(A) 0.77 V
(B) 1.57 V
(C) 3.17 V
(D) 3.97 V
(E) 0.38 V

Practice Test Two

59. What could cause this change in the energy diagram of a reaction?

From: to:

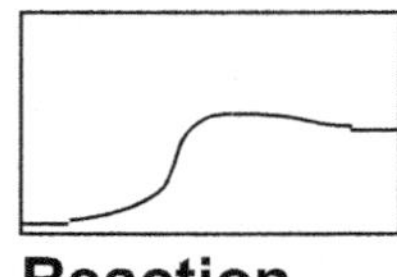

Reaction pathway →

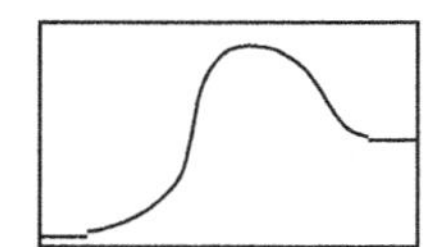

Reaction pathway →

(A) Adding a catalyst to an endothermic reaction
(B) Removing a catalyst from an endothermic reaction
(C) Adding a catalyst to an exothermic reaction
(D) Removing a catalyst from an exothermic reaction
(E) Adding heat to an endothermic reaction.

60. In the following phase diagram, _______ occurs as pressure is decreased from A to B at constant temperature and _______ occurs as temperature is increased from C to D at constant pressure.

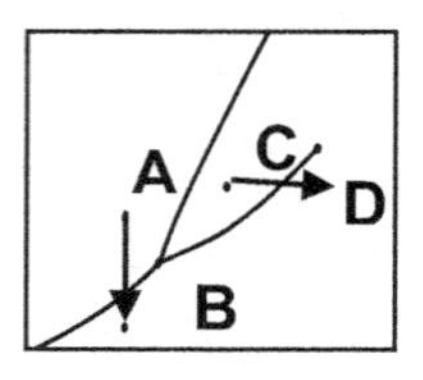

(A) deposition, melting
(B) sublimation, melting
(C) deposition, vaporization
(D) sublimation, vaporization
(E) melting, vaporization

61. Heat is added to a pure solid at its melting point until it all becomes liquid at its freezing point. Which of the following occur(s)?

(A) Intermolecular attractions are weakened.
(B) The kinetic energy of the molecules does not change.
(C) The freedom of the molecules to move about increases.
(D) The temperature of the system remains constant.
(E) All of the above.

62. **This compound**

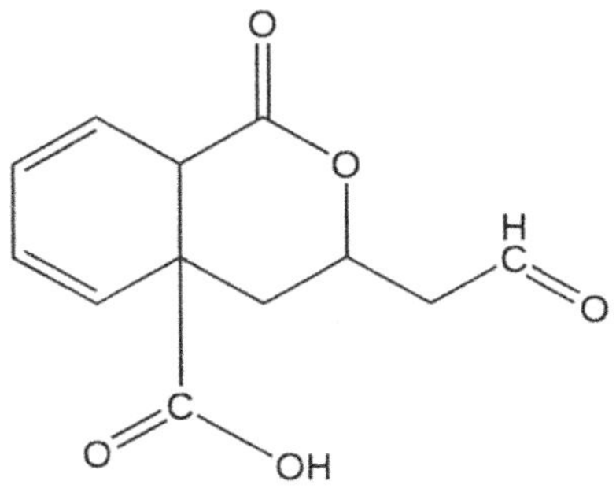

contains which of the following?

(A) alkene, carboxylic acid, ester, and ketone
(B) B. aldehyde, alkyne, ester, and ketone
(C) aldehyde, alkene, carboxylic acid, and ester
(D) acid anhydride, aldehyde, alkene, and amine
(E) aldehyde, amine, ketone, alcohol

63. **Which of the following pairs are isomers?**

I.

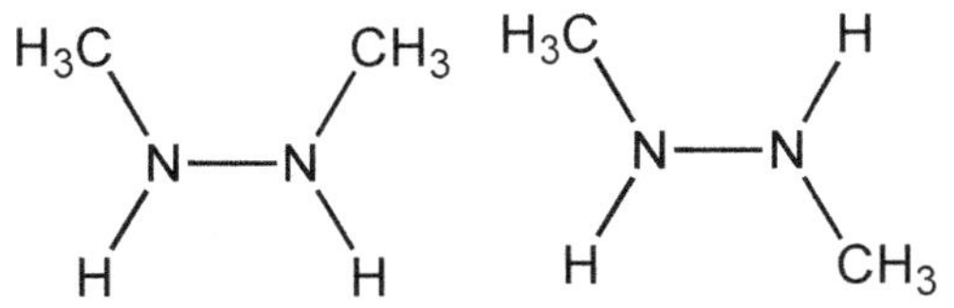

II. pentanal, 2-pentanone

III.

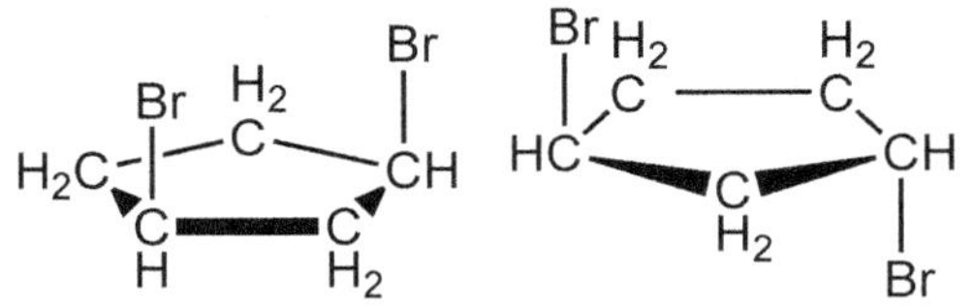

IV.

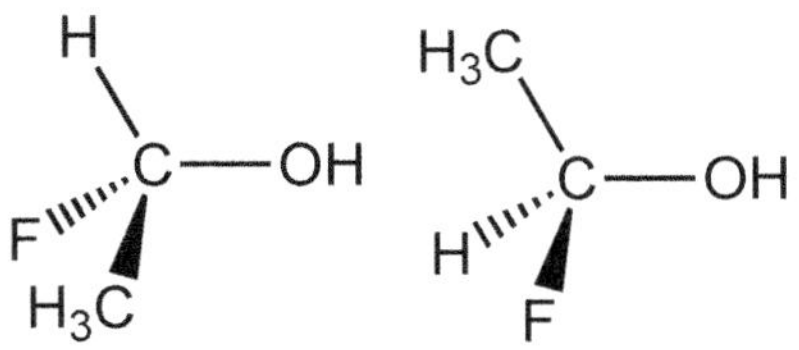

(A) I and IV
(B) II and III
(C) I, II, and III
(D) I, II, III, and IV
(E) None of the above

Practice Test Two

64. Which of the following is *not* a colligative property?

(A) Viscosity lowering
(B) Freezing point lowering
(C) Boiling point elevation
(D) Vapor pressure lowering
(E) All of the above

65. A sample of 50.0 ml KOH is titrated with 0.100 M $HClO_4$. The initial buret reading is 1.6 ml and the reading at the endpoint is 22.4 ml. What is [KOH]?

(A) 0.0416 M
(B) 0.0481 M
(C) 0.0832 M
(D) 0.0962 mM
(E) 0.0962 M

66. When KNO_3 dissolves in water, the water grows slightly colder. An increase in temperature will _______ the solubility of KNO_3.

(A) increase
(B) decrease
(C) double
(D) have no effect on
(E) have an unknown effect with the information given on the solubility.

67. Classify these biochemicals:

I.

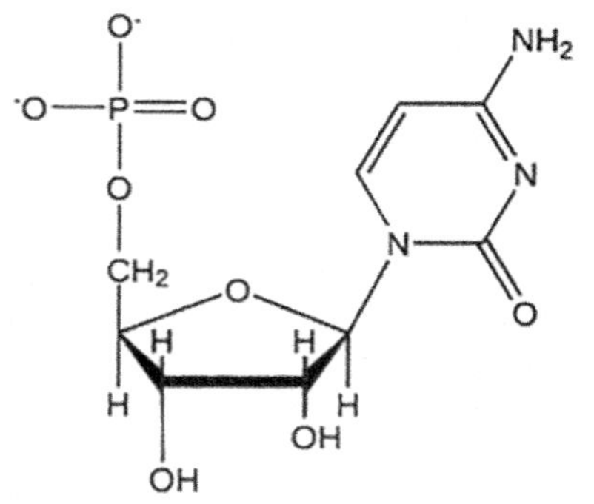

II.

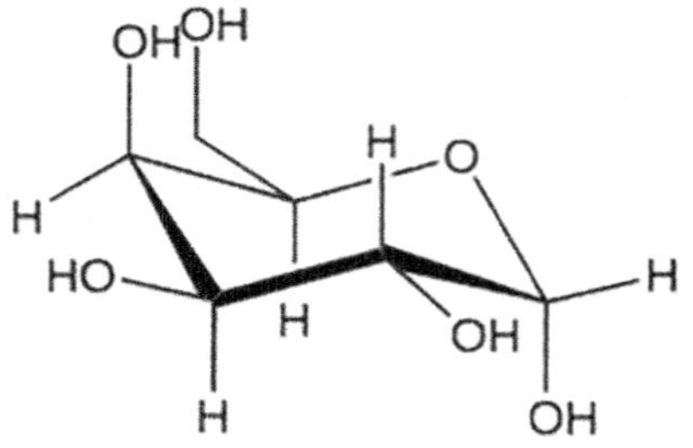

III.

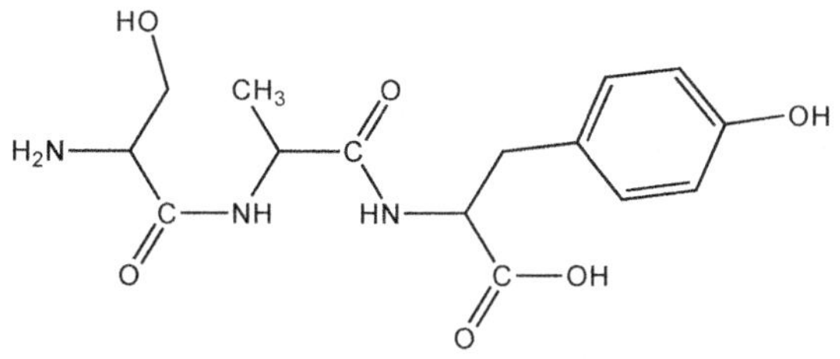

IV.

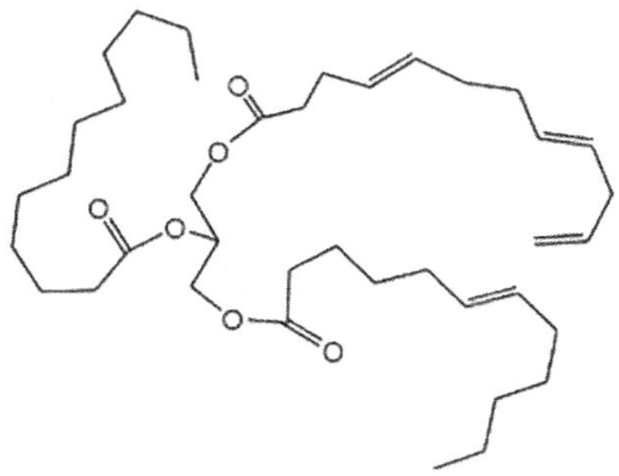

(A) I - nucleotide, II - sugar, III - peptide, IV – fat
(B) I – DNA, II sugar, III- peptide, IV lipid
(C) I - disaccharide, II - sugar, III - fatty acid, IV-polypeptide
(D) I - disaccharide, II - amino acid, III - fatty acid, IV - polysaccharide
(E) I - nucleotide, II - sugar, III - triglyceride, IV – DNA

68. Which of the following can be determined from the periodic table?

I. The number of protons
II. The number of neutrons
III. The number of isotopes of that atom
IV. The number of valence electrons

(A) I only
(B) I and II
(C) I, II, and III
(D) I, II, III, IV
(E) I, II and IV only

69. Which of the following quantum numbers are needed to define the position of the electrons in an element:

(A) principal, angular momentum, magnetic, and spin
(B) principal, circular, magnetic and electromagnetic
(C) angular, magnetic, electronic, spin
(D) primary, angular momentum, magnetic, and spin
(E) principal only

70. **Which of the following are true:**

(A) Anions are larger than their corresponding atom
(B) Second Ionization energy is greater than the first ionization energy
(C) As you move down a group on the periodic table the atomic radius increases.
(D) Atoms with completed shells are more stable.
(E) All of the above are true.

71. **The solubility of $CoCl_2$ is 54 g per 100 g of ethanol. Three flasks each contain 100 g of ethanol. Flask #1 also contains 40 g $CoCl_2$ in solution. Flask #2 contains 56 g $CoCl_2$ in solution. Flask #3 contains 5 g of solid $CoCl_2$ in equilibrium with 54 g $CoCl_2$ in solution. Which of the following describes the solutions present in the liquid phase of the flasks?**

(A) #1 - saturated, #2 - supersaturated, #3 - unsaturated.
(B) #1 - unsaturated, #2 - miscible, #3 - saturated.
(C) #1 - unsaturated, #2 - supersaturated, #3 - saturated.
(D) #1 - unsaturated, #2 - not at equilibrium, #3 – miscible.
(E) #1 unsaturated, #2 saturated, #3 miscible.

72. **An experiment requires 100. mL of a 0.500 M solution of $MgBr_2$. How many grams of $MgBr_2$ will be present in this solution?**

(A) 9.2 g
(B) 18.4 g
(C) 11.7 g
(D) 12.4 g
(E) 15.6 g

73. **Which of the following is most likely to dissolve in water?**

(A) H_2
(B) CCl_4
(C) SiO_2
(D) CH_3OH
(E) CH_4

74. **10. kJ of heat are added to one kilogram of iron at 10.° C. What is its final temperature? The specific heat of iron is 0.45 J/g•° C.**

(A) 22° C
(B) 27° C
(C) 32° C
(D) 37° C
(E) 14.5° C

Practice Test Two

75. A student wishes to prepare 4.0 liters of a 0.500 M KIO_3 (molar mass 214 g). The proper procedure is to weigh out:

(A) 42.8 grams of KIO_3, and add 4 kilograms of H_2O
(B) 428 grams of KIO_3 and add H_2O until the final homogenous solution has a volume of 4.0 L
(C) 21.4grams of KIO_3, added to 4L of water
(D) 428 g of KIO_3 added to 4 L of water.
(E) 214 g of KIO_3 added to 4.0 L of H_2O

76. A 50.0 mL sample of 0.50 M HCl is titrated with 0.50 M NaOH. What is the pH of the solution after 28.0 mL of NaOH have been added to the acid?

(A) A. 14
(B) B 8.5
(C) C 0.85
(D) D. 1.70
(E) E. 1.85

Part II. Classification Questions

Questions 77-81 refer to the following:

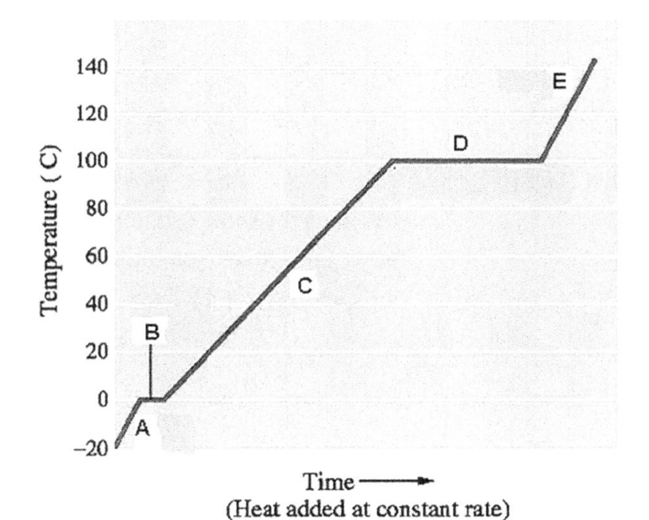

77. Represents the phase change from liquid to gas-liquid

78. Represents the heating of a liquid

79. Represents the system in the solid state

80. Represents the melting of a liquid

81. The time when there is no temperature change.

Practice Test Two

Part III: Relationship Questions

Determine if each statement is TRUE or FALSE.

If BOTH statements are true – decide if the first statement is TRUE BECAUSE of the second statement.

82. When an Al atom combines with 3 Cl atoms to form $AlCl_3$, the aluminum atom is reduced and loses electrons and reduces in size	BECAUSE	In this reaction the chlorine atoms are oxidized and gain electrons and increase in size
83. Among the halogens, fluorine is a gas, chlorine is a heavy gas moving on the ground, bromine is a liquid and iodine is a solid	BECAUSE	The intermolecular attraction between nonpolar compounds is London dispersion forces, which increase with the size of compounds. For example, boiling points in straight-chain alkanes increase with increasing the number of carbon in the chain.
84. A compressed gas in a sealed piston at room temperature is relieved to expand and the piston stops at a point where the volume of the gas is twice that in the initial position. The pressure is reduced to half of what it was.	BECAUSE	At the new position, the average speed of the molecules increases and the molecular collisions with the container walls per second decreases
85. In the reaction of hydrogen with oxygen $2\ H_2 + O_2 \longrightarrow 2\ H_2O$ 22.4 L of oxygen will react exactly with 44.8 L of hydrogen to produce water	BECAUSE	At STP (standard temperature and pressure) the molar volume of any gas, the volume that one mole of any gas occupies, is 22.4 L.

Practice Test Two

SAT Chemistry Practice Test 2

Answer Key

Question Number	Correct Answer	Your Answer
1	C	
2	B	
3	E	
4	A	
5	C	
6	D	
7	C	
8	B	
9	B	
10	C	
11	D	
12	A	
13	A	
14	E	
15	C	
16	B	
17	C	
18	A	
19	D	
20	D	
21	B	
22	C	
23	A	
24	A	
25	E	
26	C	
27	A	
28	D	
29	A	
30	B	

Question Number	Correct Answer	Your Answer
31	A	
32	C	
33	E	
34	A	
35	C	
36	D	
37	A	
38	C	
39	D	
40	D	
41	C	
42	B	
43	A	
44	C	
45	A	
46	D	
47	A	
48	D	
49	D	
50	A	
51	A	
52	B	
53	B	
54	D	
55	A	
56	C	
57	B	
58	C	
59	B	
60	D	

Question Number	Correct Answer	Your Answer
61	E	
62	C	
63	B	
64	A	
65	A	
66	A	
67	A	
68	E	
69	A	
70	E	
71	C	
72	A	
73	D	
74	C	
75	B	
76	C	
77	D	
78	C	
79	A	
80	B	
81	B or D	
82	F, F	
83	T, T	
84	T, F	
85	T, T	

SAT Chemistry Practice Test 2

Explanations

1. A piston compresses a gas at constant temperature. Which gas properties increase?

I. Average speed of molecules
II. Pressure
III. Molecular collisions with container walls per second

(A) I and II
(B) I and III
(C) II and III
(D) I, II, and III
(E) None of the above

The correct answer is C

A decrease in volume (V) occurs at constant temperature (T). Average molecular speed is determined only by temperature and will be constant. V and P are inversely related, so pressure will increase. With less wall area and at higher pressure, more collisions occur per second.

2. The temperature of a liquid is raised at atmospheric pressure. Which property of liquids increases?

(A) Critical pressure
(B) Vapor pressure
(C) Surface tension
(D) Viscosity
(E) Boiling Point

The correct answer is B

The critical pressure of a liquid is its vapor pressure at the critical temperature and is always a constant value. A rising temperature increases the kinetic energy of molecules and decreases the importance of intermolecular attraction. More molecules will be free to escape to the vapor phase (vapor pressure increases), but the effect of attractions at the liquid-gas interface will fall (surface tension decreases) and molecules will flow against each other more easily (viscosity decreases).

3. **Potassium crystallizes with two atoms contained in each unit cell. What is the mass of potassium found in a lattice $1.0 \bullet 10^6$ unit cells wide, $2.0 \bullet 10^6$ unit cells high, and $5.0 \bullet 10^5$ unit cells deep?**

(A) A. 85 ng
(B) B. 32.5 µg
(C) C. 64.9 µg
(D) D. 13 µg
(E) E. 130 µg

The correct answer is E

First we find the number of unit cells in the lattice by multiplying the number in each row, stack, and column:

$1.00 \bullet 10^6 \text{ x } 2.00 \bullet 10^6 \text{ x } .500 \bullet 10^6 = 1 \bullet 10^{18}$ unit cells.

However, there are two potassium atoms contained in each unit cell, so there are $2 \bullet 10^{18}$ potassium atoms in the described crystal.

Avogadro's number (6.022•1023) and the molecular weight of potassium (K) (39 g/mol) are used to calculate the mass in the crystal:

$2 \bullet 10^{18}$ K atoms x 1 mol / $6.022 \bullet 10^{23}$ atoms x 39 g / 1 mol = $1.29 \bullet 10^{-4}$ grams

This is equal to $129 \bullet 10^{-6}$ grams, or 129 µg

Since there are only 2 significant digits in this problem, this makes the answer 130 µg.

4. **A gas is heated in a sealed container. Which of the following occur(s)?**

(A) Gas pressure rises
(B) Gas density decreases
(C) The average distance between molecules increases
(D) The volume increases
(E) All of the above

The correct answer is A

The same material is kept in a constant volume, so neither density nor the distance between molecules will change. Pressure will rise because of increasing molecular kinetic energy impacting container walls.

5. **How many molecules are in 2.20 pg of a protein with a molecular weight of 150 kDa?**

(A) $8.83 \bullet 10^9$
(B) $1.82 \bullet 10^9$
(C) $8.83 \bullet 10^6$
(D) $1.82 \bullet 10^6$
(E) $8.83 \text{ x } 10^{15}$

The correct answer is C

The prefix "p" for "pico-" indicates 10^{-12}. A kilodalton is 1000 atomic mass units, so 150 kDa = 1.5×10^5 amu, so the molecular weight is 1.5×10^5 grams/mol.

Therefore, 2.20×10^{-12} grams x 1 mole protein / 1.5×10^5 g x 6.022×10^{23} molecules / 1 mole protein = $8.83 \cdot 10^6$ molecules

6. At STP, 20 μL of O2 contain $5.4 \cdot 10^{16}$ molecules. According to Avogadro's hypothesis, how many molecules are in 20 μL of Ne?

(A) $5.4 \cdot 10^{15}$
(B) $1.0 \cdot 10^{16}$
(C) $2.7 \cdot 10^{16}$
(D) $5.4 \cdot 10^{16}$
(E) $1.3 \cdot 10^{6}$

The correct answer is D

Avogadro's hypothesis states that equal volumes of different gases at the same temperature and pressure contain equal numbers of molecules.

7. An ideal gas at 50.0° C and 3.00 atm is in a 300 cm^3 cylinder. The cylinder volume changes by moving a piston until the gas is at 50.0° C and 1.00 atm. What is the final volume?

(A) 100 cm^3
(B) 450 cm^3
(C) 900 cm^3
(D) 1.20 dm^3
(E) 150.0 cm^3

The correct answer is C

A three-fold decrease in pressure of a constant quantity of gas at constant temperature will cause a three-fold increase in gas volume.

8. 1-butanol, ethanol, methanol, and 1-propanol are all liquids at room temperature. Rank them in order of increasing boiling point.

(A) 1-butanol < 1-propanol < ethanol < methanol
(B) methanol < ethanol < 1-propanol < 1-butanol
(C) methanol < ethanol < 1-butanol < 1-propanol
(D) 1-propanol < 1-butanol < ethanol < methanol
(E) ethanol < methanol < 1-butanol < 1-propanol

The correct answer is B

Higher boiling points result from stronger intermolecular attractive forces. The molecules listed are all alcohols with the -OH functional group attached to the end of a straight-chain alkane. In other words, they all have the formula $CH_3(CH_2)_{n-1}OH$. The only difference between the molecules is the length of the alkane corresponding to the value of *n*. With all else identical, larger molecules have greater intermolecular attractive forces due to a greater molecular surface for the attractions. Therefore the boiling points are ranked: methanol (CH_3OH) < ethanol (CH_3CH_2OH) < 1-propanol ($CH_3CH_2CH_2OH$) < 1-butanol ($CH_3CH_2CH_2CH_2OH$).

9. One mole of an ideal gas at STP occupies 22.4 L. At what temperature will one mole of an ideal gas at 1 atm occupy 31.0 L?

(A) 34.6° C
(B) 139° C
(C) 378° C
(D) 442° C
(E) 28 ° C

The correct answer is B

Either Charles' law, the combined gas law, or the ideal gas law may be used with temperature in Kelvin. Charles' law or the combined gas law with $P_1 = P_2$ may be manipulated to equate a ratio between temperature and volume when *P* and *n* are constant. First, realize the STP means the temperature is 25°C or 298.15 K.

Using the ideal gas law and realizing the constant and n (number of moles) are the same on both sides:

P1 * V1 / nR * T1 = P2 * V2 / nR * T2

1 atm * 22.4 L / 298.15 K = 1 atm * 31.0 L / T2

T2 = 31 L * 298.15 K / 22.4 L

T2 = 412.6 K = 139 ° C in 3 significant digits

You could also think of this as:

V1 / T1 = V2 / T2

22.4 L / 298.15 K = 31.0 L / T2

T2 = 31.0 L * 298.15 K / 22.4 L

T2 = 412.61 = 139.47 °C = 139 ° C in 3 significant digits

10. What pressure is exerted by a mixture of 2.7 g of H_2 and 59 g of Xe at STP on a 50. L container?

(A) 0.69 atm
(B) 0.76 atm
(C) 0.88 atm
(D) 0.97 atm
(E) 27.0 atm

The answer is C

Grams of gas are first converted to moles:

$2.7 \text{ g } H_2 \text{ x } 1 \text{ mol } H_2 / 2 \text{ g } H_2 = 1.35 \text{ mol } H_2$

$59 \text{ g Xe x } 1 \text{ mol Xe} / 131 \text{ g Xe} = 0.45 \text{ mol Xe}$

Dalton's law of partial pressures for an ideal gas is used to find the pressure of the mixture:

$P_{total} = P1 + P2$

$P_{total} = (n1 * RT / V) + (n2 * RT / V)$

$R = 0.082 \text{ L} * \text{atm} / \text{K} * \text{mol}$

$T = 298.15 \text{ K}$

$P_{total} = [(1.35 \text{ mol H2}) * (0.082 \text{ L} * \text{atm} / \text{K} * \text{mol}) * (298.15 \text{ K}) / (50.0 \text{ L})] + [(0.45 \text{ mol Xe}) * (0.082 \text{ L} * \text{atm} / \text{K} * \text{mol}) * (298.15 \text{ K}) / (50.0 \text{ L})]$

$P_{total} = [.66 \text{ atm}] + [.22 \text{ atm}] = 0.88 \text{ atm}$

alternatively, you could bypass Dalton's law of partial pressures and simply add the moles of gas, and then solve using the ideal gas equation:

$P_{total} = (1.8 \text{ moles gas}) * (0.082 \text{ L} * \text{atm} / \text{K} * \text{mol}) * (298.15 \text{ K}) / 50.0 \text{ L}$

$P_{total} = 0.88 \text{ atm}$

11. The boiling point of water at sea level on the Kelvin scale is closest to:

(A) 112 K
(B) 212 K
(C) 273 K
(D) 373 K
(E) 298 K

The correct answer is D

Temperature in Kelvin is equal to the temperature in Celsius plus 273.15. Since the boiling point of water at 1 atm pressure (that is the pressure at sea level) is 100° C, it will boil at 373.15 K, corresponding to answer D.

12. Which phases may be present at the triple point of a substance?

I. Gas
II. Liquid
III. Solid
IV. Supercritical fluid

(A) I, II, and III
(B) I, II, and IV
(C) II, III, and IV
(D) I, II, III, and IV
(E) I, III, IV

The correct answer is A

Gas, liquid, and solid may exist together at the triple point.

13. Carbonated water is bottled at 25° C under pure CO_2 at 4.0 atm. Later the bottle is opened at 4° C under air at 1.0 atm that has a partial pressure of $3 \cdot 10^{-4}$ atm CO_2. Why do CO_2 bubbles form when the bottle is opened?

(A) CO_2 falls out of solution due to a drop in solubility at the lower total pressure.
(B) CO_2 falls out of solution due to an increase in solubility at the lower CO_2 pressure.
(C) CO_2 falls out of solution due to a drop in solubility at the lower temperature.
(D) CO_2 falls out of solution due to the decreased concentration of water.
(E) CO_2 is formed by the decomposition of carbonic acid.

The correct answer is A

Henry's law states that CO_2 solubility in M (mol/L) will be proportional to the partial pressure of the gas. A four-fold increase in pressure from 1.0 atm to 4.0 atm will increase solubility four-fold from 0.034 M to 0.14 M. Opening the bottle at the lower pressure will cause a decrease in solubility, causing the gas to come out of solution.

14. Which statement about molecular structures is false?

(A) $H_2C{=}CH{-}CH{=}CH_2$ is a conjugated molecule.
(B) A bonding σ orbital connects two atoms in a double bond.
(C) A bonding π orbital connects two atoms via a single bond in a separate region from the straight line between them.
(D) The anion with resonance forms

$[\,O{=}CH{-}O\,]^{-} \longleftrightarrow [\,O{-}CH{=}O\,]^{-}$ will always exist in one form or the other.

(E) They are all false

The correct answer is E

All of these statements are false.

15. What is the chemical composition of magnesium nitrate?

(A) 11.1% Mg, 22.2% N, 66.7% O
(B) 50.1% Mg, 22.2% N, 33.0% O
(C) 16.4% Mg, 18.9% N, 64.7% O
(D) 20.9% Mg, 24.1% N, 55.0% O
(E) 28.2% Mg, 16.2% N, 55.7% O

The correct answer is C

First find the formula for magnesium nitrate. Mg is an alkali earth metal and will always have a 2+ charge. The nitrate ion is NO_3^-. Two nitrate ions are required for each Mg^{2+} ion. Therefore the formula is $Mg(NO_3)_2$

Determine the chemical composition:

Determine the number of atoms for each element in $Mg(NO_3)_2$:

1 Mg, 2 N, 6 O.

Multiply by the molecular weight of the elements to determine the grams of each in one mole of the formula:

$$\frac{1 \text{ mol Mg}}{\text{mol Mg(NO}_3)_2} \times \frac{24.3 \text{ g Mg}}{\text{mol Mg}} = 24.3 \text{ g Mg/mol Mg(NO}_3)_2$$

$$2(14.0) = 28.0 \text{ g N/mol Mg(NO}_3)_2$$

$$6(16.0) = 96.0 \text{ g O/mol Mg(NO}_3)_2$$

$$148.3 \text{ g Mg(NO}_3)_2\text{/mol Mg(NO}_3)_2$$

Divide to determine % composition:

$$\%\text{Mg} = \frac{24.3 \text{ g Mg/mol Mg(NO)}}{148.3 \text{ g Mg(NO) /mol Mg(NO)}} = 0.164 \text{ g Mg/g Mg(NO)} \times 100\% = 16.4\%$$

$$\%\text{N} = \frac{28.0}{148.3} \times 100\% = 18.9\% \qquad \%\text{O} = \frac{96.0}{148.3} \times 100\% = 64.7\%$$

16. How many neutrons are there in the radioactive isotope Cobalt-60 (atomic mass = 60)?

(A) 27
(B) 33
(C) 60
(D) 87
(E) 14

The correct answer is B

The number of neutrons is found by subtracting the atomic number (27) from the mass number (60).

17. Select the list of atoms that is arranged in order of increasing size:

(A) Mg, Na, Si, Cl
(B) Si, Cl, Mg, Na
(C) Cl, Si, Mg, Na
(D) Na, Mg, Si, Cl
(E) Mg, Si, Cl, Na

The correct answer is C

These atoms are all in the same row of the periodic table. Size increases further to the left for atoms in the same row.

18. Based on trends in the periodic table, which of the following properties would you expect to be greater for Rb than for K?

I. Density
II. Melting point
III. Ionization energy
IV. Oxidation number in a compound with chlorine

(A) I only
(B) I, II, and III
(C) II and III
(D) I, II, III, and IV
(E) None of the above

The correct answer is A

Rb is underneath K in the alkali metal column (group 1) of the periodic table. There is a general trend for density to increase as one moves down a group; therefore, we select choice I. Rb and K experience metallic bonds for intermolecular forces, and the strength of metallic bonds decreases for larger atoms further down the periodic table resulting in a lower melting point for Rb, so we do not choose II. Ionization energy decreases for larger atoms further down the periodic table, so we do not choose III. Both Rb and K would be expected to have a charge of +1 and therefore an oxidation number of +1 in a compound with chlorine, so we do not choose IV.

19. Why does $CaCl_2$ have a higher normal melting point than NH_3?

(A) London dispersion forces in $CaCl_2$ are stronger than covalent bonds in NH_3.
(B) Covalent bonds in NH_3 are stronger than dipole-dipole bonds in $CaCl_2$.
(C) Ionic bonds in $CaCl_2$ are stronger than London dispersion forces, the strongest intermolecular forces in NH_3.
(D) Ionic bonds in $CaCl_2$ are stronger than hydrogen bonds, the strongest intermolecular forces in NH_3.
(E) None of the above

Practice Test Two

The correct answer is D

London dispersion forces are weaker than covalent bonds, eliminating choice A. A higher melting point will result from stronger intermolecular bonds, eliminating choice B. $CaCl_2$ is an ionic solid resulting from a cation on the left and an anion on the right of the periodic table. The dominant attractive forces between NH_3 molecules are hydrogen bonds. (All molecules have London forces, but these are weak intermolecular forces.)

20. Rank the following bonds from least to most polar:

C-H, C-Cl, H-H, C-F

(A) C-H < H-H < C-F < C-Cl
(B) H-H < C-H < C-F < C-Cl
(C) C-F < C-Cl < C-H < H-H
(D) H-H < C-H < C-Cl < C-F
(E) H-H < C-CL < C-H< C-F

The correct answer is D

Bonds between atoms of the same element are completely non-polar, so H-H is the least polar bond in the list, eliminating choices A and C. The C-H bond is considered to be non-polar even though the electrons of the bond are slightly unequally shared. C-Cl and C-F are both polar covalent bonds, but C-F is more strongly polar because F has a greater electronegativity.

21. In C_2H_2, each carbon atom contains the following valence orbitals:

(A) p only
(B) p and sp hybrids
(C) p and sp^2 hybrids
(D) sp^3 hybrids only
(E) s, p, and sp^2 hybrids

The correct answer is B

An isolated C has the valence electron configuration $2s^22p^2$. Before bonding, one *s* electron is promoted to an empty *p* orbital. In C_2H_2, each C atom bonds to 2 other atoms. Bonding to two other atoms is achieved by combination into two *p* orbitals and two *sp* hybrids.

22. The boiling points of N_2, O_2 and Cl_2 are -196, -182, and -34 respectively. Cl_2 boils at much higher temperature than the others. This is explained by

(A) Single bonds are easier to break than double or triple bonds
(B) Cl_2 has a longer bond length than the others
(C) Cl_2 has greater London dispersion forces because it has more electrons.
(D) Hydrogen bonding is stronger in Cl_2
(E) None of the above

The correct answer is C

Liquid oxygen and liquid nitrogen and liquid chlorine are all non-polar substances that experience only London dispersion forces of attraction. These forces are greater for Cl_2 because it has more electrons, so it has the highest boiling point.

23. Which of the following statements regarding molecular geometries are true?

I. Double bonds are planar
II. A central atom with sp2 hybridization is trigonal
III. A central atom with sp2 hybridization is bipyramidal
IV. Single bonds are always planar

(A) I and II are true
(B) I and IV are true
(C) II and IV are true
(D) II and III are true
(E) I, II and IV are true

The correct answer is A

I is true, double bonds are planar; single bonds are never planar. II is true sp2 hybridization is trigonal. Therefore only I and II are true.

24. The electronic configuration or Iron (Fe) is

(A) $1s^2\ 2s^2\ 2p^6\ 3s^2\ 3p^6\ 4s^2\ 3d^6$
(B) $1s^2\ 2s^2\ 2p^6\ 3s^2\ 3p^6\ 3d^8$
(C) $1s^2\ 2s^2\ 2p^6\ 3s^2\ 3p^6\ 4s^2$
(D) $1s^2\ 2s^2\ 2p^6\ 3s^2\ 3p^6\ 4s^2\ 4p^6$
(E) $1s^2\ 2s^2\ 2p^6\ 3s^2\ 3p^6\ 4s^2\ 4p^3\ 4d^3$

The correct answer is A

Fe has 26 electrons. Filling each shell from 1s 2s 2p 3s 3p ... then remember to fill 4s before 3d.

25. Which of the following is a correct electron arrangement for oxygen?

(A) 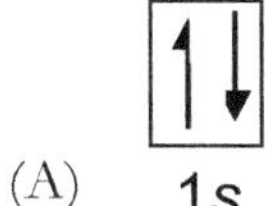 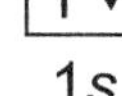 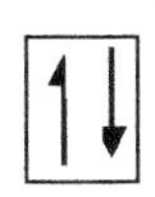 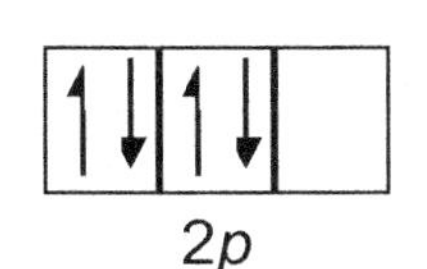

(B) $1s^2 2s^2, 2s^2, 2p^2$
(C) 2, 2, 4
(D) 2, 2, 4, 1/2
(E) None of the above

The correct answer is E

Choice A violates Hund's rule. The two electrons on the far right should occupy the final two orbitals. B should be $1s^22s^22p^4$. There is no $1p$ subshell. C should be 2, 6. Number lists indicate electrons in shells. There is no value 1/2

26. Which of the following is a proper Lewis dot structure of CHClO?

(A)

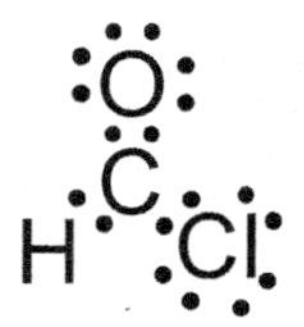

(B)

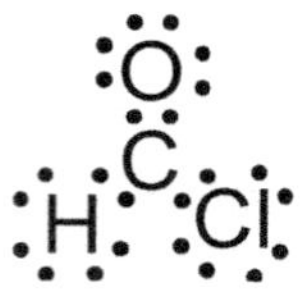

(C)

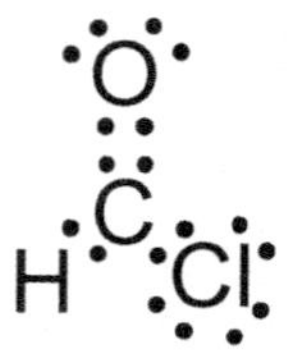

(D)

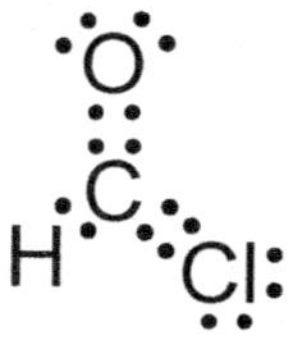

(E) None of the above

The correct answer is C

C has 4 valence shell electrons, H has 1, Cl has 7, and O has 6. The molecule has a total of 18 valence shell electrons. This eliminates choice B which has 24. Choice B is also incorrect because has an octet around a hydrogen atom instead of 2 electrons and because there are only six electrons surrounding the central carbon. A single bond connecting all atoms would give choice A. This is incorrect because there are only 6 electrons surrounding the central carbon. A double bond between C and O gives the correct answer, C. A double bond between C and O and also between C and Cl would give choice D. This is incorrect because there are 10 electrons surrounding the central carbon.

27. Which intermolecular attraction explains the following trend in straight-chain alkanes?

Condensed structural formula	Boiling point (° C)
CH_4	-161.5
CH_3CH_3	-88.6
$CH_3CH_2CH_3$	-42.1
$CH_3CH_2CH_2CH_3$	-0.5
$CH_3CH_2CH_2CH_2CH_3$	36.0
$CH_3CH_2CH_2CH_2CH_2CH_3$	68.7

(A) London dispersion forces
(B) hydrophobic interactions
(C) Dipole-dipole interactions
(D) Hydrogen bonding
(E) Ion-induced dipole interactions

The correct answer is A

Alkanes are composed entirely of non-polar C-C and C-H bonds, resulting in no dipole interactions or hydrogen bonding. London dispersion forces increase with the size of the molecule, resulting in a higher temperature requirement to break these bonds and a higher boiling point.

28. Match the theory with the scientist who first proposed it:

I. Electrons, atoms, and all objects with momentum also exist as waves.
II. Electron density may be accurately described by a single mathematical equation.
III. There is an inherent indeterminacy in the position and momentum of particles.
IV. Radiant energy is transferred between particles in exact multiples of a discrete unit.

(A) I - de Broglie, II - Planck, III - Schrödinger, IV - Thomson
(B) I - Dalton, II - Bohr, III - Planck, IV - de Broglie
(C) I - Henry, II - Bohr, III - Heisenberg, IV - Schrödinger
(D) I - de Broglie, II - Schrödinger, III - Heisenberg, IV - Planck
(E) I - Schrödinger, II - de Broglie, III - Plank, IV - Heisenberg

The correct answer is D

Henry's law relates gas partial pressure to liquid solubility. de Broglie theorized that particles such as **electrons, atoms, and all objects with momentum also exist as waves.** Schrödinger's equation is the equation that describes electron density, Heisenberg developed the Uncertainty Principle and Plank theorized discrete energy levels. So D is the correct answer.

29. **The terrestrial composition of an element is: 50% as a stable isotope with an atomic mass of 78 amu and 50% as a stable isotope with an atomic mass of 80 amu. Calculate the atomic mass of the element.**

(A) 79 amu
(B) 78 amu
(C) 77 amu
(D) 81 amu
(E) 80 amu

The correct answer is A

[(78) + (80)] / 2 = 79 amu

30. $^{3}_{1}H$ **decays with a half-life of 12 years. 3.0 g of pure** $^{3}_{1}H$ **were placed in a sealed container 24 years ago. How many grams of** $^{3}_{1}H$ **remain?**

(A) 0.38 g
(B) 0.75 g
(C) 1.5 g
(D) 0.125 g
(E) 3.0 g

The correct answer is B

Every 12 years, the amount remaining is cut in half. After 12 years, 1.5 g will remain. After another 12 years, 0.75 g will remain.

31. **Moving down a column in the Periodic Table:**

I. the atomic radius increase
II. Ionization energy increase
III. Protons are added
IV. metallic characteristics increase

(A) I and III
(B) I and II only
(C) III only
(D) III and IV only
(E) None of the above

The correct answer is A

Moving down a column the atomic radii increases, the ionization energy decreases, and protons are added. The metallic characteristics do not increase. So, only I and III are correct.

32. Moving from left to right on a Periodic Table (Li – to Ne) which of the following are true:

I. the atomic radius decreases
II. electrons are added
III. ionization energies increase
IV. electronegativity decreases

(A) I only
(B) I, II
(C) I, II, III
(D) I, II, III, IV
(E) None of the above

The correct answer is C

Moving from left to right along a row of the periodic table the atomic radius decreases, electrons are added, ionization energies increase and electronegativities increase.

33. NH_4F is dissolved in water. Which of the following are conjugate acid/base pairs present in the solution?

I. NH_4^+/NH_4OH
II. HF/F^-
III. H_3O^+/H_2O
IV. H_2O/OH^-

(A) I only
(B) I, II, and III
(C) I, III, and IV
(D) II and IV
(E) II, III, and IV

The correct answer is E

NH_4F is soluble in water and completely dissociates to NH_4^+ and F^-. F^- is a weak base with HF as its conjugate acid (II). NH_4^+ is a weak acid with NH_3 as its conjugate base. A conjugate acid/base pair must have the form HX/X (where X is one lower charge than HX). NH_4^+/NH_4OH (I) is *not* a conjugate acid/base pair, eliminating Choices A and B. H_3O^+/H_2O and $H_2O/OH-$ (III and IV) are always present in water and in all aqueous solutions as conjugate acid/base pairs. All of the following equilibrium reactions occur in $NH_4F(aq)$:

$$NH_4^+ (aq) + OH^- (aq) \longleftrightarrow NH_3 (aq) + H_2O (l)$$

$$F^- (aq) + H_3O^+ (aq) \longleftrightarrow HF (aq) + H_2O (l)$$

$$2H_2O (l) \longleftrightarrow H_3O^+ (aq) + OH^- (aq)$$

Practice Test Two

34. Rank the following from lowest to highest pH. Assume a small volume for the component given in moles:

I. 0.01 mol HCl added to 1 L H_2O
II. 0.01 mol HI added to 1 L of an acetic acid/sodium acetate solution at pH 4.0
III. 0.01 mol NH_3 added to 1 L H_2O
IV. 0.1 mol HNO_3 added to 1 L of a 0.1 M $Ca(OH)_2$ solution

(A) I < II < III < IV
(B) I < II < IV < III
(C) II < I < III < IV
(D) II < I < IV < III
(E) IV < III < II < I

The correct answer is A

HCl is a strong acid.

HI is also a strong acid and would have a pH of 2 at this concentration in water, but the buffer will prevent the pH from dropping this low. Solution II will have a pH above 2 and below 4, eliminating choices C and D.

If a strong base were in Solution III, its pOH would be 2. Using the equation pH + pOH = 14, its pH would be 12. Because NH_3 is a weak base, the pH of Solution III will be greater than 7 and less than 12.

A neutralization reaction occurs in Solution IV between 0.1 mol of H^+ from the strong acid HNO_3 and 0.2 mol of OH^- from the strong base $Ca(OH)_2$. Each mole of $Ca(OH)_2$ contributes two base equivalents for the neutralization reaction. The base is the excess reagent and 0.1 mol of OH^- remains after the reaction. This resulting solution will have a pOH of 1 and a pH of 13.

A is correct because: 2 < between 2 and 4 < between 7 and 12 < 13.

35. Which statement about acids and bases is *not* true?

(A) All strong acids ionize in water.
(B) All Lewis acids accept an electron pair.
(C) All Brønsted bases use OH^- as a proton acceptor
(D) All Arrhenius acids form H^+ ions in water.
(E) Water can act as either an acid or a base.

The correct answer is C

Choice A is the definition of a strong acid, Choice B is the definition of a Lewis acid, and Choice D is the definition of an Arrhenius acid. By definition, all Arrhenius bases form OH^- ions in water, and all Brønsted bases are proton acceptors. But not all Brønsted bases use OH^- as a proton acceptor. For example, NH_3 is a Brønsted base.

36. What is the pH of a buffer solution made of 0.128 M sodium formate (NaCOOH) and 0.072 M formic acid (HCOOH)? The pK_a of formic acid is 3.75.

(A) 2.0
(B) 3.0
(C) 3.75
(D) 4.0
(E) 5 .0

The correct answer is D

From the pK_a, we may find the K_a of formic acid: 1.8×10^{-4}

$$K_a = 10^{-pK_a} = 10^{-3.75} = 1.78 \times 10^{-4}$$

$$K_a = \frac{[H^+][HCOO^-]}{[HCOOH]} = 1.78 \times 10^{-4}$$ for the dissociation: $HCOOH \text{ ƒ } H^+ + HCOO^-$

e pH is found by solving for the H^+ concentration:

$$[H^+] = K_a \frac{[HCOOH]}{[HCOO^-]} = (1.78 \times 10^{-4}) \frac{0.072}{0.128} = 1.0 \times 10^{-4} \text{ M}$$

$$pH = -\log_{10}[H^+] = -\log_{10}(1.0 \times 10^{-4}) = 4.0 \text{ (choice C)}$$

37. A 100. L vessel of pure O_2 at 500. kPa and 20.° C is used for the combustion of butane:

$$2\,C_4H_{10} + 13\,O_2 \longrightarrow 8\,CO_2 + 10\,H_20$$

Find the mass of butane that consumes all the O_2 in the vessel. Assume O_2 is an ideal gas and use a value of R = 8.314 J/(mol•K).

(A) 183 g
(B) 467 g
(C) 1.83 kg
(D) 2.6 kg
(E) 7.75 kg

The correct answer is A

We are given a volume and asked for a mass. The steps will be "volume to moles to moles to mass." "Volume to moles..." requires the ideal gas law, but first several units must be altered:

Units of joules are identical to m^3•Pa
500 kPa is 500•10^3 Pa
100 L is 0.100 m^3
20° C is 293.*15* K

$PV=nRT$ is rearranged to give:

$$n = \frac{PV}{RT} = \frac{(500\times10^3\ PA)(0.100m^3\ O_2)}{\left(8.314\frac{m^3\ g\ Pa}{mol\ g\ K}\right)(293.15K)} = 20.5\mathit{1}\ mol\ O_2$$

"...to moles to mass" utilizes stoichiometry. The molecular weight of butane is 58.1 u:

$$20.5\mathit{1}\ mol\ O_2 \times \frac{2\ mol\ C_4H_{10}}{13\ mol\ O_2} \times \frac{58.1\ g\ C_4H_{10}}{1\ mol\ C_4H_{10}} = 183\ g\ C_4H_{10}$$

38. **32.0 g of hydrogen and 32.0 grams of oxygen react to form water until the limiting reagent is consumed. What is present in the vessel after the reaction is complete?**

(A) 16.0 g O_2 and 48.0 g H_2O
(B) 24.0 g H_2 and 40.0 g H_2O
(C) 28.0 g H_2 and 36.0 g H_2O
(D) 28.0 g H_2 and 34.0 g H_2O
(E) 28.0 g H_2 and 16.0 g O_2

The correct answer is C

First the equation must be constructed:

$$2H_2 + O_2 \longrightarrow 2H_2O$$

A fast and intuitive solution would be to recognize that:

1. One mole of H_2 is about 2.0 g, so about 16 moles of H_2 are present.
2. One mole of O_2 is 32.0 g, so one mole of is O_2 is present.
3. Imagine the 16 moles of H_2 reacting with one mole of O_2. 2 moles of H_2 will be consumed before the one mole of O_2 is gone. O_2 is limiting. (Eliminate choice A.)
4. 16 moles less 2 leaves 14 moles of H_2 or about 28 g. (Eliminate choice B.)
5. The reaction began with 64.0 g total. Conservation of mass for chemical reactions forces the total final mass to be 64.0 g also. (Eliminate choice D.)
6. Of course, choice E is incorrect because it does not include any product, H_2O.

39. Which reaction is *not* a redox process?

(A) Combustion of octane: $2C_8H_{18} + 25O_2 \longrightarrow 16CO_2 + 18H_2O$
(B) Depletion of a lithium battery: $Li + MnO_2 \longrightarrow LiMnO_2$
(C) Corrosion of aluminum by acid: $2Al + 6HCl \longrightarrow 2AlCl_3 + 3H_2$
(D) Taking an antacid for heartburn:
$CaCO_3 + 2HCl \longrightarrow CaCl_2 + H_2CO_3 \longrightarrow CaCl_2 + CO_2 + H_2O$
(E) None of the above

The correct answer is D

The oxidation state of atoms is altered in a redox process. During combustion (Choice A), the carbon atoms are oxidized from an oxidation number of -4 to +4. Oxygen atoms are reduced from an oxidation number of 0 to -2. All batteries (Choice B) generate electricity by forcing electrons from a redox process through a circuit. Li is oxidized from 0 in the metal to +1 in the $LiMnO_2$ salt. Mn is reduced from +4 in manganese (IV) oxide to +3 in lithium manganese (III) oxide salt. Corrosion (Choice C) is due to oxidation. Al is oxidized from 0 to +3. H is reduced from +1 to 0. Acid-base neutralization (Choice D) transfers a proton (an H atom with an oxidation state of +1) from an acid to a base. The oxidation state of all atoms remains unchanged (Ca at +2, C at +4, O at -2, H at +1, and Cl at -1), so D is correct. Note that Choices C and D both involve an acid. The availability of electrons in aluminum metal favors electron transfer but the availability of CO_3^{2-} as a proton acceptor favors proton transfer.

40. Which of the following are true about Galvanic cells?

I. two half reactions take place in the two chambers
II. oxidation takes place at the anode
III. reduction takes place at the cathode
IV. if the salt bridge is removed the voltage drops to zero
V. La Châtelier's principle can be applied to the systematic

(A) I , II, III ARE TRUE
(B) ONLY II AND III ARE TRUE
(C) ONLY I AND IV ARE TRUE
(D) All are true
(E) none are true

The correct answer is D

All of the statements are true when learning about Galvanic cells.

41. Balance the equation for the neutralization reaction between phosphoric acid and calcium hydroxide by filling in the blank stoichiometric coefficients.

$H_3PO_4 + \quad Ca(OH)_2 \longrightarrow Ca_3(PO_4)_2 + H_2O$

Practice Test Two

The correct answer is C

We are given the unbalanced equation. Next we determine the number of atoms on each side. For reactants (left of the arrow): 5 H, 1 P, 6 O, and 1 Ca. For products: 2 H, 2 P, 9 O, and 3 Ca.

We assume that the molecule with the most atoms – $Ca_3(PO_4)_2$ – has a coefficient of one, and find the other coefficients required to have the same number of atoms on each side of the equation. Assuming $Ca_3(PO_4)_2$ has a coefficient of one means that there will be 3 Ca and 2 P on the right because H_2O has no Ca or P. A balanced equation would also have 3 Ca and 2 P on the left. This is achieved with a coefficient of 2 for H_3PO_4 and 3 for $Ca(OH)_2$. Now we have:

$$2H_3PO_4 + 3Ca(OH)_2 \longrightarrow Ca_3(PO_4)_2 + H_2O$$

The coefficient for H_2O is found by balancing H or O. Whichever one is chosen, the other atom should be checked to confirm that a balance actually occurs. For H, there are 6 H from $2H_3PO_4$ and 6 from $3Ca(OH)_2$ for a total of 12 H on the left. There must be 12 H on the right for balance. None are accounted for by $Ca_3(PO_4)_2$, so all 12 H must be associated with H_2O. It has a coefficient of 6:

$$2H_3PO_4 + 3Ca(OH)_2 \longrightarrow Ca_3(PO_4)_2 + 6H_2O$$

This is choice C, but if time is available, it is best to check that the remaining atoms are balanced. There are 8 O from $2H_3PO_4$ and 6 from $3Ca(OH)_2$ for a total of 14 on the left, and 8 O from $Ca_3(PO_4)_2$ and 6 from $6H_2O$ for a total of 14 on the right. The equation is balanced.

Multiplication by a whole number is not required because the stoichiometric coefficients from step 3 already are whole numbers.

An alternative method would be to try the coefficients given for answer A, answer B, etc. until we recognize a properly balanced equation.

42. Write an equation showing the reaction between calcium nitrate and lithium sulfate in aqueous solution. Include all products.

(A) $Ca(NO_3)_2 + 2H_2O + Li_2SO_4 \longrightarrow CaSO_4\ (s) + 2\ LiNO_3 + 2OH- + H_2(g)$
(B) $Ca(NO_3)_2 + Li_2SO_4 \longrightarrow CaSO_4\ (s) + 2\ LiNO_3$
(C) $Ca(NO_3)_2$ and $Li_2SO_2 \longrightarrow CaSO_2\ (s) + 2\ LiNO_3$
(D) $Ca(NO_3)$ and $Li_2SO_4 \longrightarrow CaSO_4\ (s) + Li_2NO_3$
(E) None of the Above

The correct answer is B

When two ionic compounds are in solution, a precipitation reaction should be considered. We can determine from their names that the two reactants are the ionic compounds $Ca(NO_3)_2$ and Li_2SO_4. The compounds are present in aqueous solution as their four component ions Ca^{2+}, NO_3^-, Li^+, and SO_4^{2-}. Solubility rules indicate that nitrates are always soluble but sulfate will form a solid precipitate with Ca^{2+}, forming $CaSO_4\ (s)$. B is correct.

43. Find the mass of CO_2 produced by the combustion of 15 kg of isopropyl alcohol in the reaction:

$$2\ C_3H_7OH\ (l) + 9\ O_2\ (g) \longrightarrow 6\ CO_2\ (g) + 8\ H_2O\ (g)$$

(A) 33 kg
(B) 44 kg
(C) 50 kg
(D) 60 kg
(E) 66 kg

The correct answer is A

Remember "grams to moles to moles to grams." Step 1 converts mass to moles for the known value. In this case, kg and kmol are used. Step 2 relates moles of the known value to moles of the unknown value by their stoichiometry coefficients. Step 3 converts moles of the unknown value to a mass.

$$\underbrace{}_{}\quad \text{step 1} \qquad \text{step 2} \qquad \text{step 3}$$

$$15\times10^3\ g\ C_4H_8O \times \frac{1\ mol\ C_4H_8O}{60\ g\ C_4H_8O} \times \frac{6\ mol\ CO_2}{2\ mol\ C_4H_8O} \times \frac{44\ g\ CO_2}{1\ mol\ CO_2} = 33\times10^3\ g\ CO_2$$

$$= 33\ kg\ CO_2$$

44. What is the density of nitrogen gas at STP? Assume an ideal gas and a value of 0.08206 L•atm/(mol•K) for the gas constant.

(A) 0.62 g/L
(B) 1.14 g/L
(C) 1.25 g/L
(D) 2.03 g/L
(E) 3.38 g/L

The correct answer is C

The molecular mass *M* of N_2 is 28.0 g/mol.
Recall that one mole of an ideal gas at STP occupies 22.4 L.

1 mole = 28.0 grams. One mole occupies 22.4 L. Therefore, 28.0 grams occupies 22.4 L.

28.0 grams / 22.4 L = 1.25 g/L

45. Find the volume of methane that will produce 12 m^3 of hydrogen in the reaction:

$$CH_4(g) + H_2O(g) \rightarrow CO(g) + 3H_2(g)$$

Assume that the temperature and pressure remain constant.

(A) 4.0 m^3
(B) 32 m^3
(C) 36 m^3
(D) 64 m^3
(E) Cannot be determined

The correct answer is A

Stoichiometric coefficients may be used directly for ideal gas volumes at constant T and P because of Avogadro's Law.

12 m^3 of H_2 will be produced from 4 m^3 of CH_4 .

46. Household "chlorine bleach" is sodium hypochlorite. Which of the following best represents the production of sodium hypochlorite, sodium chloride, and water by bubbling chlorine gas through aqueous sodium hydroxide?

(A) $Cl + NaOH \longrightarrow NaCl + H_2O$
(B) $Cl_2 + 2NaOH \longrightarrow 2NaClO_2 + H_2O$
(C) $Cl + 2NaOH \longrightarrow Na^+ + NaOCl + H_2O$
(D) $Cl_2 + 2NaOH \longrightarrow NaCl + NaOCl + H_2O$
(E) None of the above.

The correct answer is D

$$Cl_2 + 2NaOH \longrightarrow NaCl + NaOCl + H_2O$$

Chlorine gas is a diatomic molecule, eliminating choices A and C. The hypochlorite ion is ClO^- eliminating choices A and B. All of the equations are properly balanced.

47. For the reaction $M(s) + Cd^{2+} \longrightarrow M^+ + Cd(s)$ in an electrochemical cell at 25 °C Which of the following are true ?

(A) $M(s) \longrightarrow M^+ + e-$ takes place at the anode
(B) As the reaction proceeds the $[M^+]$ decreases
(C) $Cd(2^+)$ looses electrons
(D) $Cd(2^+) \longrightarrow Cd\ (s)$ takes place at the anode
(E) There is not enough information to determine

Practice Test Two

The correct answer is A

A is true because M(s) loses electrons, therefore this is oxidation (LEO) and oxidation occurs at the anode.

As the reaction proceeds the $[M^+]$ increase so B is false. Cd(2+) gains the electrons (GER) , so this is the reduction reaction which takes place at the cathode, so C and D are also false.

48. Write the equilibrium expression K_{eq} for the reaction:

$$CO_2 (g) + H_2 (g) \longleftrightarrow CO (g) + H_2O (g)$$

(A) $K_{eq} = [CO_2] [H_2]^3 / [H_2O] [CO]$
(B) $K_{eq} = [H_2O] [CO] / [CO_2] [H_2]$
(C) $K_{eq} = [CO_2] [H_2]^2 / [H_2O] [CO]$
(D) $K_{eq} = [CO_2] [H_2] / [H_2O] [CO]$
(E) None of the above

The correct answer is D

Product concentrations are multiplied together in the numerator and reactant concentrations in the denominator, eliminating choice B. The stoichiometric coefficient of H_2 is one, eliminating choice A and C. D is correct.

49. The exothermic reaction $2NO (g) + Br_2 (g) \longleftrightarrow 2NOBr (g)$ is at equilibrium. According to Le Châtelier's principle:

(A) Adding Br_2 will increase [NO].
(B) Adding [NO] will increase Br_2.
(C) An increase in container volume (with T constant) will increase [NOBr].
(D) An increase in pressure (with T constant) will increase [NOBr].
(E) An increase in temperature (with P constant) will increase [NOBr].

The correct answer is D

Le Châtelier's principle predicts that equilibrium will shift to partially offset any change. Adding Br_2 will be partially offset by reducing $[Br_2]$ and [NO] via a shift to the right (not Choice A). Adding [NO] will have the same effect. For the remaining possibilities, we may write the reaction as: 3 moles $\longrightarrow$ 2 moles + heat. An increase in container volume will decrease pressure. This change will be partially offset by an increase in the number of moles present, shifting the reaction to the left (not Choice B). An increase in pressure will be offset by a decrease in the number of moles present, shifting the reaction to the right (Choice D, correct). Raising the temperature by adding heat will shift the reaction to the left (not Choice C).

50. At a certain temperature, T, the equilibrium constant for the reaction $2NO\ (g) \longleftrightarrow N_2\ (g) + O_2\ (g)$ is $K_{eq} = 2x10^3$. If a 1.0 L container at this temperature contains 90 mM N_2, 20 mM O_2, and 5 mM NO, what will occur?

(A) The reaction will make more N_2 and O_2.
(B) The reaction is at equilibrium.
(C) The reaction will make more NO.
(D) The temperature, T, is required to solve this problem.
(E) None of the above

The correct answer is A

Calculate the reaction quotient at the actual conditions:

$$Q = \frac{[N_2][O_2]}{[NO]^2} = \frac{(0.090\ M)(0.020\ M)}{(0.005\ M)^2} = 72$$

This value is less than K_{eq} ($72 < 2X10^3$), therefore $Q < K_{eq}$. To achieve equilibrium, the numerator of Q must be larger relative to the denominator. This occurs when products turn into reactants. Therefore NO will react to make more N_2 and O_2.

51. $BaSO_4$ ($K_{sp} = 1x10^{-10}$) is added to pure H_2O. How much is dissolved in 1 L of saturated solution?

(A) 2mg
(B) 10 μg
(C) 2 μg
(D) 100 pg
(E) Cannot determine from the information given

The correct answer is A

$BaSO_4\ (s) \longrightarrow Ba^{2+}\ (aq) + SO_4^{2-}\ (aq)$, therefore, $K_{sp} = [Ba^{2+}][SO_4^{2-}]$.

In a saturated solution: $[Ba^{2+}] = [SO_4^{2-}] = \sqrt{1\times10^{-10}} = 1\times10^{-5}\ M$

The mass in one liter is found from the molarity:

$$1\times10^{-5}\ \frac{\text{mol } Ba^{2+} \text{ or } SO_4^{2-}}{L} \times \frac{1 \text{ mol dissolved } BaSO_4}{1 \text{ mol } Ba^{2+} \text{ or } SO_4^{2-}} \times \frac{(137+32+4\times16)\text{ g } BaSO_4}{1 \text{ mol } BaSO_4}$$

$$= 0.002\ \frac{g}{L}\ BaSO_4 \times 1\text{ L solution} \times \frac{1000\text{ mg}}{g} = 2\text{ mg } BaSO_4$$

Practice Test Two

52. What are the pH and the pOH of 0.010 M HNO_3 (*aq*)?

(A) pH = 1.0, pOH = 9.0
(B) pH = 2.0, pOH = 12.0
(C) pH = 2.0, pOH = 8.0
(D) pH = 8.0, pOH = 6.0
(E) pH = 0.1, pOH = 0.9

The correct answer is B

HNO_3 is a strong acid, so it completely dissociates:

$$HNO_3 \longrightarrow H+ \; + \; NO_3^-$$

Therefore, at 0.010 moles/L, there will be 0.010 moles of H+. That is, 1×10^{-2} . The negative log is the pH, which is 2. pOH = 14 – pH, so pOH = 12.

53. Which statements about reaction rates are true?

I. Catalysts shift an equilibrium to favor product formation.
II. Catalysts increase the rate of forward and reverse reactions.
III. A greater temperature increases the chance that a molecular collision will overcome a reaction'sactivation energy.
IV. A catalytic converter contains a homogeneous catalyst.

(A) I and II
(B) II and III
(C) I, II, and III
(D) II, III, and IV
(E) I, III, and IV

The correct answer is B

Catalysts provide an alternate mechanism in both directions, but do not alter equilibrium (I is false, II is true). The kinetic energy of molecules increases with temperature, so the energy of their collisions increases also (III is true). Catalytic converters contain a heterogeneous catalyst (IV is false).

54. Consider the reaction between iron and hydrogen chloride gas:

$$2\ Fe(s) + 6\ HCl(aq) = 2\ FeCl_3(aq) + 3\ H_2(g)$$

7 moles of iron and 10 moles of HCl react until the limiting reagent is consumed. Which statements are true?

I. HCl is the excess reagent
II. HCl is the limiting reagent
III. 7 moles of H_2 are produced
IV. 2 moles of the excess reagent remain

(A) I and III
(B) I and IV
(C) II and III
(D) II and IV
(E) I only

The correct answer is D

The limiting reagent is found by dividing the number of moles of each reactant by its stoichiometric coefficient. The lowest result is the limiting reagent.

Therefore, HCl is the limiting reagent (II is true) and Fe is the excess reagent.

Knowing that 6 moles of HCl are required for the reaction, and only 10 moles are supplied, it would not be possible for 7 moles of H_2 to be produced, so III cannot be correct.

55. The reaction:

$$(CH_3)_3CBr(aq) + OH^-(aq) \longrightarrow (CH_3)_3COH(aq) + Br$$

occurs in three elementary steps:

$(CH_3)_3CBr \longrightarrow (CH_3)_3C^+ + Br^-$ is slow

$(CH_3)_3C^+ + H_2O \longrightarrow (CH_3)_3COH_2^+$ is fast

$(CH_3)_3COH_2^+ + OH^- \longrightarrow (CH_3)_3COH + H_2O$ is fast

What is the rate law for this reaction?

(A) Rate = $k[(CH_3)_3CBr]$
(B) Rate = $k[OH^-]$
(C) Rate = $k[(CH_3)_3CBr][OH^-]$
(D) Rate = $k[(CH_3)_3CBr]^2$
(E) Rate = k[(CH3)3 C Br]

Practice Test Two

The correct answer is A

Rate = $k[(CH_3)_3CBr]$

The first step will be rate-limiting. It will determine the rate for the entire reaction because it is slower than the other steps. This step is a unimolecular process with the rate given by answer A. Choice C would be correct if the reaction as a whole were one elementary step instead of three, but the stoichiometry of a reaction composed of multiple elementary steps cannot be used to predict a rate law.

56. Which statement about thermochemistry is true?

(A) Particles in a system move about less freely at high entropy.
(B) Water at 100° C has the same internal energy as water vapor at 100° C.
(C) A decrease in the order of a system corresponds to an increase in entropy.
(D) The Heat of Fusion is the energy needed to turn from a gas into a solid.
(E) At its sublimation temperature, dry ice has higher entropy than gaseous CO_2.

The correct answer is C

At high entropy, particles have a large freedom of molecular motion (A is false). Water and water vapor at 100° C contain the same translational kinetic energy, but water vapor has additional internal energy in the form of resisting the intermolecular attractions between molecules (B is false). We also know water vapor has a higher internal energy because heat must be added to boil water. Entropy may be thought of as the disorder in a system (C is correct). Sublimation is the phase change from solid to gas, and there is less freedom of motion for particles in solids than in gases. Solid CO_2 (dry ice) has a lower entropy than gaseous CO_2 because entropy decreases during a phase change that prevents molecular motion (E is false).

57. Which statement about reactions is true?

(A) All spontaneous reactions are exothermic and cause an increase in entropy.
(B) An endothermic reaction that increases the order of the system cannot be spontaneous.
(C) A reaction can be non-spontaneous in one direction and also non-spontaneous in the opposite direction.
(D) Melting snow is an exothermic process.
(E) Thermodynamic functions are dependent on the reaction pathway.

The correct answer is B

All reactions that are both exothermic and cause an increase in entropy will be spontaneous, but the converse (Choice A) is not true. Some spontaneous reactions are exothermic but decrease entropy and some are endothermic and increase entropy. Choice B is correct. The reverse reaction of a non-spontaneous reaction (Choice C) will be spontaneous. Melting snow (Choice D) requires heat. Therefore it is an endothermic process.

58. **Given:**

$E° = -2.37V$ for

Mg^{2+} (*aq*) +2e⁻ ⟶ Mg (*s*)

and

$E°=0.80$ V for

Ag^{+} (*aq*) + e⁻ ⟶ Ag (*s*)

what is the standard potential of a voltaic cell composed of a piece of magnesium dipped in a 1 M Ag^{+} solution and a piece of silver dipped in 1 M Mg^{2+}?

(A) 0.77 V
(B) 1.57 V
(C) 3.17 V
(D) 3.97 V
(E) 0.38 V

The correct answer is C

Ag^{+} (*aq*) + e^{-} ⟶ Ag (*s*) has a larger value for $E°$ (reduction potential) than Mg^{2+} (*aq*) + $2e^{-}$ ⟶ Mg (*s*). Therefore, in the cell described, reduction will occur at the Ag electrode and it will be the cathode. Using the equation:

$$E^{o}_{cell} = E^{o}(\text{cathode}) - E^{o}(\text{anode}),$$

we obtain:

$$E^{o}_{cell} = 0.80\text{ V} - (-2.37\text{ V}) = 3.17\text{ V (Answer C)}$$

Choice D results from the incorrect assumption that electrode potentials depend on the amount of material present. The balanced net reaction for the cell is:

$Mg(s) \rightarrow Mg^{2+}(aq) + 2e^{-}$	$E^{o}_{ox} = 2.37\text{ V}$
$2Ag^{+}(aq) + 2e^{-} \rightarrow 2Ag(s)$	$E^{o}_{red} = 0.80\text{ V}$ **(not 1.60 V)**
$Mg(s) + 2Ag^{+}(aq) \rightarrow 2Ag(s) + Mg^{2+}(aq)$	$E^{o}_{cell} = 3.17\text{ V}$ **(not 3.97 V)**

59. What could cause this change in the energy diagram of a reaction?

From: **to:**

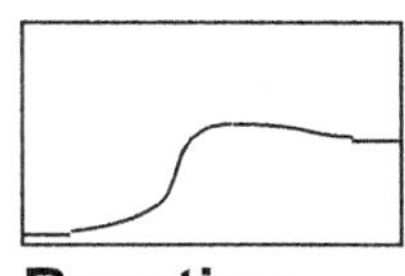

Reaction pathway →

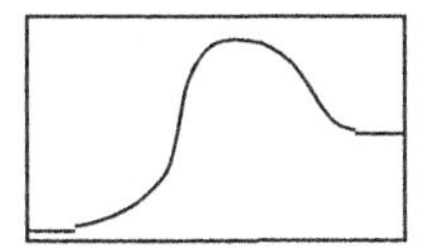

Reaction pathway →

(A) Adding a catalyst to an endothermic reaction
(B) Removing a catalyst from an endothermic reaction
(C) Adding a catalyst to an exothermic reaction
(D) Removing a catalyst from an exothermic reaction
(E) Adding heat to an endothermic reaction.

The correct answer is B

The products at the end of the reaction pathway are at a greater energy than the reactants, so the reaction is endothermic (narrowing down the answer to A or B). The maximum height on the diagram corresponds to activation energy. An increase in activation energy could be caused by removing a heterogeneous catalyst.

60. In the following phase diagram, _______ occurs as pressure is decreased from A to B at constant temperature and _______ occurs as temperature is increased from C to D at constant pressure.

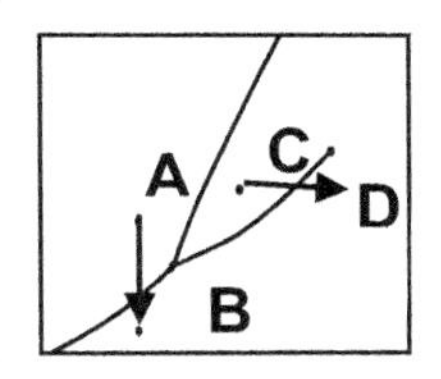

(A) deposition, melting
(B) sublimation, melting
(C) deposition, vaporization
(D) sublimation, vaporization
(E) melting, vaporization

The correct answer is D

Point A is located in the solid phase; point C is located in the liquid phase. Points B and D are located in the gas phase. The transition from solid to gas is sublimation and the transition from liquid to gas is vaporization.

61. Heat is added to a pure solid at its melting point until it all becomes liquid at its freezing point. Which of the following occur(s)?

(A) Intermolecular attractions are weakened.
(B) The kinetic energy of the molecules does not change.
(C) The freedom of the molecules to move about increases.
(D) The temperature of the system remains constant.
(E) All of the above.

The correct answer is E

Intermolecular attractions are lessened during melting. This permits molecules to move about more freely, but there is no change in the kinetic energy of the molecules because the temperature has remained the same.

62. This compound

contains which of the following?

(A) alkene, carboxylic acid, ester, and ketone
(B) B. aldehyde, alkyne, ester, and ketone
(C) aldehyde, alkene, carboxylic acid, and ester
(D) acid anhydride, aldehyde, alkene, and amine
(E) aldehyde, amine, ketone, alcohol

The correct answer is C

Choice A is wrong because there are no ketones in the molecule. A ketone has a carbonyl group linked to two hydrocarbons as shown to the right. All the carbonyls in the molecule are linked to at least one oxygen atom. Choice B is wrong because there are no ketones *and* no alkynes in the molecule. An alkyne contains a $C \equiv C$ triple bond. Choice D is wrong because there are no acid anhydrides (shown below to the left) *and* no amines (shown below bto the right). Amines require at least one N-C bond and there are no nitrogen atoms in the molecule.

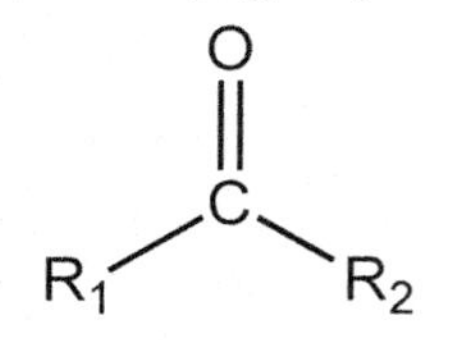

R_1—N-R_3 or H
R_2 or H

63. Which of the following pairs are isomers?

I.

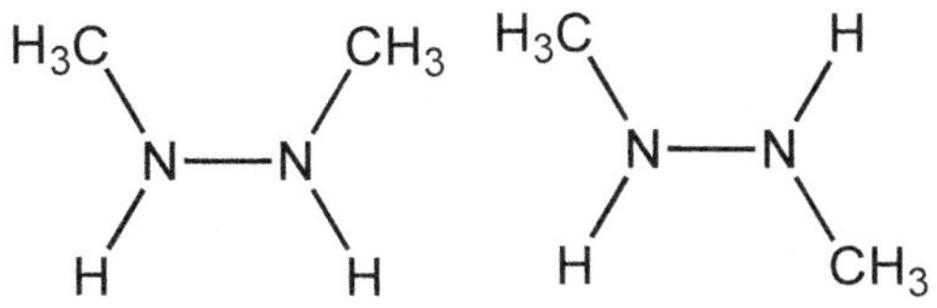

H_3C CH_3 H_3C H
N—N N—N
H H H CH_3

II. pentanal, 2-pentanone

III.

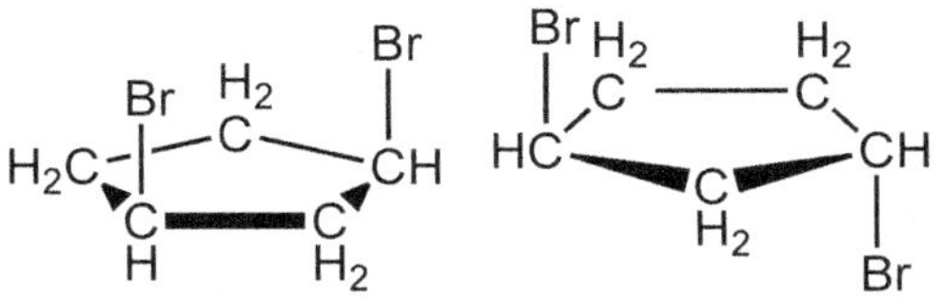

Br H_2 Br Br H_2 H_2
H_2C C CH HC C C CH
C C C
H H_2 H_2 Br

IV.

H H_3C
C—OH C—OH
F H
H_3C F

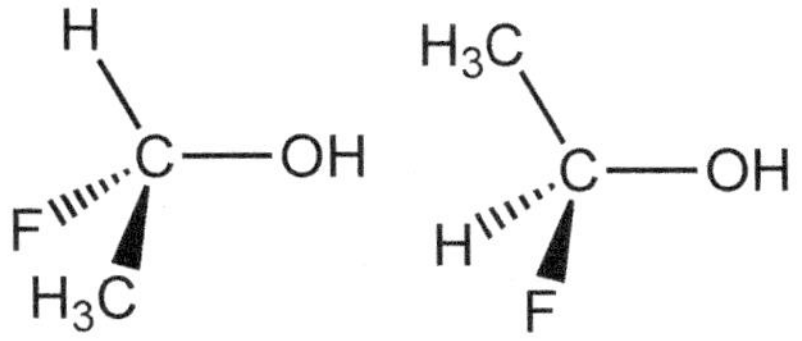

(A) I and IV
(B) II and III
(C) I, II, and III
(D) I, II, III, and IV
(E) None of the above

The correct answer is B

In Pair I, the N—N bond may freely rotate in the molecule because it is not a double bond. The identical molecule is represented twice.

For Pair II, pentanal is

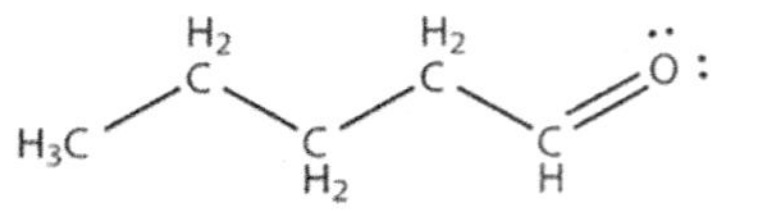

and 2-pentanone is:

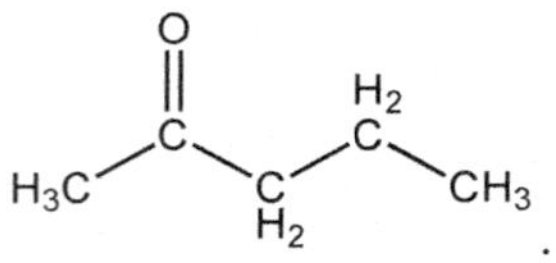

Both molecules are $C_5H_{10}O$, and they are isomers because they have the same formula with a different arrangement of atoms.

In Pair III, both molecules are 1,3-dibromocyclopentane, $C_5H_8Br_2$. In the first molecule, the bromines are in a *trans* configuration, and in the second molecule, they are *cis*. The two molecules are also viewed from different perspectives. Unlike Pair I, no bond rotation may occur because the intervening atoms are locked into place by the ring, so they are different arrangements and are isomers.

In Pair IV (1-fluoroethanol), there is a chiral center, so stereoisomers are possible, but as in Pair I, the same molecule is represented twice. Rotating the C-O bond indicates that the two structures are superimposable. The molecule shown to the right is a stereoisomer to the molecule represented in IV. The answer is B (Pairs II and III).

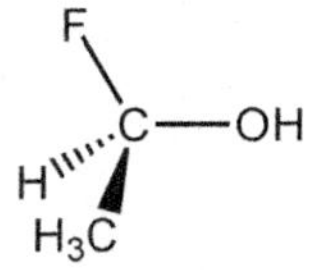

64. Which of the following is *not* a colligative property?

(A) Viscosity lowering
(B) Freezing point lowering
(C) Boiling point elevation
(D) Vapor pressure lowering
(E) All of the above

The correct answer is A

Vapor pressure lowering, boiling point elevation, and freezing point lowering may all be visualized as a result of solute particles interfering with the interface between phases in a consistent way. This is not the case for viscosity.

65. A sample of 50.0 ml KOH is titrated with 0.100 M $HClO_4$. The initial buret reading is 1.6 ml and the reading at the endpoint is 22.4 ml. What is [KOH]?

(A) 0.0416 M
(B) 0.0481 M
(C) 0.0832 M
(D) 0.0962 mM
(E) 0.0962 M

The correct answer is A

$HClO_4$ and KOH are both strong electrolytes. If you are good at memorizing formulas, solve the problem this way:

$$C_{unknown} = \frac{C_{known}\left(V_{final} - V_{initial}\right)}{V_{unknown}} = \frac{0.100\ M\ \left(22.4\ ml - 1.6\ ml\right)}{50.0\ ml} = 0.0416\ M$$

66. When KNO_3 dissolves in water, the water grows slightly colder. An increase in temperature will _______ the solubility of KNO_3.

(A) increase
(B) decrease
(C) double
(D) have no effect on
(E) have an unknown effect with the information given on the solubility.

The correct answer is A

The decline in water temperature indicates that the net solution process is endothermic (requiring heat). A temperature increase supplying more heat will favor the solution and increase solubility according to Le Châtelier's principle.

67. Classify these biochemicals:

I.

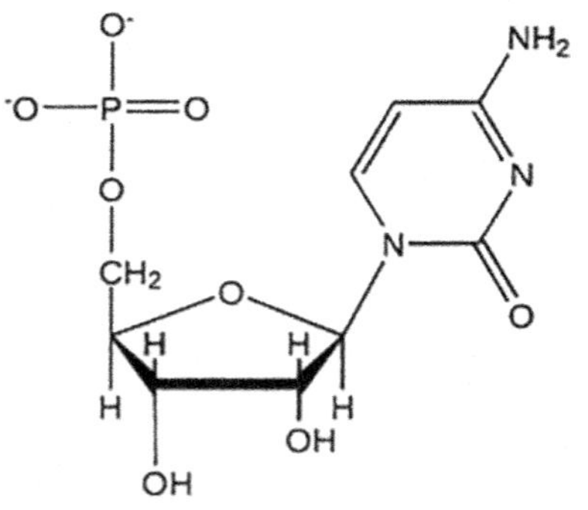

II.

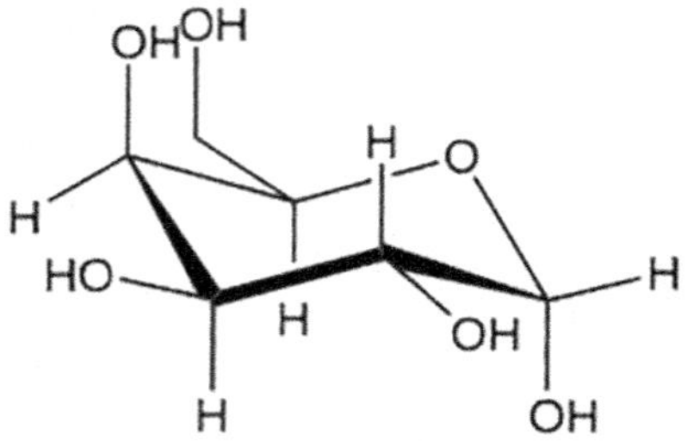

III.

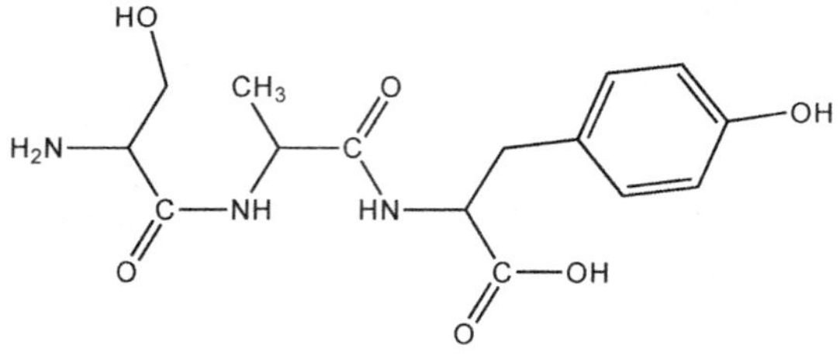

IV.

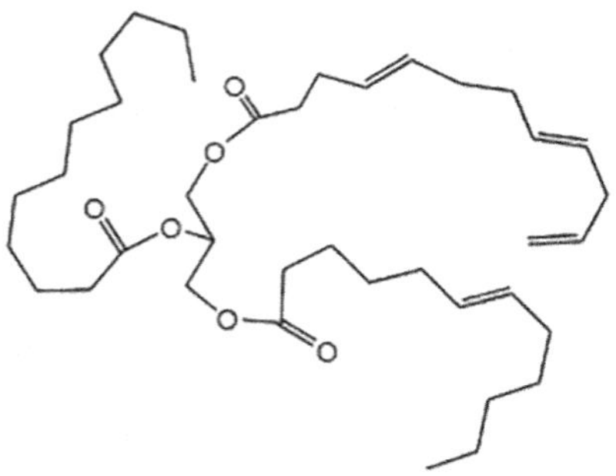

(A) I - nucleotide, II - sugar, III - peptide, IV – fat
(B) I – DNA, II sugar, III- peptide, IV lipid
(C) I - disaccharide, II - sugar, III - fatty acid, IV-polypeptide
(D) I - disaccharide, II - amino acid, III - fatty acid, IV - polysaccharide
(E) I - nucleotide, II - sugar, III - triglyceride, IV – DNA

The correct answer is A

I is a phosphate (PO_4) linked to a sugar and an amine: a nucleotide. II has the formula $C_nH_{2n}O_n$, indicative of a sugar. III contains three amino acids linked with peptide bonds. It is a tripeptide. IV is a triglyceride, a fat molecule.

68. Which of the following can be determined from the periodic table?

I. The number of protons
II. The number of neutrons
III. The number of isotopes of that atom
IV. The number of valence electrons

(A) I only
(B) I and II
(C) I, II, and III
(D) I, II, III, IV
(E) I, II and IV only

The correct answer is E

The number of protons, neutrons and electrons can be determined from the periodic table. An isotope of any atom varies only in the number of extra neutrons – which can vary from zero to infinity – even if we haven't found them all yet!

69. Which of the following quantum numbers are needed to define the position of the electrons in an element:

(A) principal, angular momentum, magnetic, and spin
(B) principal, circular, magnetic and electromagnetic
(C) angular, magnetic, electronic, spin
(D) primary, angular momentum, magnetic, and spin
(E) principal only

The correct answer is A

All 4 quantum numbers, principal, angular, magnetic and spin are required to completely define the position of an electron in an element.

70. Which of the following are true:

(A) Anions are larger than their corresponding atom
(B) Second Ionization energy is greater than the first ionization energy
(C) As you move down a group on the periodic table the atomic radius increases.
(D) Atoms with completed shells are more stable.
(E) All of the above are true.

The correct answer is E

All of the statements are true.

71. The solubility of $CoCl_2$ is 54 g per 100 g of ethanol. Three flasks each contain 100 g of ethanol. Flask #1 also contains 40 g $CoCl_2$ in solution. Flask #2 contains 56 g $CoCl_2$ in solution. Flask #3 contains 5 g of solid $CoCl_2$ in equilibrium with 54 g $CoCl_2$ in solution. Which of the following describes the solutions present in the liquid phase of the flasks?

(A) #1 - saturated, #2 - supersaturated, #3 - unsaturated.
(B) #1 - unsaturated, #2 - miscible, #3 - saturated.
(C) #1 - unsaturated, #2 - supersaturated, #3 - saturated.
(D) #1 - unsaturated, #2 - not at equilibrium, #3 – miscible.
(E) #1 unsaturated, #2 saturated, #3 miscible.

The correct answer is C

Flask #1 contains less solute than the solubility limit, and is unsaturated. Flask #2 contains more solute than the solubility limit, and is supersaturated and also not at equilibrium. Flask #3 contains the solubility limit and is a saturated solution. The term "miscible" applies only to liquids that mix together in all proportions.

72. An experiment requires 100. mL of a 0.500 M solution of $MgBr_2$. How many grams of $MgBr_2$ will be present in this solution?

(A) 9.2 g
(B) 18.4 g
(C) 11.7 g
(D) 12.4 g
(E) 15.6 g

The correct answer is A

100 mL = .1 L

.1 L x 0.500 moles $MgBr_2$ / 1 L x (24 + 80 + 80) g $MgBr_2$ / 1 mole $MgBr_2$ = 9.2 g

73. Which of the following is most likely to dissolve in water?

(A) H_2
(B) CCl_4
(C) SiO_2
(D) CH_3OH
(E) CH_4

The correct answer is D

The best solutes for a solvent have intermolecular bonds of similar strength to the solvent. H_2O molecules are connected by fairly strong hydrogen bonds. H_2 and CCl_4 are molecules with intermolecular attractions due to weak London dispersion forces. $(SiO_2)_n$ is a covalent network solid and is essentially one large molecule with bonds that are much stronger than hydrogen bonds. CH_3OH (methanol) is miscible with water because it contains hydrogen bonds between molecules.

74. **10. kJ of heat are added to one kilogram of iron at 10.° C. What is its final temperature? The specific heat of iron is 0.45 J/g•° C.**

(A) 22° C
(B) 27° C
(C) 32° C
(D) 37° C
(E) 14.5° C

The correct answer is C

The expression for heat as a function of temperature change:

$$q = n \times C \times \Delta T$$

may be rearranged to solve for the temperature change:

$$\Delta T = \frac{q}{n \times C}$$

In this case, *n* is a mass and *C* is the specific heat of iron:

$$\Delta T = \frac{10000\ \text{J}}{1000\ \text{g} \times 0.45\ \frac{\text{J}}{\text{g}\ ^\circ\text{C}}} = 22\ ^\circ\text{C}$$

This is not the final temperature (choice A is incorrect). It is the temperature difference between the initial and final temperature:

$$\Delta T = T_{final} - T_{initial} = 22\ ^\circ\text{C}$$

Solving for the final temperature gives us:

$$T_{final} = \Delta T + T_{initial} = 22\ ^\circ\text{C} + 10\ ^\circ\text{C} = 32\ ^\circ\text{C (Choice C)}$$

Practice Test Two

75. A student wishes to prepare 4.0 liters of a 0.500 M KIO_3 (molar mass 214 g). The proper procedure is to weigh out:

(A) 42.8 grams of KIO_3, and add 4 kilograms of H_2O
(B) 428 grams of KIO_3 and add H_2O until the final homogenous solution has a volume of 4.0 L
(C) 21.4grams of KIO_3, added to 4L of water
(D) 428 g of KIO_3 added to 4 L of water.
(E) 214 g of KIO_3 added to 4.0 L of H_2O

The correct answer is B

4.0 L * 0.5 moles/liter * 214g/mole = 428g

Dissolve 428 g of KIO_3 in water, then add more water until the final homogenous solution has a volume of 4.0 L. You will not necessarily add 4 L of water.

76. A 50.0 mL sample of 0.50 M HCl is titrated with 0.50 M NaOH. What is the pH of the solution after 28.0 mL of NaOH have been added to the acid?

(A) A. 14
(B) B 8.5
(C) C 0.85
(D) D. 1.70
(E) E. 1.85

The correct answer is C

First calculate the moles of HCl and NaOH:

moles HCl : (0.50 mol/L) (0.050 L) = 0.025 mol
moles NaOH : (0.50 mol/L) (0.028 L) = 0.014 mol

then calculate the moles of HCl remaining:

Since HCl and NaOH react in a 1:1 molar ratio:
0.025 mol - 0.014 mol = 0.011 mol HCl remaining

then calculate [HCl] of new solution:

0.011 mol / 0.078 L = 0.141 M
Note volume of 78 mL, derived from 50 + 28.

Finally calculate pH:

pH = -log $[H^+]$
Since HCl dissociates 100%:
pH = -log 0.141 = 0.85

Part II. Classification Questions

Questions 77-81 refer to the following:

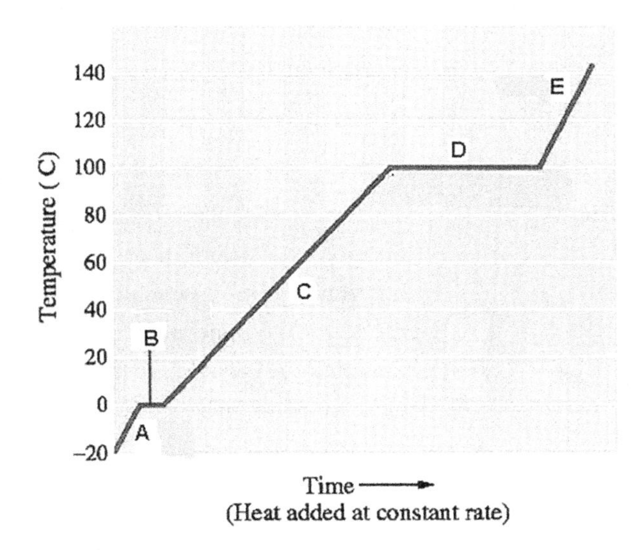

77. **Represents the phase change from liquid to gas-liquid:** The correct answer is D.

78. **Represents the heating of a liquid:** The correct answer is C.

79. **Represents the system in the solid state:** The correct answer is A.

80. **Represents the melting of a liquid:** The correct answer is B.

81. **The time when there is no temperature change:** The correct answer is B or D.

Part III: Relationship Questions

Determine if each statement is TRUE or FALSE.

If BOTH statements are true – decide if the first statement is TRUE BECAUSE of the second statement.

82. When an Al atom combines with 3 Cl atoms to form $AlCl_3$, the aluminum atom is reduced and loses electrons and reduces in size	BECAUSE	In this reaction the chlorine atoms are oxidized and gain electrons and increase in size
83. Among the halogens, fluorine is a gas, chlorine is a heavy gas moving on the ground, bromine is a liquid and iodine is a solid	BECAUSE	The intermolecular attraction between nonpolar compounds is London dispersion forces, which increase with the size of compounds. For example, boiling points in straight-chain alkanes increase with increasing the number of carbon in the chain.
84. A compressed gas in a sealed piston at room temperature is relieved to expand and the piston stops at a point where the volume of the gas is twice that in the initial position. The pressure is reduced to half of what it was.	BECAUSE	At the new position, the average speed of the molecules increases and the molecular collisions with the container walls per second decreases
85. In the reaction of hydrogen with oxygen $2\,H_2 + O_2 \longrightarrow 2\,H_2O$ 22.4 L of oxygen will react exactly with 44.8 L of hydrogen to produce water	BECAUSE	At STP (standard temperature and pressure) the molar volume of any gas, the volume that one mole of any gas occupies, is 22.4 L.

Practice Test Two

Explanations for Questions 81-84:

82. False, False

Metals like Fe, Cu, Al, and Ni..., tend to lose electrons to be oxidized and become cations with smaller sizes. Here , aluminum is losing electrons; therefore, it is oxidized, whereas statement "I" says it is reduced. So statement I is false. On the other hand, non-metals such as Cl, Fe, Be ..., tend to gain the electrons and become reduced, while statement II mentions chlorine is oxidized, which is false. Therefore, the correct answer is false, false.

83. True, True

Non-polar compounds have no dipole interactions or hydrogen bonding. London dispersion forces increase with the size of the molecule, resulting in a higher temperature requirement to break these bonds and a higher boiling point. This will match with statement I. Also, the halogens, which are binuclear molecules (F_2, Cl_2, Br_2, I_2) do not have polar bonds between F-F, Cl-Cl, Br-Br, and I-I. They therefore have no dipole interactions or hydrogen bonding. The molecules are attracted to each other by London dispersion forces, which are weakest in fluorine, so it is a gas; stronger in chlorine because in the period trend this force increases by the size of atom; chlorine is a heavy gas creeping on the ground. London forces are even stronger in bromine, which is a liquid, and the strongest London dispersion forces of the group are in iodine which is the largest halogen and is solid. Statement II is correct as an explanation of statement I. **Therefore the correct answer is true, true, correct explanation.**

84. True, False

According to the Gas Law equation at constant temperature $P_1V_1=P_2 V_2$. The volume of a gas at constant temperature is doubled and its pressure is reduced to half. Thus the statement I is correct. At the new position the average speed of the molecules is unchanged because the average speed is related to temperature and the temperature is constant. But the molecular collision with container walls per second decreases because it is related to pressure and the pressure has decreased by expansion of the piston. Thus statement II is false because it says the average speed of molecules increases. Therefore the correct answer is true, false

85. True, True

The stoichiometry of this reaction indicates that two moles (2 x 22.4 Lit = 44.8 Lit) of hydrogen reacts with one mole (1 x 22.4 = 22.4 Lit) of oxygen to produce 2 moles of water. Therefore, the Statement I is correct.

Statement II is also TRUE – for any gas, one mole of the gas takes up 22.4 liters at STP. This is a corrected explanation for Statement 1.

CPSIA information can be obtained
at www.ICGtesting.com
Printed in the USA
BVOW09s0603261016
465955BV00022B/69/P

9 781607 875680